mm Universe @ NIKA2 - Observing the mm Universe with the NIKA2 Camera

Held online due to COVID-19

EPJ Web of Conferences Volume 257 (2022)

Rome, Italy
28 June - 2 July 2021

Editors:

M. De Petris
A Ferragamo
F. Mayet

ISBN: 978-1-7138-4245-3

Printed from e-media with permission by:

Curran Associates, Inc.
57 Morehouse Lane
Red Hook, NY 12571

Some format issues inherent in the e-media version may also appear in this print version.

For additional information, please contact EDP Sciences – Web of Conferences
at the address below.

EDP Sciences – Web of Conferences
17, Avenue du Hoggar
Parc d'Activité de Courtabœuf
BP 112
F-91944 Les Ulis Cedex A
France

Phone: +33 (0) 1 69 18 75 75
Fax: +33 (0) 1 69 28 84 91

contact-edps@webofconferences.org

Additional copies of this publication are available from:

Curran Associates, Inc.
57 Morehouse Lane
Red Hook, NY 12571 USA
Phone: 845-758-0400
Fax: 845-758-2633
Email: curran@proceedings.com
Web: www.proceedings.com

TABLE OF CONTENTS

tSZ NIKA2 map of Coliseum

Observing the millimeter Universe with the NIKA2 camera

Sapienza University, Department of Physics - Rome

28 June - 2 July 2021

The "mm Universe @NIKA2" Conference is organised by the Department of Physics at Sapienza University in Rome and held online via Zoom due to the current situation. We hope that the virtual format will allow all the interested participants to be able to take part! "mm Universe @NIKA2" is the second edition in a series of workshops dedicated to the scientific exploitation of the NIKA2 camera installed at the IRAM 30-m telescope (Pico Veleta, Spain) and opened to the scientific community to report about related targets, such as:

- Cosmology with clusters
- Galaxy cluster science (SZ, X, visible)
- Galaxy formation in the early Universe
- Nearby galaxies
- Galactic star formation
- From dust to planets
- Instrumentation, other mm-instruments.

Registration will open on 1st March 2021 until 16th May 2021. The mm Universe 2021 conference is funded by Sapienza University.

https://agenda.infn.it/event/25056/

International Scientific Committee

M. De Petris (Univ. Sapienza Rome) - *co-chair*
J. M. Girart (IEEC Barcelona)
G. Lagache (LAM Marseille)
J. F. Lestrade (LERMA Paris)
F. Mayet (LPSC Grenoble) - *chair*
A. Monfardini (I. Neel Grenoble)
N. Ponthieu (IPAG Grenoble)
R. Sunyaev (MPIA Garching)
G. Yepes (UAM Madrid)

Local Organizing Committee

M. De Petris - *chair*
V. Capalbo
F. De Luca
A. Ferragamo - *co-chair*
G. Gianfagna
A. Paliwal

International Scientific Committee

- M. De Petris (Univ. Sapienza Rome) - *co-chair*
- J. M. Girart (IEEC Barcelona)
- G. Lagache (LAM Marseille)
- J. F. Lestrade (LERMA Paris)
- F. Mayet (LPSC Grenoble) - *chair*
- A. Monfardini (I. Neel Grenoble)
- N. Ponthieu (IPAG Grenoble)
- R. Sunyaev (MPIA Garching)
- G. Yepes (UAM Madrid)

Local Organizing Committee

- Marco De Petris - *chair*
- Valentina Capalbo
- Federico De Luca
- Antonio Ferragamo - *co-chair*
- Giulia Gianfagna
- Aishwarya Paliwal

Acknowledgements

The Local Organizing Committee would like to thank the Department of Physics at Sapienza University in Rome for administrative support and for *virtually* hosting the Conference.

The organization of the conference was made possible thanks to funding from Sapienza University: Finanziamenti per convegni, seminari, workshop – Classe A 2019 n. CC11916B7A19FBC7 and Progetti di Ricerca Medi 2019 n. RM11916B7540DD8D.

All authors and anonymous referees should be thanked for their contribution to the editing process of the mm Universe conference proceedings.

Cover Picture

The cover picture is a mock image of tSZ signal of Coliseum as it could be mapped with NIKA2.

Foreword

Observing the mm Universe with the NIKA2 camera

The "mm Universe @NIKA2" conference is the second edition in a series of biennial workshops. The first edition was held in Grenoble and organised by the LPSC (CNRS, Université Grenoble Alpes) in June 2019.
This international conference is dedicated to the scientific exploitation of the NIKA2 camera installed at the 30-m telescope of IRAM (Institut de Radio Astronomie Millimétrique) located in Pico Veleta (Spain) but it is open to the whole scientific community working on related science. NIKA2 is a millimetre camera, capable of surveying large areas of the sky at a high-angular resolution, with a high sensitivity and a large field of view. It was designed and built by the NIKA2 collaboration.

It has been installed in October 2015 at the IRAM 30-m telescope and commissioned in 2017. NIKA2 is open to the scientific community for observations for the next decade and provides key observations from the galactic scale to the high-redshift Universe. It allows observers to address questions, such as the environmental impact on dust properties, the star formation processes at low and high redshifts, the evolution of large-scale structures and the use of galaxy clusters for precision cosmology.
The conference covered theoretical, observational and instrumental topics related to the mm Universe, *i.e.* from stellar to cosmological scales.
The Conference was organised by the Department of Physics at Sapienza University in Rome and held online via Zoom due to the current situation. Luckily the virtual format allowed all the interested participants to be able to take part, even if the useful exchange of discussions was limited to dedicated Slack channels. The number of

participants was 145, from several countries. The conference lasted 5 days with 64 presentations in plenary sessions[1].

We believe that it was a great success thanks to the feedback received, the enthusiasm of the participants and the high quality of their presentations. All the remote presentations and discussions went smoothly thanks to the great involvement of the chairpersons and the invaluable help of all the members of the Local Organising Committee. The organization of the conference was made possible thanks to funding from the Sapienza University: Finanziamenti per convegni, seminari, workshop – Classe A 2019 n. CC11916B7A19FBC7 and Progetti di Ricerca Medi 2019 n. RM11916B7540DD8D.

All proceedings articles in this volume have been peer-reviewed. Thanks are due to all referees and authors for their contributions to the editing process of the "mm Universe @NIKA2" conference proceedings.

M. De Petris (chair), A. Ferragamo and F. Mayet
Scientific editors of the 2nd edition of the conference "mm Universe @ NIKA2"

---

[1] see https://agenda.infn.it/event/25056/

Statement of Peer review

In submitting conference proceedings to *Web of Conferences*, the editors of the proceedings certify to the Publisher that

1. They adhere to its **Policy on Publishing Integrity** in order to safeguard good scientific practice in publishing.
2. All articles have been subjected to peer review administered by the proceedings editors.
3. Reviews have been conducted by expert referees, who have been requested to provide unbiased and constructive comments aimed, whenever possible, at improving the work.
4. Proceedings editors have taken all reasonable steps to ensure the quality of the materials they publish and their decision to accept or reject a paper for publication has been based only on the merits of the work and the relevance to the journal.

Title, date and place of the conference

mm Universe @ NIKA2
28 June - 2 July 2021
Rome (Italy)

Proceedings editor(s):

M. De Petris, A. Ferragamo and F. Mayet

Date and editor's signature

15/11/2021

Probing the role of magnetic fields in star-forming filaments: NIKA2-Pol commissioning results toward OMC-1

H. Ajeddig[1,*], *R.* Adam[2], *P.* Ade[3], *P.* André[1], *E.* Artis[4], *H.* Aussel[1], *A.* Beelen[5], *A.* Benoît[6], *S.* Berta[7], *L.* Bing[8], *O.* Bourrion[4], *M.* Calvo[6], *A.* Catalano[4], *M.* De Petris[9], *F.-X.* Désert[10], *S.* Doyle[3], *E. F. C.* Driessen[7], *A.* Gomez[11], *J.* Goupy[6], *F.* Kéruzoré[4], *C.* Kramer[12], *B.* Ladjelate[12], *G.* Lagache[8], *S.* Leclercq[7], *J.-F.* Lestrade[13], *J.-F.* Macías-Pérez[4], *A.* Maury[1], *P.* Mauskopf[3,14], *F.* Mayet[4], *A.* Monfardini[6], *M.* Muñoz-Echeverría[4], *L.* Perotto[4], *G.* Pisano[3], *N.* Ponthieu[10], *V.* Revéret[1], *A. J.* Rigby[3], *A.* Ritacco[5,15], *C.* Romero[16], *H.* Roussel[17], *F.* Ruppin[18], *K.* Schuster[7], *S.* Shu[19], *A.* Sievers[12], *C.* Tucker[3], *R.* Zylka[7], and *Y.* Shimajiri[20,21]

[1]AIM, CEA, CNRS, Université Paris-Saclay, Université Paris Diderot, Sorbonne Paris Cité, 91191 Gif-sur-Yvette, France

[2]LLR (Laboratoire Leprince-Ringuet), CNRS, École Polytechnique, Institut Polytechnique de Paris, Palaiseau, France

[3]School of Physics and Astronomy, Cardiff University, Queen's Buildings, The Parade, Cardiff, CF24 3AA, UK

[4]Univ. Grenoble Alpes, CNRS, Grenoble INP, LPSC-IN2P3, 53, avenue des Martyrs, 38000 Grenoble, France

[5]Institut d'Astrophysique Spatiale (IAS), CNRS, Université Paris Sud, Orsay, France

[6]Institut Néel, CNRS, Université Grenoble Alpes, France

[7]Institut de RadioAstronomie Millimétrique (IRAM), Grenoble, France

[8]Aix Marseille Univ, CNRS, CNES, LAM (Laboratoire d'Astrophysique de Marseille), Marseille, France

[9]Dipartimento di Fisica, Sapienza Università di Roma, Piazzale Aldo Moro 5, I-00185 Roma, Italy

[10]Univ. Grenoble Alpes, CNRS, IPAG, 38000 Grenoble, France

[11]Centro de Astrobiología (CSIC-INTA), Torrejón de Ardoz, 28850 Madrid, Spain

[12]Instituto de Radioastronomía Milimétrica (IRAM), Granada, Spain

[13]LERMA, Observatoire de Paris, PSL Research University, CNRS, Sorbonne Université, UPMC, 75014 Paris, France

[14]School of Earth and Space Exploration and Department of Physics, Arizona State University, Tempe, AZ 85287, USA

[15]Laboratoire de Physique de l'École Normale Supérieure, ENS, PSL Research University, CNRS, Sorbonne Université, Université de Paris, 75005 Paris, France

[16]Department of Physics and Astronomy, University of Pennsylvania, 209 South 33rd Street, Philadelphia, PA, 19104, USA

[17]Institut d'Astrophysique de Paris, CNRS (UMR7095), 98 bis boulevard Arago, 75014 Paris, France

[18]Kavli Institute for Astrophysics and Space Research, Massachusetts Institute of Technology, Cambridge, MA 02139, USA

[19]Caltech, Pasadena, CA 91125, USA

[20]National Astronomical Observatory of Japan, National Institutes of Natural Sciences, Osawa, Mitaka, Tokyo 181-8588, Japan

[21]Department of Physics and Astronomy, Graduate School of Science and Engineering, Kagoshima University, 1-21-35 Korimoto, Kagoshima, Kagoshima 890-0065, Japan

*e-mail: hamza.ajeddig@cea.fr

Abstract. Dust polarization observations are a powerful, practical tool to probe the geometry (and to some extent, the strength) of magnetic fields in star-forming regions. In particular, Planck polarization data have revealed the importance of magnetic fields on large scales in molecular clouds. However, due to insufficient resolution, Planck observations are unable to constrain the B-field geometry on prestellar and protostellar scales. The high angular resolution of 11.7 arcsec provided by NIKA2-Pol 1.15 mm polarimetric imaging, corresponding to 0.02 pc at the distance of the Orion molecular cloud (OMC), makes it possible to advance our understanding of the B-field morphology in star-forming filaments and dense cores (IRAM 30m large program B-FUN). The commissioning of the NIKA2-Pol instrument has led to several challenging issues, in particular, the instrumental polarization or intensity-to-polarization "leakage" effect. In the present paper, we illustrate how this effect can be corrected for, leading to reliable exploitable data in a structured, extended source such as OMC-1. We present a statistical comparison between NIKA2-Pol and SCUBA2-Pol2 results in the OMC-1 region. We also present tentative evidence of local pinching of the B-field lines near Orion-KL, in the form of a new small-scale hourglass pattern, in addition to the larger-scale hourglass already seen by other instruments such as Pol2.

1 Introduction

Observations of nearby molecular clouds by the *Herschel Gould Belt Survey* have shown the crucial role of filaments in the star formation process [3, 4], but the detailed fragmentation manner of star-forming filaments remains under debate. In parallel to *Herschel*, dust polarization observations from the *Planck satellite* have highlighted the role of magnetic fields in interstellar clouds [15, 16]. The magnetic field inferred from *Planck* polarization observations appears to have a regular, structured geometry on large scales (>> 0.1 pc), suggesting it is dynamically important. Stars form inside molecular filaments characterized by a common inner width of ~0.1 pc [7] and the angular resolution of *Planck* is insufficient to probe the geometry of the B-field in the <0.1 pc interior of such filaments. The IRAM 30m large program B-FUN will perform high-resolution polarization observations of a broad sample of nearby star-forming filaments imaged by *Herschel* [3], in an effort to improve our understanding of the role of magnetic fields in core/star formation along filaments. It is important to characterize the B-field at different wavelengths and high angular resolution to understand how it is involved in the star formation process [5, 9, 14]. NIKA2-Pol, the polarization channel of the NIKA2 camera [1] on the IRAM 30m telescope (Pico Veleta, Spain), is under commissioning and will provide 1.15 mm continuum polarization observations with high angular resolution (11.7 arcsec). Polarimetric imaging observations are challenging due, in particular, to the intensity-to-polarization leakage (hereafter "leakage"), which needs to be characterized and corrected for before proper scientific observations can be made. Qualitatively, the leakage affecting NIKA2-Pol data has the shape of a cloverleaf with a central positive lobe surrounded by a negative quadrupolar pattern [2]. The leakage effect can be characterized by observing a strong unpolarized point-like source such as the planet Uranus, and then subtracted from the Stokes Q, U data using a "deconvolution" technique described in detail in [6, 19, 25]. A similar instrumental polarization effect is also present in the case of other polarimeters such as XPOL, SCUBA2-POL2, and NIKA1 [11, 18, 25].

Here, we report on the results of NIKA2-Pol commissioning observations of Orion OMC-1 that were used to characterize the leakage effect in an extended source. OMC-1 is a well documented source that has been observed by all polarimetric instruments (see, e.g., Sect. 4

below). As part of our attempts to finalize the NIKA2-Pol commissioning, we discuss the leakage effect in OMC-1 and compare our polarization findings with recently published data (here from SCUBA2-POL2 [24]). As OMC-1 is one of the targets of the B-FUN large program, we also present a preliminary scientific analysis of the results (see Sect. 4), illustrating the great potential of NIKA2-Pol.

2 Correcting NIKA2-Pol data for instrumental polarization

To test the repeatability of NIKA2-Pol results, OMC-1 was observed on several nights and under good weather conditions in Nov. 2020. Figure 1 shows the difference between data taken on Nov. 12 and 15, 2020 before and after leakage correction, in panels b) and c), respectively. On the same night, the leakage pattern was observed on Uranus under similar conditions of focus and elevation, which guarantee an optimal correction of instrumental polarization (IP). Without IP correction, the OMC-1 polarization data taken on different days are in good agreement (e.g., with a root mean square difference in polarization angles of $< 3°$), except at the location of the compact source Orion-KL, which is strong in Stokes I but weakly polarized ($P \sim 1\%$), and thus more strongly affected by IP. Prior to IP correction, a very significant difference in polarization angles ($\sim 50°$) is observed near Orion-KL between data sets taken on two different nights (see Fig. 1b). Indeed, the impact of IP is more pronounced in areas of the sky where the signal in Stokes I is strong and structured on small angular scales rather than extended and slowly varying. After subtracting the IP estimated using Uranus data observed before OMC-1, however, the polarization data from different nights agree well with each other, with a maximum difference in polarization angles reduced to $< 20°$ near Orion-KL (see Fig. 1c) and $\leq 2°$ outside of Orion-KL. This analysis illustrates the ability of NIKA2-Pol to provide robust (repeatable), high-quality polarization maps for scientific purposes (see Fig. 3a below).

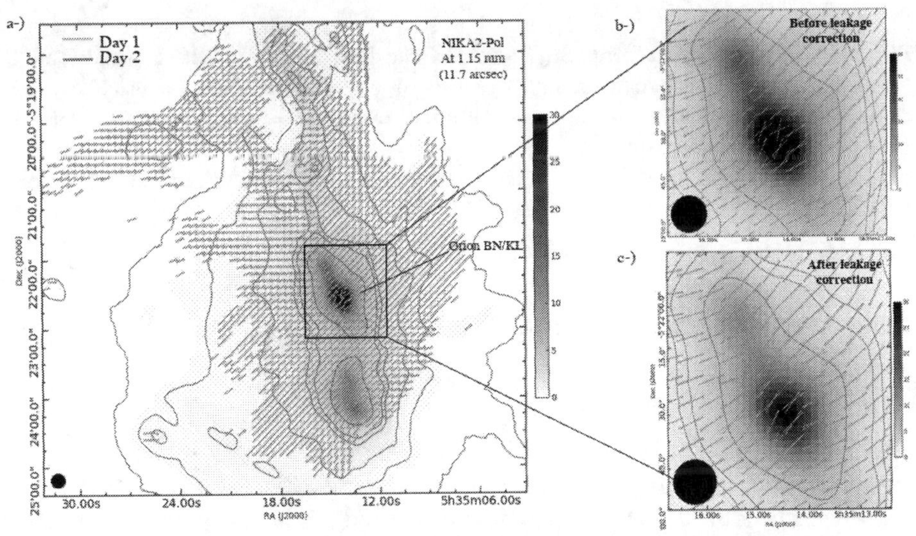

Figure 1. (*a*): B-field vectors from NIKA2-Pol polarization data at positions of significant detections in polarized intensity (i.e., $I_{pol}/\sigma_{I_{pol}}$ >3) on two different days, day 1 and day 2 (before leakage correction), corresponding to the green and red colors, respectively. (*b*): Closeup view of the vicinity of Orion-KL position overlaid with B-field vectors before correcting for instrumental polarization. (*c*) similar to panel *b* but after correcting for instrumental polarization.

3 Comparison of NIKA2-POL vs SCUBA2-POL2 results

In order to compare NIKA2-Pol polarization observations at 1.15 mm with published SCUBA2-POL2 data at 850 μm [24], we re-projected the POL2 and NIKA2-Pol data to the same grid, and smoothed the NIKA2-Pol data to the POL2 angular resolution (14 arcsec). Figure 2a shows map of the difference in B-field angles between POL2 and NIKA2-Pol data, where vectors represent the B-field angle difference each 14″. Where the signal-to-noise ratio in polarized intensity is high ($I_{pol}/\sigma_{I_{pol}}$ >5) for both instruments, the NIKA2-Pol and POL2 data are in excellent agreement, within a typical 1° uncertainty (standard deviation of the mean difference in polarization angles – see Fig. 2*b*). This result has been confirmed using three independent data sets from the NIKA2-Pol commissioning campaigns of December 2018, February 2020, and November 2020.

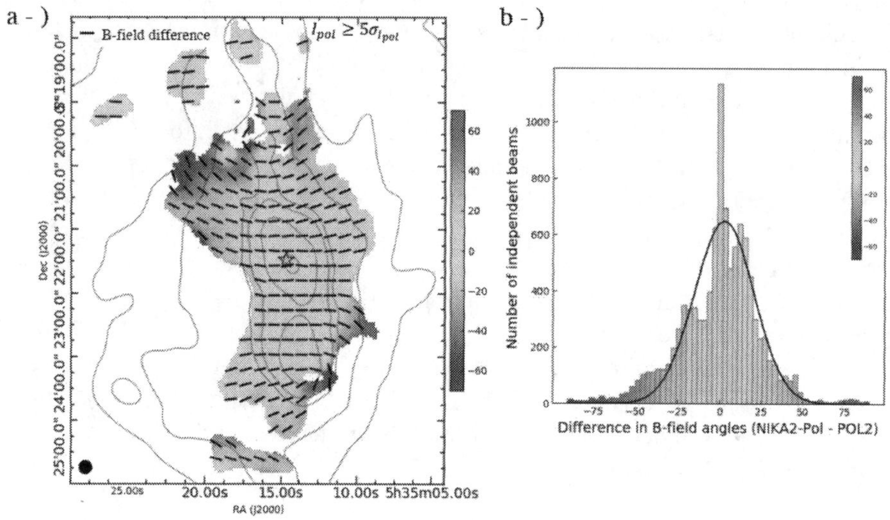

Figure 2. Differences in B-field angle between NIKA2-Pol and POL2 data in OMC-1 at the same angular resolution (14″) and for positions with high signal-to-noise ratio in polarized intensity (I_{pol} > $5\sigma_{I_{pol}}$ for both NIKA2-Pol and POL2). *a-)*: Map of the differences in B-field angle (NIKA2-Pol – POL2). The overlaid black segments display difference vectors plotted every 14″ beam. *b-)*: Histogram of the differences in B-field angle (NIKA2-Pol – POL2). The amplitude of the difference in B-field angle between NIKA2-Pol and POL2 is less than 10° for 67% of the positions.

4 A possible new local hourglass at the Orion-KL position

The magnetic field line geometry in OMC-1 has been under investigation for more than two decades. Schleuning (1998) [20] reported an hourglass shape of the B-field geometry based on a 8′ × 8′ mapping of OMC-1 at 100 μm and 35 arcsec resolution. More recent submillimeter polarization observations with the SOFIA/HAWC+ (214 μm), SCUBA2-POL2 (850 μm) and POLKA (870 μm) polarimeters [10, 11, 21] have confirmed this large-scale hourglass B-field geometry in OMC-1 and the weak polarization fraction around Orion-KL (∼1–2%). In particular, based on their SCUBA-POL2 study, Pattle et al. (2019) [14] proposed a scenario in which the evolution of the OMC-1 region is dynamically regulated by the magnetic field. In this scenario, the large-scale gravitational collapse of the OMC-1 region and/or a central explosive outflow are responsible for the large-scale hourglass. At the Orion-KL position, however, the B-field geometry was unclear. For instance, Vaillancourt et al. (2008) [23]

pointed out that B-field angles inferred from SHARP data at 350 μm and 450 μm differ by 25° and reported a drop in 450/350 μm polarization ratio by a factor of 2 toward Orion-KL. In addition, the B-field lines seemed to be distorted at the Orion-KL position (see Fig. 1 of [23]; see also the ALMA observations of Cortes et al. 2021 [8]). Thanks to the high angular resolution and sensitivity provided by NIKA2-Pol at 1.15 mm, this distortion seems to be confirmed and even looks like another hourglass, on smaller scales, centered on Orion KL, which is suggestive of local gravitational collapse at this location (see Fig. 3a). The large-scale hourglass observed earlier is sketched in Fig. 3b from [14], while the possible new local hourglass revealed by the NIKA2-Pol polarization data is shown and contrasted with the other hourglass in Fig. 3c.

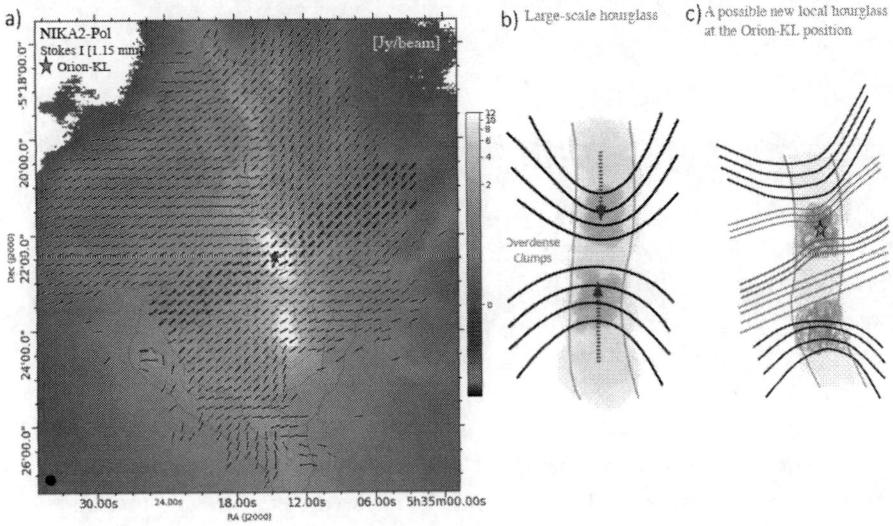

Figure 3. *a-)* Stokes I 1.15 mm map of OMC-1 region obtained from NIKA2-Pol polarization commissioning data, with magnetic field vectors at positions where $I_{pol}/\sigma_{I_{pol}} > 3$. The red vectors emphasize the possible new local hourglass revealed by NIKA2-Pol. *b)*: Illustration of the large-scale magnetic field as seen by POL2 from Pattle et al. (2019). *c)* Schematic illustration of the large-scale hourglass geometry as seen with NIKA2-Pol (in black lines) and the new possible local hourglass at the position of Orion-KL revealed by the high sensitivity and resolution of NIKA2-Pol (in blue lines).

5 Concluding remarks

The NIKA2-Pol commissioning data have led to a better characterization of the reliability of the NIKA2 instrument in polarization mode. In this paper, we showed how NIKA2-Pol observations taken toward a region with extended and structured emission such as Orion OMC-1 can be corrected for instrumental polarization in practice. We achieved consistency between NIKA2-Pol data taken two years apart and also with independent observations from POL2. The NIKA2-Pol observations confirm the presence of a large-scale hourglass pattern seen by previous polarimeters in OMC-1. In addition, the NIKA2-Pol data reveal a possible new local hourglass centered at the Orion-KL position. The NIKA2-Pol commissioning results obtained on OMC-1 are very promising and suggest that high-quality scientific data will be obtained as part of the IRAM 30m large program B-FUN.

Acknowledgements

We would like to thank the IRAM staff for their support during the campaigns. The NIKA2 dilution cryostat has been designed and built at the Institut Néel. In particular, we acknowledge the crucial contribution of the Cryogenics Group, and in particular Gregory Garde, Henri Rodenas, Jean Paul Leggeri, Philippe Camus. This work has been partially funded by the Foundation Nanoscience Grenoble and the LabEx FOCUS ANR-11-LABX-0013. This work is supported by the French National Research Agency under the contracts "MKIDS", "NIKA" and ANR-15-CE31-0017 and in the framework of the "Investissements d'avenir" program (ANR-15-IDEX-02). This work has benefited from the support of the European Research Council Advanced Grant ORISTARS under the European Union's Seventh Framework Programme (Grant Agreement no. 291294). F.R. acknowledges financial supports provided by NASA through SAO Award Number SV2-82023 issued by the Chandra X-Ray Observatory Center, which is operated by the Smithsonian Astrophysical Observatory for and on behalf of NASA under contract NAS8-03060.

References

[1] Adam, R *et al.*, Astron. Astrophys. **609**, A115 (2018)
[2] Ajeddig, H *et al.*, EPJWC. **228**, 00002 (2020)
[3] André, Ph *et al.*, Astron. Astrophys. **518**, L102 (2010)
[4] André, Ph *et al.*, in *Protostars and Planets VI*, 27–51 (2014)
[5] André, Ph *et al.*, PASA. **36**, e029 (2019)
[6] Andrianasolo, Aina, http://www.theses.fr/2019GREAY077, (2019)
[7] Arzoumanian, D *et al.*, Astron. Astrophys. **529**, L6 (2011)
[8] Cortes P. C.*et al.*, 2021, ApJ, 907, 94. doi:10.3847/1538-4357/abcafb
[9] Crutcher, R. M. 2012, , 50, 29. doi:10.1146/annurev-astro-081811-125514
[10] Dowell, C. D., HAWC+ Instrument Team, & HAWC+ Science Team 2018,
[11] Friberg, P *et al.*, Millimeter, Submillimeter, and Far-Infrared Detectors and Instrumentation for Astronomy IX **10708** 3M (2018)
[12] Houde, M., Dowell, C. D., Hildebrand, R. H., et al. 2004, , 604, 717. doi:10.1086/382067
[13] Gonatas, D. P., Engargiola, G. A., Hildebrand, R. H., et al. 1990, , 357, 132. doi:10.1086/168898
[14] Pattle, K. & Fissel, L. 2019, Frontiers in Astronomy and Space Sciences, 6, 15. doi:10.3389/fspas.2019.00015
[15] Planck Collaboration, Ade, P. A. R., Aghanim, N., et al. 2013, , 557, A53.
[16] Planck collaboration *et al.*, Astron. Astrophys. **594**, A1 (2016)
[17] Rao, R., Crutcher, R. M., Plambeck, R. L., et al. 1998, , 502, L75. doi:10.1086/311485
[18] Ritacco, A *et al.*, Astron. Astrophys. **599**, A34 (2017)
[19] Ritacco, A , http://www.theses.fr/2016GREAY018, (2016)
[20] Schleuning, D. A. 1998, , 493, 811. doi:10.1086/305139
[21] Siringo, G., Kreysa, E., Reichertz, L. A., et al. 2004, The Magnetized Interstellar Medium, 215
[22] Thum, C *et al.*, PASP. **120**, 777–790 (2008)
[23] Vaillancourt, J. E., Dowell, C. D., Hildebrand, R. H., et al. 2008, , 679, L25. doi:10.1086/589152
[24] Ward-Thompson, D., Pattle, K., Bastien, P., et al. 2017, , 842, 66. doi:10.3847/1538-4357/aa70a0
[25] Wiesemeyer, H *et al.*, PASP. **126**, 1027 (2014)

EPJ Web of Conferences 257, 00003 (2022) https://doi.org/10.1051/epjconf/202225700003
mm Universe @ NIKA2

PSZ2 G091: A massive double cluster at $z \sim 0.822$ observed by the NIKA2 camera

E. Artis[1,*], *R.* Adam[2], *P.* Ade[3], *H.* Ajeddig[4], *P.* André[4], *M.* Arnaud[4], *H.* Aussel[4], *I.* Bartalucci[7], *A.* Beelen[6], *A.* Benoît[8], *S.* Berta[9], *L.* Bing[10], *O.* Bourrion[1], *M.* Calvo[8], *A.* Catalano[1], *M.* De Petris[11], *F.-X.* Désert[12], *S.* Doyle[3], *E. F. C.* Driessen[9], *A.* Ferragamo[11], *A.* Gomez[13], *J.* Goupy[8], *F.* Kéruzoré[1], *C.* Kramer[14], *B.* Ladjelate[14], *G.* Lagache[10], *S.* Leclercq[9], *J.-F.* Lestrade[15], *J.-F.* Macías-Pérez[1], *A.* Maury[4], *P.* Mauskopf[3,16], *F.* Mayet[1], *A.* Monfardini[8], *M.* Muñoz-Echeverría[1], *A.* Paliwal[11], *L.* Perotto[1], *G.* Pisano[3], *E.* Pointecouteau[5], *N.* Ponthieu[12], *G. W.* Pratt[4], *V.* Revéret[4], *A. J.* Rigby[3], *A.* Ritacco[6,17], *C.* Romero[18], *H.* Roussel[19], *F.* Ruppin[21], *K.* Schuster[9], *S.* Shu[22], *A.* Sievers[14], *C.* Tucker[3], and *G.* Yepes[20]

[1]Univ. Grenoble Alpes, CNRS, Grenoble INP, LPSC-IN2P3, 53, avenue des Martyrs, 38000 Grenoble, France
[2]LLR (Laboratoire Leprince-Ringuet), CNRS, École Polytechnique, Institut Polytechnique de Paris, Palaiseau, France
[3]School of Physics and Astronomy, Cardiff University, Queen's Buildings, The Parade, Cardiff, CF24 3AA, UK
[4]AIM, CEA, CNRS, Université Paris-Saclay, Université Paris Diderot, Sorbonne Paris Cité, 91191 Gif-sur-Yvette, France
[5]Univ. de Toulouse, UPS-OMP, Institut de Recherche en Astrophysique et Planétologie, Toulouse, France CNRS, IRAP, 9 Av. colonel Roche, BP 44346, F-31028 Toulouse cedex 4, France
[6]Institut d'Astrophysique Spatiale (IAS), CNRS, Université Paris Sud, Orsay, France
[7]INAF, IASF-Milano, Via A. Corti 12, 20133 Milano, Italy
[8]Institut Néel, CNRS, Université Grenoble Alpes, France
[9]Institut de RadioAstronomie Millimétrique (IRAM), Grenoble, France
[10]Aix Marseille Univ, CNRS, CNES, LAM (Laboratoire d'Astrophysique de Marseille), Marseille, France
[11]Dipartimento di Fisica, Sapienza Università di Roma, Piazzale Aldo Moro 5, I-00185 Roma, Italy
[12]Univ. Grenoble Alpes, CNRS, IPAG, 38000 Grenoble, France
[13]Centro de Astrobiología (CSIC-INTA), Torrejón de Ardoz, 28850 Madrid, Spain
[14]Instituto de Radioastronomía Milimétrica (IRAM), Granada, Spain
[15]LERMA, Observatoire de Paris, PSL Research University, CNRS, Sorbonne Université, UPMC, 75014 Paris, France
[16]School of Earth and Space Exploration and Department of Physics, Arizona State University, Tempe, AZ 85287, USA
[17]Laboratoire de Physique de l'École Normale Supérieure, ENS, PSL Research University, CNRS, Sorbonne Université, Université de Paris, 75005 Paris, France
[18]Department of Physics and Astronomy, University of Pennsylvania, 209 South 33rd Street, Philadelphia, PA, 19104, USA
[19]Institut d'Astrophysique de Paris, CNRS (UMR7095), 98 bis boulevard Arago, 75014 Paris, France
[20]Departamento de Física Teórica and CIAFF, Facultad de Ciencias, Modulo 8, Universidad Autánoma de Madrid, 28049 Madrid, Spain

*e-mail: emmanuel.artis@lpsc.in2p3.fr

[21]Kavli Institute for Astrophysics and Space Research, Massachusetts Institute of Technology, Cambridge, MA 02139, USA
[22]Caltech, Pasadena, CA 91125, USA

Abstract. PSZ2 G091.83+26.11 is a massive galaxy cluster with $M_{500} = 7.43 \times 10^{14} M_\odot$ at $z = 0.822$. This object exhibits a complex morphology with a clear bimodality observed in X-rays. However, it was detected and analysed in the *Planck* sample as a single, spherical cluster following a universal profile [1]. This model can lead to miscalculations of thermodynamical quantities, like the pressure profile. As future multiwavelength cluster experiments will detect more and more objects at higher redshifts (where we expect the fraction of merging objects to be higher), it is crucial to quantify this systematic effect. In this work, we use high-resolution observations of PSZ2 G091.83+26.11 by the NIKA2 camera to integrate the morphological characteristics of the cluster in our modelling. This is achieved by fitting a two-halo model to the SZ image and then by reconstruction of the resulting projected pressure profile. We then compare these results with the spherical assumption.

1 Introduction

Galaxy clusters are the largest virialized objects in the universe. As such, they can constrain cosmological models, fondamental physics, and astrophysics [2, 3]. Precisely, the number of clusters per unit of mass and volume, modelled as the halo mass function (historically from [4]), constitutes a robust cosmological probe, measuring among others the matter density parameter Ω_m, the rms linear fluctuations on scales of 8 h^{-1}Mpc, σ_8, the dark energy equation of state parameters, as well as the sum of the mass of the neutrinos [5, 6]. Cluster counts have been used by many surveys like *Planck* [7], *DES* [8], and *XXL* [9] to cite a few examples.
However, the total mass of dark matter halos hosting clusters is not an observable quantity, and must be inferred from different physical phenomena [10], like the thermal Sunyaev-Zel'dovich effect (tSZ, [11]). As a consequence, astrophysical systematic effects are biasing our measurements [12], and their effects must be integrated in any cosmological analysis. Moreover, given the fact that the size of the catalogs will be increased by several orders of magnitude in the future, effects that are now neglected will play a crucial role [13, 14].
The Large program SZ of the NIKA2 experiment (LPSZ, [15]) aims at investigating these systematic effects, taking advantage of the spatial resolution and FoV of the NIKA2 camera. In this work, we focus on the impact of cluster morphology on the reconstruction of observable quantities in the case of the massive cluster PSZ2G091.

2 The case of PSZ2G091

Table 1: Characteristics of PSZ2 G091.83+26.11 as described in [16], with the observation length and SNR at SZ peak.

z	M_{500}	θ_{500}	t_{obs}/t_{LPSZ}	tSZ decrement peak
0.822	$7.43 \times 10^{14} M_\odot$	2.2 arcmin	2.5h/2.5h=1	14.9σ

As part of the LPSZ, PSZ2G091 was observed in October, 2017, with an average elevation of 58.5° and an average atmospheric opacity at 225 GHz of 0.243. These conditions are standard

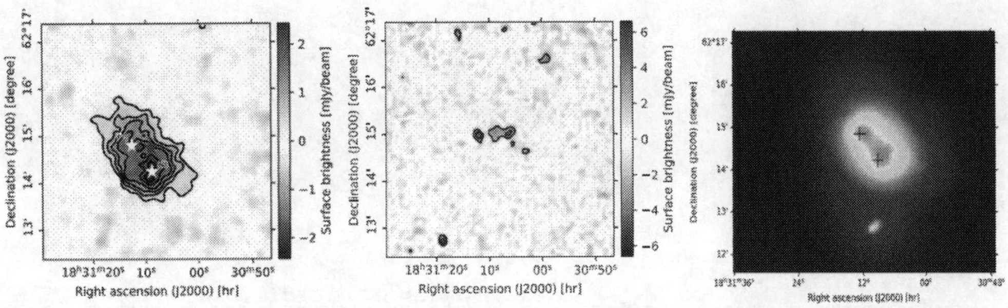

Figure 1. Left: Surface brightness at 2 mm from NIKA2 observations. The black contours represent the SNR levels starting at 3σ increasing by 2σ at each step. The northern (N) and southern (S) X-ray peaks are shown as white stars, the X-ray centroid being in purple. **Middle**: Surface brightness at 1 mm. The signal exhibits several point sources. Their treatment is explained in section 3. **Right**: X-ray surface brightness of PSZ2G091 from *XMM-Newton*. The cluster exhibits two peaks at the position $(\alpha_N, \delta_N) = (277.80, 62.24)$, and $(\alpha_S, \delta_S) = (277.79, 62.237)$, shown with the black crosses.

for observations at the IRAM 30 m telescope, at this season.
The data from NIKA2 were reduced following the pipeline described in [17]. In figure 1, we show the results of the data reduction at 1 and 2 mm. The cluster can clearly be seen in the 2 mm image, and it is clearly elongated in the NW-SW direction. There is a clear departure from sphericity, and a hint of bimodality later confirmed in the X-ray surface brightness. Conversely, in the 1 mm map, we only detect point sources, since the SZ signal is completely dominated by the noise as expected. The peaks in the X-ray map are in good agreement with the one observed in the NIKA2 2 mm map. This would imply the presence of two well-defined sub-halos in the first stages of a major merger.

3 Point sources treatment

The 1 mm map of PSZ2G091 exhibits 7 point sources (see figure 1). Their fluxes must be inferred at 2 mm, in order to fit the ICM signal. If external information on these sources is available, we can usually use SED models to infer a PDF of the flux at 2 mm for each source, and use it as a prior for the model reconstruction. However, in this case, no additional data was available. The point sources observed at this wavelength are either millimetric sources, or radio objects. For the latter, the flux expected at 2 mm is higher than the one seen in the 1 mm map. Given the value of the SZ signal in the 2 mm map, we were consequently able to identify the sources as millimetric sources.Thus we decided to used flat priors for the fluxes of the sources at 2 mm, with a conservative upper limit of $F_{2,\text{upper}} = F_1/3$. The fluxes of the point sources were then fitted jointly with the parameters of the profiles, using these flat priors.

4 Imaging analysis

4.1 Single spherical model

As a first test case, we consider a single spherical halo centred on the X-ray centroid coordinates. A forward modelling approach is incorporated in an MCMC sampling framework to

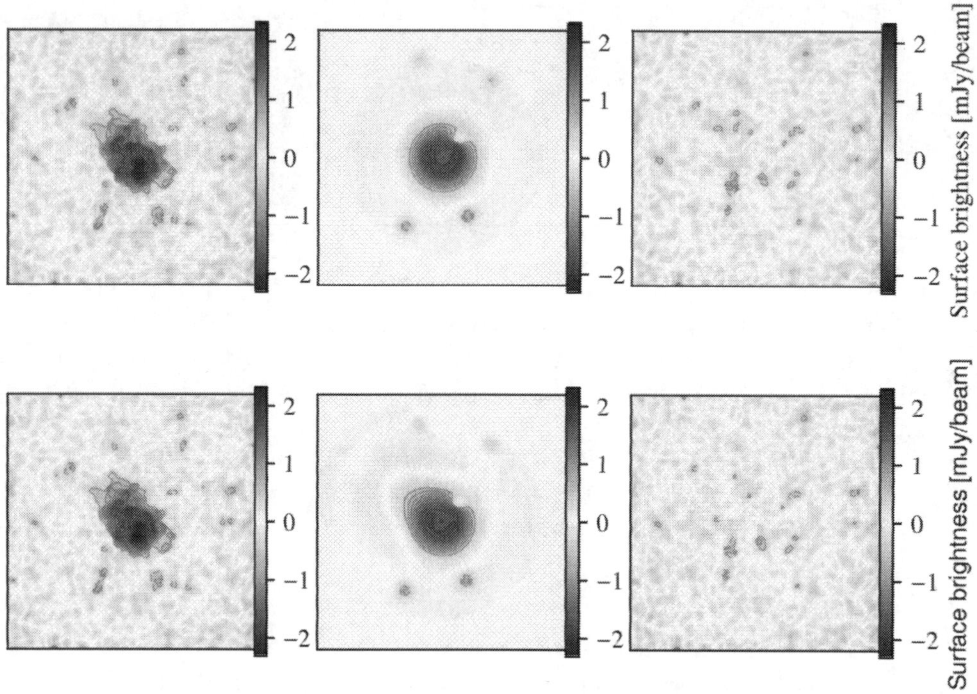

Figure 2. Single spherical model fit of the NIKA2 2 mm map for PSZ2G091 centered on the X-ray centroid coordinates (**top**), and 2-halo model (**bottom**). From **left** to **right**, we display the data, the single halo model, with the point sources treated as described in section 3, and the residuals. The maps are given in a $5' \times 5'$ area, and for display purposes, each map is smoothed with a gaussian kernel. The contours are showing the SNR level sets, starting at 3σ and increasing with a step of 1σ.

fit the parameters of the pressure profile, as well as the point sources, as described in section 3, with the collaboration software PANCO2. Instead of considering a generalised Navarro-Frenk-and-White profile [18], we use a power law model, where the pressure is divided into 6 bins, equally distributed in *log*-space. Each one of these bins follow the identity

$$P(r) = P_i (r/r_i)^{-\alpha_i}. \qquad (1)$$

The top row of figure 2 shows the results of the fitting procedure. Although returning reasonable residuals, the spherical symmetry clearly does not encapsulate the bimodal nature of the cluster. It is then required to improve our modelling.

4.2 Two-halo model

Instead of considering a single pressure profile, we jointly fit two halos at the positions of the X-ray peaks. The top row of figure 2 shows the results of the MCMC fitting. The results of the fits are shown on the bottom row of figure 2. It is clear that the two halo model yields a more realistic representation of the dynamical state of the cluster. Additionally, the residuals are slightly improved in the region of the northern subhalo. Of course, due to the non-spherical nature of this cluster, it is not possible to consider a radial pressure profile. However, in section 5, we describe how we recover an average radial profile for the two-halo model.

5 Pressure profile reconstruction

Our model takes into account the specific morphology of this cluster. However, the thermodynamical profiles are usually considered with the goal of reconstructing a mass profile, using the hydrostatic equilibrium equation. This requires the presence of a 1D pressure profile, when we previously fitted a 2D map. Thus, we use the following procedure to recover an average 1D profile. We consider both the profiles of the two halos fitted in section 4.2. With the assumption of the two sub-halos lying in the same plane perpendicular to the line of sight, we recover a mean radial pressure profile by integrating both the profiles in annuli centered around the X-ray centroid coordinates. In this framework, the pressure in the i-th bin reads:

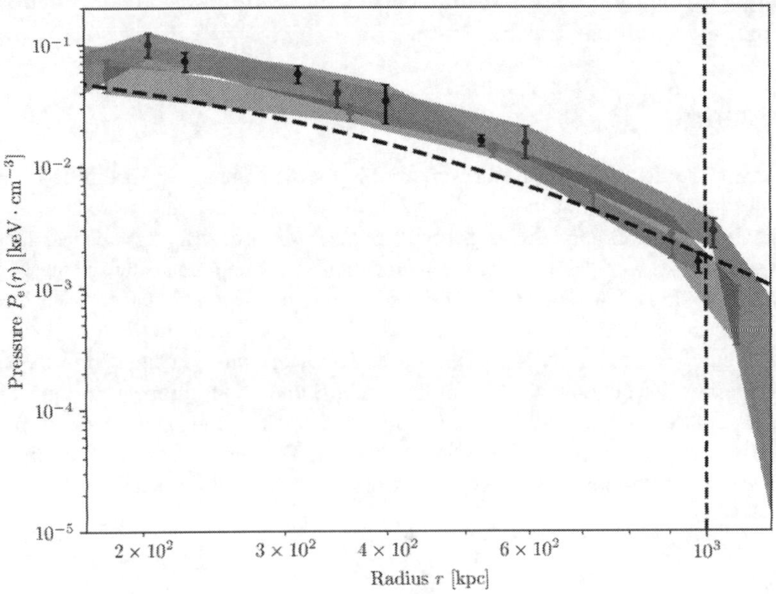

Figure 3. In red, composite pressure profile obtained by combining the northern (N) and southern (S) profiles, as described in equation 2. In blue, profile obtained with the single spherical model. In black, we represent the contours obtained from XMM X-ray data only. The black dashed line represents the universal profile from [19]. The vertical dashed line represents the radius R_{500} obtained from X-ray data assuming spherical symmetry. Although the single spherical model seems to slightly overestimate the pressure, both profile look compatible for the considered radius range.

$$P_i = \left(\int_{r_i}^{r_{i+1}} (P_N(r) + P_S(r)) 2\pi r \mathrm{d}r \right) / \left(\pi (r_{i+1}^2 - r_i^2) \right), \tag{2}$$

where P_i is the value of the pressure, P_N and P_S are respectively the profiles of the northern and southern subhalos, r_i and r_{i+1} being the inner and outer radii of the annulus. This allows us to reconstruct the profile available in figure 3. The integration is performed up to the radius R_{500} that is found with the X-ray data. The 1D and average 2D models seem to be slightly compatible, which tends to show that even in the case of an highly disturbed cluster, the spherical model still yields robust results.

6 Conclusions

This analysis shows the challenges related to the complexity of cluster morphologies. In this work, we showed that for the case of the highly disturbed cluster PSZ2 G091, taking into account the merging state of the cluster yields results compatible with the spherical profile. This is a promising result, as the pressure profile impacts the $Y_{500} - M$ relation. To complete this analysis, we plan to perform a full thermodynamical analysis of this cluster, recovering the 2D maps of physical quantities like the temperature and the entropy. This a precondition to assess the impact of the morphology of this cluster on its full mass reconstruction, and generally on cosmological inference using clusters.

Acknowledgements

We would like to thank the IRAM staff for their support during the campaigns. The NIKA2 dilution cryostat has been designed and built at the Institut Néel. In particular, we acknowledge the crucial contribution of the Cryogenics Group, and in particular Gregory Garde, Henri Rodenas, Jean Paul Leggeri, Philippe Camus. This work has been partially funded by the Foundation Nanoscience Grenoble and the LabEx FOCUS ANR-11-LABX-0013. This work is supported by the French National Research Agency under the contracts "MKIDS", "NIKA" and ANR-15-CE31-0017 and in the framework of the "Investissements d'avenir" program (ANR-15-IDEX-02). This work has benefited from the support of the European Research Council Advanced Grant ORISTARS under the European Union's Seventh Framework Programme (Grant Agreement no. 291294). F.R. acknowledges financial supports provided by NASA through SAO Award Number SV2-82023 issued by the Chandra X-Ray Observatory Center, which is operated by the Smithsonian Astrophysical Observatory for and on behalf of NASA under contract NAS8-03060.

References

[1] M. Arnaud et al., A&A**517**, A92 (2010), 0910.1234
[2] M. Cataneo, D. Rapetti, International Journal of Modern Physics D **27**, 1848006 (2018)
[3] M. Cataneo et al., Phys. Rev. D**92**, 044009 (2015), 1412.0133
[4] W.H. Press, P. Schechter, ApJ**187**, 425 (1974)
[5] L. Salvati et al., A&A**614**, A13 (2018), 1708.00697
[6] Í. Zubeldia, A. Challinor, MNRAS**489**, 401 (2019), 1904.07887
[7] Planck Collaboration et al., A&A**594**, A24 (2016), 1502.01597
[8] T.M.C. Abbott et al., Phys. Rev. D**102**, 023509 (2020), 2002.11124
[9] F. Pacaud et al., A&A**620**, A10 (2018), 1810.01624
[10] G.W. Pratt et al., Space Sci. Rev.**215**, 25 (2019), 1902.10837
[11] R.A. Sunyaev, Y.B. Zeldovich, Ap&SS**7**, 3 (1970)
[12] S. Bocquet et al. (2015), 1502.07357
[13] E. Artis et al., A&A**649**, A47 (2021), 2101.02501
[14] A. Fumagalli et al., A&A**652**, A21 (2021), 2102.08914
[15] F. Mayet et al., in *EPJ Web of Conferences 2228* (2020), p. 00017, 1911.03145
[16] Planck Collaboration et al., A&A**594**, A27 (2016), 1502.01598
[17] L. Perotto et al., A&A**637**, A71 (2020), 1910.02038
[18] D. Nagai et al., ApJ**668**, 1 (2007), astro-ph/0703661
[19] Planck Collaboration et al., A&A**550**, A131 (2013), 1207.4061

Galaxy cluster number counts with individual lensing mass estimates: Forecasts for *Euclid*

E. Artis[1,*], J.-B. Melin[2], J. G. Bartlett[2,3], and Calum Murray[1], on behalf of the *Euclid* Consortium

[1]Univ. Grenoble Alpes, CNRS, Grenoble INP, LPSC-IN2P3, 53, avenue des Martyrs, 38000 Grenoble, France
[2]IRFU, CEA, Université Paris-Saclay, F-91191 Gif-sur-Yvette, France
[3]Université de Paris, CNRS, Astroparticule et Cosmologie, F-75006 Paris, France
[4]Jet Propulsion Laboratory, California Institute of Technology, Pasadena, California, USA

Abstract. The ΛCDM model is gradually becoming challenged by observational data. Upcoming cosmological surveys will increase the number of detected galaxy clusters by several orders of magnitude. Therefore, clusters will shortly provide precise cosmological constraints and improve our understanding of structure formation in the Universe. In the following, we present a cluster likelihood based on individual weak lensing mass estimates and forecast *Euclid*'s performances within this framework. We use a matched filter for weak lensing mass estimation and model its characteristics with a set of simulations. We use the *Flagship* N-body simulation to emulate the expected cluster mass distribution of a *Euclid*-like sample and test our statistical framework against it. Finally, we simultaneously constrain the observable-mass relation and the cosmological parameters.

1 Introduction

Galaxy clusters are the most massive virialized structures in the universe. Among the various ways of testing cosmological models with clusters [e.g., 1], their abundance as a function of mass and redshift is one of the primary probes. The precision of current measurements is limited by our understanding of the observable-mass relation [see 2, for a review]. The upcoming years will transform our understanding of cluster cosmology, as the number of detected structures will increase by several orders of magnitude. Next-generation large surveys like *Euclid* [3] will provide an unprecedented quantity of data. In this proceeding, we propose an approach where a mass proxy is estimated for each cluster in the catalog to calibrate the observable mass relation. Individual masses have been introduced in cluster surveys [4], and has been applied in the context of SZ observations by [5]. The match filtering method extracting the lensing masses has been developed in Murray et al., in prep. In this work, we analyze the output of the matched filter and parameterize it to inject the results in a Poisson unbinned likelihood. We show that, in the context of a *Euclid*-like survey, one can use this match filtering approach to constrain the observable-mass relation and the cosmological parameters jointly.

*e-mail: emmanuel.artis@lpsc.in2p3.fr

2 Mass estimation

Mass estimation is the current bottleneck of cluster abundance cosmology. For future surveys, we need to construct robust mass proxies with well understood properties. In the case of *Euclid*, the weak lensing measurements will provide the data required to estimate the dark matter halo mass. Clusters are primarily detected (in the optical) as overdensities of galaxies. One of the algorithms selected by the *Euclid* consortium [6] is *AMICO* [7]. It provides a natural mass proxy as an output, which is the quantity related to the selection function. The weak lensing masses serve to calibrate its relation with the true halo mass.
In this section we briefly describe the two mass proxies that we use to forecast the performance of our framework.

2.1 Lensing mass estimates

From Murray et al., in prep., the approach of the matched filter is the following: assuming a Navarro-Frenk-White (NFW) profile [8], we want to build an estimator of the scale radius r_s of a given cluster in the form

$$\tilde{r}_s = \sum \Delta\Sigma_i \, h_i \,, \tag{1}$$

where $\Delta\Sigma_i$ is the measured excess surface density at a distance r_i from the cluster center and h_i are weights to be determined. From [9], the theoretical excess surface density is proportional to the scale radius of the cluster $\Delta\Sigma_{i,\mathrm{th}} = r_s \tau_i$. We thus have that $\Delta\Sigma_i = \Delta\Sigma_{i,\mathrm{th}} + n_i = r_s \tau_i + n_i$, where n_i is the noise, with the covariance matrix $\mathbf{S} = \mathbf{n}\mathbf{n}^\intercal$. If we want the filter to be unbiased and the noise to be minimal, a good choice of coefficients is

$$\mathbf{h} = \frac{\boldsymbol{\tau}^\intercal \mathbf{S}^{-1}}{\boldsymbol{\tau}^\intercal \mathbf{S}^{-1} \boldsymbol{\tau}} \,. \tag{2}$$

We can then recover the mass $M_L = 4/3\pi \, \Delta \, (c_\Delta \tilde{r}_s)^3 \, \rho_c(z)$, where c_Δ is the concentration and $\rho_c(z)$ is the critical overdensity of the universe at redshift z. This estimator provides a robust summary statistic. Note that the lensing masses explicitly depend on the cosmological model. From the calculation of [9], it can be shown that, within the framework of the matched filter and assuming that the clusters are at low to moderate redshifts ($z < 1$ on average)

$$M_L(\Theta) \propto D_A^{-3}(z|\Theta) \, \rho_c^{-2}(z|\Theta) \,, \tag{3}$$

where D_A is the angular diameter distance and Θ is the set of parameters of the background cosmological model. For readability, we define the function $c(\Theta) \equiv D_A^{-3}(z|\Theta) \, \rho_c^{-2}(z|\Theta)$. Assuming that the masses where computed at a fiducial cosmology Θ_{fid}, we therefore have that the masses at a given cosmology Θ follow $M_L(\Theta) = (c(\Theta)/c(\Theta_{\mathrm{fid}})) \, M_{L,\mathrm{fid}}$.

2.2 Amplitude

Clusters are selected through their signature on the distribution of galaxies in the sky. In this work we focus on results from *AMICO* [7, 10].
This algorithm detects clusters as overdensities of galaxies in the sky, with no colour assumptions. The proxy of the mass is the normalization of the cluster galaxy distribution, called the amplitude A. We assume a power law relation between the amplitude and the mass, $A(M) = A_0 \, (M/M_0)^\alpha$.

3 Likelihood

In this section, we briefly describe our likelihood model. We use an unbinned Poissonian likelihood, where the abundance of observed clusters per unit of lensing mass estimates $M_{\rm L}$, amplitude A and redshift z follows:

$$\frac{\mathrm{d}\tilde{N}}{\mathrm{d}M_{\rm L}\,\mathrm{d}A\,\mathrm{d}z}(\Theta) = \Omega \int_{-\infty}^{+\infty} \mathrm{d}\ln M \,\frac{\mathrm{d}N}{\mathrm{d}\ln M\,\mathrm{d}\Omega\,\mathrm{d}z}(\Theta)\, f(A, M_{\rm L}|M, z, \Theta)\, \chi(A(M), z)\,, \tag{4}$$

where $\mathrm{d}N/\mathrm{d}\ln M\,\mathrm{d}\Omega\,\mathrm{d}z$ is the halo mass function described in [11], $\chi(A(M), z)$ is the fraction of clusters detected at a given mass and redshift, $f(A, M_{\rm L}|M, z, \Theta)$ is the probability distribution function of the two observables A and $M_{\rm L}$, at a given mass and redshift and Ω is the sky coverage of the survey. In our case, we assume that,

$$f(A, M_{\rm L}|M, z, \Theta) = \frac{1}{2\pi\,|\mathbf{C}|^{1/2}} \exp\left[-\frac{1}{2}\left(\begin{pmatrix} A \\ M_{\rm L}\end{pmatrix} - \nu(\Theta)\right)^{\rm T} \mathbf{C}^{-1} \left(\begin{pmatrix} A \\ M_{\rm L}\end{pmatrix} - \nu(\Theta)\right)\right], \tag{5}$$

with the correlation matrix,

$$\mathbf{C}(M, z) = \begin{pmatrix} \sigma_{\rm A}^2(M, z) & \rho\,\sigma_{\rm A}(M, z)\,\sigma_{M_{\rm L}}(M, z) \\ \rho\,\sigma_{\rm A}(M, z)\,\sigma_{M_{\rm L}}(M, z) & \sigma_{M_{\rm L}}^2(M, z)\end{pmatrix}, \tag{6}$$

where ρ is our correlation coefficient assumed to be constant and

$$\nu(\Theta) = \begin{pmatrix} A_0 \left(\frac{(1-b_{\rm A})\,M}{M_0}\right)^{\alpha} \\ \frac{c(\Theta_{\rm fid})}{c(\Theta)}\,(1 - b_{\rm L})\,M \end{pmatrix}, \tag{7}$$

where $b_{\rm A}$ and $b_{\rm L}$ are potential biases, respectively on the amplitude and the lensing proxies. The noise related to the amplitude $\sigma_{\rm A}(M, z)$ is parameterised following [6], while we assume that,

$$\sigma_{M_{\rm L}}^2(M, z) = A_{\rm L}\,\overline{\theta_\Delta}^{\beta}(M, z)\,/\,n_{\rm gal}(z) + \sigma_{\rm int}^2(M, z)\,/\,n_{\rm gal}(z)\,, \tag{8}$$

where $n_{\rm gal}(z)$ is the number of background galaxies at redshift z, $\sigma_{\rm int}$ is the intrinsic scatter, and $\overline{\theta_\Delta}$ is the mean angular size of a cluster. The fraction of detected objects is assumed to follow,

$$\chi(A, z) = \mathbb{P}(q > q_{\rm obs}) = \frac{1}{\sqrt{2\pi}} \int_{q_{\rm obs}}^{+\infty} \mathrm{d}q\, \exp\left(-\frac{(q - A/\sigma_{\rm A}^2(A, z))}{2}\right), \tag{9}$$

where q is the signal to noise ratio (S/N), and $q_{\rm obs}$ is the detection treshold.

4 Simulation and mock generation

The impact of the systematics that we will have to face must be measured at the precision of a *Euclid*-like survey, which means $15,000$ deg^2, $z \in [0, 2]$, reliable lensing information and baryonic physics correctly incorporated. This represents an enormous volume, and currently, no data set possesses these characteristics. To tackle this issue, we have chosen to use three different simulations, each one having the required properties that we are looking for. The *RayGalGroupSims* [12], are used to derive the lensing properties. For the selection function, and the performance of the detection algorithm, we refer to the the results of [6]. Finally, we used the *Flagship* dark matter halo catalog, the official *Euclid* simulation, as a baseline to reconstruct a mock survey that is as close as possible from a *Euclid*-like data set. The *RayGalGroupSims* are a set of N-body simulations that are dedicated to ray-tracing. The

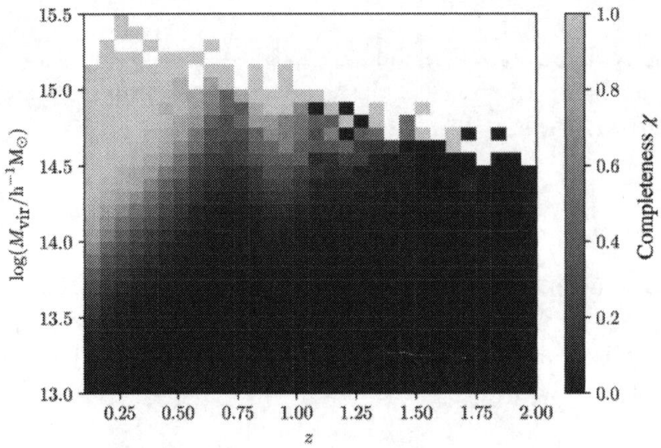

Figure 1: Distribution of the clusters selected in the Flagship dark matter halo catalog. We set the S/N (q_{obs} = 6, from equation 9), so that we remove most of the low mass systems for which the uncertainties are high, especially regarding the selection function. The color scale represents the fraction of detected clusters.

main interest of these simulations is their spatial resolution (5 h^{-1}kpc), and the excellence of the lensing calculations. We use these simulations to test and constrain the relation on the scatter of the lensing mass estimates introduced with equation 8. They extend up to $z = 0.5$, so we assume that our results are valid for higher redshifts. Precisely, we fit the normalization and the power law of the relation. In this study, we fix the value of the intrinsic scatter σ_{int} to 0.3. We do verify that our fiducial value corresponds to a reasonable fit of the masses. The results are presented in table 1.
In order to create a catalog of objects similar to the one expected from *Euclid*, we select a spherical region of 15,000 deg^2, in which we select all the halos down to a virial mass of $M_{vir,cut} = 10^{13}\ h^{-1}M_{\odot}$. Starting from these masses, we assign to each halo an amplitude, A, and a lensing mass M_L using the distribution presented in section 3. From our dark matter halo catalog, we select the objects that would be detected by *AMICO* according to equation 9. We fix the S/N threshold in agreement with the expected number of objects to be detected with *Euclid*. We chose an S/N threshold of q_{obs} = 6, which gives provides us with a catalog of 56,549 clusters. The clusters' distribution is represented in figure 1.

5 Performance forecasts

We first consider a standard ΛCDM cosmology. The full set of fitted parameters is described in table 1. Figure 2 shows the $\Omega_m - \sigma_8$ plane for the cases when all the systematics are entirely known and when we freed all the nuisance parameters. As illustrated, it is clear that a joint fitting of the mass-observable relation and cosmological parameters is possible and provides competitive results, providing that one has statistically robust lensing mass estimates. Then we can compare these results to the one presented by [3]. Our contours are comparable to what they are presenting for their number counts only analysis. Figure 2 shows the results obtained for a $w_0 w_a$CDM cosmology.

Table 1: True values adopted for the parameters, priors and results with 1σ uncertainties. Some parameters are left blank in the posterior section when they are not constrained. The dark energy equation of state parameters, are in blue, with the following parameterization: $w(a) = w_0 + w_a(1 - a)$ [13]. The true values, for the lensing mass estimates are the best fit values introduced in equation 8. Note that although it is present in our table, the intrinsic scatter is not fitted, but fixed. As expected, the value of the power β is close to the intuitive value -2, as the errors should be proportional to the inverse of the angular area covered by the cluster.

Parameter	Meaning	True value	Prior	Posteriors
		Cosmological parameters		
Ω_m	Matter density	0.319	[0.1,0.5]	0.322 ± 0.0019
σ_8	r.m.s. matter fluctuation	0.83	[0.6,0.9]	0.835 ± 0.0056
H_0	Hubble constant ($\mathrm{km \cdot s^{-1} \cdot Mpc^{-1}}$)	67	$\mathcal{N}(67, 1000)$	65.28 ± 1.21
Ω_b	Baryonic density	0.049	$\mathcal{N}(0.049, 0.0026)$ and $\Omega_b > 0$	-
n_s	Spectral index	0.96	[0.87,1.07]	-
w_0	First parameter	-1	$[-2.5, 0]$	-0.967 ± 0.034
w_a	Second parameter	0	$[-2, 2]$	0.19 ± 0.12
		Halo Mass Function parameters		
a_0	High mass cut-off	0.938	Covariance matrix fitted on the *Flagship*	-
a_z	Redshift dependence a_0	-0.12		-
p	Shape at low masses	-0.2		-
$A_{0,\mathrm{MF}}$	Normalization	0.31		-
		Richness-mass parameters		
A_0	Normalization	0.5	[0.05,5]	0.495 ± 0.0045
α	Slope	0.5	[0.05,2]	0.5 ± 0.0012
		Lensing mass estimates uncertainties		
A_L	Normalization of the scatter	0.836	[0.001,4]	0.758 ± 0.042
β	Slope	-2.09	$[-6, 0]$	-2.2 ± 0.136
σ_{int}	Intrinsic scatter	0.3	[0.001,2]	0.35 ± 0.025
		Correlation richness-lensing mass estimates		
ρ	-	0.6	[0,1]	0.599 ± 0.0037
		Bias on the richness		
b_λ	-	0	[0,0.99]	-
		Bias on the lensing masses		
b_L	-	0	Fixed	-

6 Conclusion

In this work, we presented a method designed for the analysis of *Euclid*'s galaxy cluster sample. This framework consists in the computation of individual lensing mass estimates, for each cluster in the catalog. We use numerical simulations to reproduce the expected properties of *Euclid*'s cluster sample. Our final cluster catalog is based on the *Flagship* DM halo catalog and comprises approximately 6×10^4 objects. We model the cosmological dependence of the lensing mass estimates, which is taken into account in the final estimation. We show that it is possible to calibrate the cosmological parameters and the parameters of the observable-mass relation jointly. Our analysis demonstrates the potential of individual masses analyses in the context of large optical surveys. We assumed analytical expressions for the selection function, observable-mass relation, and cosmological dependence of the lensing mass estimates, relying on a set of numerical simulations. Each one of these elements has to be refined for the final analysis and is under investigation.

Acknowledgements

The authors acknowledge the Euclid Consortium, the European Space Agency, and a number of agencies and institutes that have supported the development of *Euclid*, in particular the Academy of Finland, the Agenzia Spaziale Italiana, the Belgian Science Policy, the Canadian

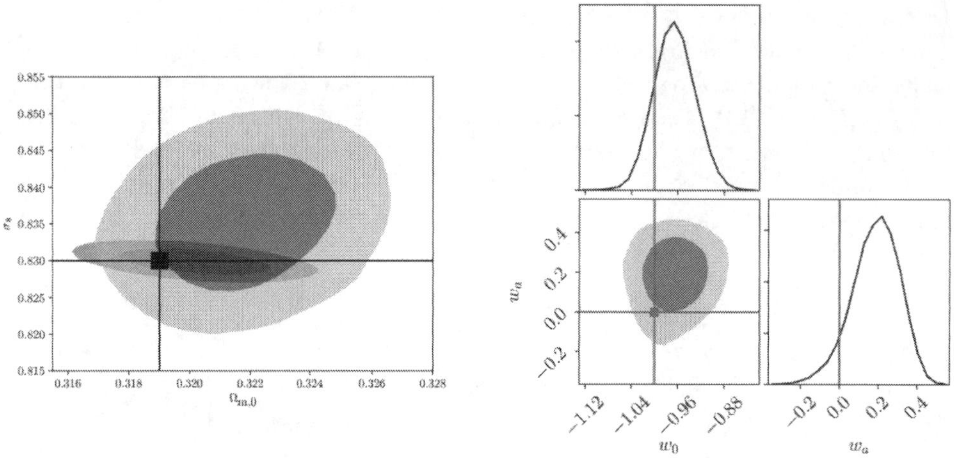

Figure 2: **Left**: 68 and 95% constraints obtained with our analysis, in case of a ΛCDM cosmology. The grey ellipse is the realistic case for which all the identified sources of systematics listed in table 1 are free. With the red ellipse, we also provide a reference case, for which all the systematics are fixed. **Right**: Constraints obtained on the dark energy equation of state parameters. The position of the best fit compared to the confidence ellipses in the grey case is an expected feature that results from the statistical production of the catalogs.

Euclid Consortium, the French Centre National d'Etudes Spatiales, the Deutsches Zentrum für Luft- und Raumfahrt, the Danish Space Research Institute, the Fundação para a Ciência e a Tecnologia, the Ministerio de Economia y Competitividad, the National Aeronautics and Space Administration, the National Astronomical Observatory of Japan, the Netherlandse Onderzoekschool Voor Astronomie, the Norwegian Space Agency, the Romanian Space Agency, the State Secretariat for Education, Research and Innovation (SERI) at the Swiss Space Office (SSO), and the United Kingdom Space Agency. A complete and detailed list is available on the *Euclid* web site (`http://www.euclid-ec.org`).

References

[1] S.W. Allen et al., ARA&A**49**, 409 (2011)
[2] G.W. Pratt et al., Space Sci. Rev.**215**, 25 (2019)
[3] B. Sartoris et al., MNRAS**459**, 1764 (2016)
[4] S. Bocquet et al., ApJ**878**, 55 (2019)
[5] Í. Zubeldia, A. Challinor, MNRAS**489**, 401 (2019)
[6] Euclid Collaboration et al., A&A**627**, A23 (2019)
[7] F. Bellagamba et al., MNRAS**473**, 5221 (2018)
[8] J.F. Navarro et al., ApJ**462**, 563 (1996)
[9] C.O. Wright, T.G. Brainerd, The Astrophysical Journal **534**, 34 (2000)
[10] M. Maturi et al., MNRAS**485**, 498 (2019)
[11] G. Despali et al., MNRAS**456**, 2486 (2016)
[12] M.A. Breton et al., MNRAS**483**, 2671 (2019)
[13] M. Chevallier, D. Polarski, International Journal of Modern Physics D **10**, 213 (2001)

The mm-to-cm SED of spiral galaxies

Synergies between NIKA2 and SRT instruments

Simone Bianchi[1,*], *Matteo* Murgia[2], *Andrea* Melis[2], *Viviana* Casasola[3], *Maud* Galametz[4], *Frédéric* Galliano[4], *Federica* Govoni[2], *Anthony* Jones[5], *Suzanne* Madden[4], *Rosita* Paladino[3], *Emmanuel* Xilouris[6], and *Nathalie* Ysard[5]

[1]INAF-Osservatorio Astrofisico di Arcetri, L. E. Fermi 50, 50125, Firenze, Italy
[2]INAF-Osservatorio Astronomico di Cagliari, Via della Scienza 5, 09047 Selargius (CA), Italy
[3]INAF-Istituto di Radioastronomia-Via P. Gobetti, 101, 40129 Bologna, Italy
[4]AIM, CEA, CNRS, Université Paris-Saclay, Université Paris Diderot, Sorbonne Paris Cité, 91191 Gif-sur-Yvette, France
[5]Université Paris-Saclay, CNRS, Institut d'Astrophysique Spatiale, 91405 Orsay, France
[6]National Observatory of Athens, Institute for Astronomy, Astrophysics, Space Applications and Remote Sensing, Ioannou Metaxa and Vasileos Pavlou, 15236 Athens, Greece

Abstract. The mm-to-cm range of the Spectral Energy Distribution of spiral galaxies remains largely unexplored. Its coverage is required to disentangle the contribution of dust emission, free-free and synchrotron radiation and can provide constraints on dust models, star-formation rates and ISM properties. We present the case for a synergy between NIKA2 observations of nearby spirals and those from planned and current instrumentation at the Sardinia Radio Telescope, and report on a pilot K-band program to search for Anomalous Microwave Emission, an elusive emission component which is presumably related to dust.

1 Galactic SEDs in the mm-to-cm range

Our knowledge of the Spectral Energy Distribution (SED) of nearby galaxies has greatly improved in recent times, thanks to multi-wavelength surveys and in particular to the *Herschel* Space Observatory. *Herschel* has extended the SED coverage to far-infrared (FIR) and submm wavelengths, a range where *dust*, i.e. solid grains in the interstellar medium (ISM), emits the radiation it has absorbed from starlight (~1/4 of the total luminosity for late type galaxies; [1]). Using a dust model and fitting tools such as, e.g., CIGALE [2] and HerBIE [3], a wealth of information can be extracted from the SED: star-formation rates (SFRs), stellar and dust masses, gas-to-dust ratios, etc. ([4–7], just to mention a few recents works). Figure 1 shows an example of the current SED knowledge: it uses data from *DustPedia* [8], a database collecting photometry and imagery for all the large ($D_{25} > 1'$) and nearby ($d < 30$ Mpc) galaxies observed by *Herschel*; and mid radio continuum flux densities from [9].

A few gaps still remain: in particular, data for $15 \leq \nu/\mathrm{GHz} \leq 300$ (mm-to-cm range) is limited to the Milky Way (MW) and a few bright objects. The high-frequency side of this gap, at $\nu \approx 200$ GHz, is still dominated by dust emission: radiation could come from

*e-mail: simone.bianchi@inaf.it

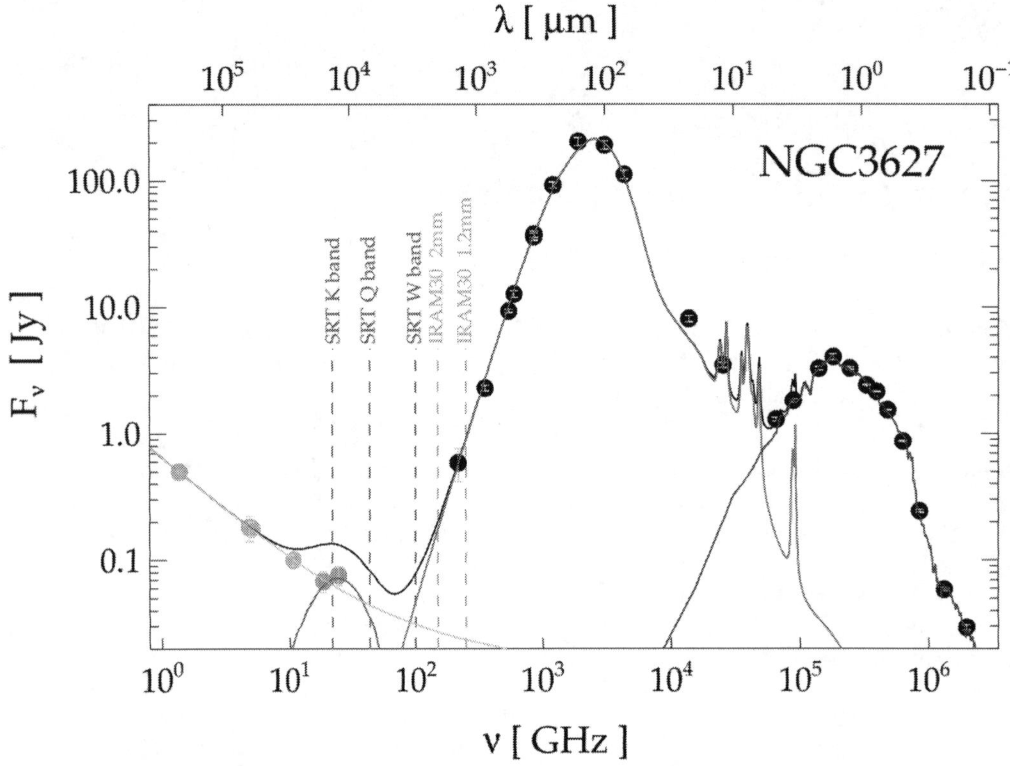

Figure 1. The SED of NGC3627. The total SED (black) results from the sum of: 1) CIGALE fits of the stellar (blue line) and dust (red line) components to the DustPedia datapoints (black dots; [4, 10]); 2) synchrotron and free-free radiation fits to the mid-radio continuum emission (green line and grey dots; [9]); 3) a prediction for AME (purple line), obtained using a model spectral template [11] normalised to the MW AME(30 GHz)/IR(100 μm) ratio [12]. NIKA2 and the SRT instrumentation can cover the mm-to-cm gap. DustPedia datapoints at mm wavelengths are from (unresolved) *Planck* observations, while the purple datapoints show the results from our pilot SRT program in the K band. DustPedia images and photometry are available at `http://dustpedia.astro.noa.gr/`.

the Rayleigh-Jeans tail of the warmer dust, or be due to colder grains undetected from FIR-submm observations. Large masses of cold grains, and/or variations of the dust absorption cross-section at longer wavelengths could be responsible for the *submm excess* detected in a few objects [13]. Thus, observations at these frequencies are important to constrain the properties of dust, which are still uncertain. Current dust models are made to reproduce the observations of extinction, emission and depletion at high galactic latitude in the MW [14]. Various dust models exist, leading to different estimates of quantities such as the dust mass (see, e.g., [4]). Furthermore, recent lab measurements are questioning the validity of the optical properties generally assumed for the grain materials [15]. Finally, there are indications that the bulk of dust in a galaxy might have different properties than those of the diffuse MW medium [16, 17]. The NIKA2 camera at the IRAM 30m radiotelescope, and in particular the IMEGIN guaranteed-time project observing large nearby galaxies at millimetre wavelengths [18, 19], will help in shedding light on these issues.

At lower frequencies, emission components other than dust need to be taken into account. Already in NIKA2 maps at 2 mm, about 10% of the SED is due to free-free and synchrotron radiation; their contribution increases, and that of dust reduces, in the mid and low-frequency side of the $15 \leq \nu/\mathrm{GHz} \leq 300$ gap (see, e.g.. [18]). Free-free emission gives the maximum contribution in the mid of the frequency gap. Being due mostly to free electrons, this radiation can be directly related to ionising photons and thus provides a direct estimate of the star-formation rate, unaffected by the dust extinction corrections needed for estimates based on the UV continuum and the hydrogen recombination lines in the optical [20]; or by the uncertain fraction of dust heating due to young stars, for estimates based on infrared dust emission [4]. The dominant emission toward the low-frequency side of the gap is due to synchrotron radiation, produced by cosmic ray electrons accelerated to relativistic velocities in supernova remnants and spiralling along the magnetic field in the ISM [20]. Since all of these emission mechanisms overlap in frequency, full SED coverage is needed to solve the degeneracies and estimate the exact contribution of each component.

An opportunity to extend the knowledge of the SED beyond the frequencies accessible to the NIKA2 camera is soon going to be offered at the Sardinia Radio telescope (SRT [21]; see Fig. 1). In the incoming year, the MISTRAL camera, operating in the W Band at $\sim$ 90 GHz, will be installed at SRT [22], as well as a 19-feed Q band receiver at $\sim$ 40 GHz [23]. Thanks to the large SRT antenna, 64m, these instruments will allow us to map nearby galaxies with HPBW=12″ and 28″, respectively, resolutions close to those of NIKA2 and the *Herschel* instruments. Already available, a 7-feed spectro-polarimetric receiver allows access to the K band at $\sim$ 22 GHz [24].

2 Anomalous Microwave Emission

A further emission component might need to be considered: the Anomalous Microwave Emission (AME). First detected in the MW in the mid-90s by Cosmic Microwave Background experiments, AME is seen as an excess over the expected free-free and synchrotron emissions, peaking at $\approx$ 30 GHz and correlating with dust emission at larger frequencies (see [25] for a review). AME is seen both in isolated clouds [26], as well as in the diffuse MW medium where its amplitude, relative to the 100 μm surface brightness, is AME(30 GHz)/I(100 μm) $\approx$ 1/3000 [12].

The most accredited theory explains AME with electric dipole emission from rapidly rotating grains [27]. For typical ISM conditions, these grains must have size of 10^{-9} m, in order to spin fast and emit at the AME frequencies: they could be associated with the macro-molecules carriers of the PAH features in the mid-infrared spectrum [28], or with nanosilicates [29]. Alternatively, AME could be due to an enhancement of the absorption/emission cross sections at microwave frequencies, either due to magnetic inclusions in grains [30] or the amorphous state of the grain material [31, 32]. AME can thus set constraints on the grain size distribution (and ISM conditions), under the spinning grain hypothesis, and/or on the grain composition, for the thermal emission hypothesis. Though these models can all explain AME, yet there is no concluding evidence for its physical mechanism: some authors find a good correlation with small-grain emission, thus favouring spinning dust [33–35]; others support the thermal hypothesis, reporting a better correlation with the total dust emission [36].

While AME is almost ubiquitous in the MW, it has been detected with high significance in three extragalactic sources only. AME has been found with pointed observations in just one star-forming region in NGC 6946 [37] and one compact radio-source in NGC 4725 ([38], even though, in this case, it does not appear to be associated with dust [39]). On the global SED, strong evidence for AME has been recently found in the Andromeda galaxy [11], with an AME component compatible with that in the MW.

3 A pilot K-band program at SRT

As a preliminary project, at the beginning of 2021 we observed four galaxies at SRT in the K band. They are: NGC 3627 (M 66), NGC 4254 (M 99), NGC4736 (M 94) and NGC5055 (M63). These galaxies are part of the NIKA2 IMEGIN sample, of the DustPedia database and of the KINGFISHER project. DustPedia provides homogenised images, full UV-to-submm global photometry and SED models [4, 10], while KINGFISHER provides radio continuum data and models at frequencies between 1 and 10 GHz [9]. The first objective of the observations was the detection of AME, by comparing the SRT data with the available fits for the free-free and synchrotron components. Ultimately, the targets could become the first nucleus for a study of the complete UV-to-radio SED, once the NIKA2 data and the future SRT instruments cover the frequency gap between the submm and the K band (Fig. 1).

The K-band receiver was used together with the SArdinia Roach2-based Digital Architecture for Radio Astronomy back end (SARDARA; [40]). We selected two 1.2 GHz sub-bands at the edges of the K-band, at 18.6 and 24.6 GHz (with HPBW = $57''$ and $45''$, respectively). These sub-bands were covered by 1.46 MHz frequency channels in full Stokes mode. Observations were done on-the-fly over an area of $16' \times 16'$, centered on each galaxy, with a scan speed $2'$/s and scan separation HPBW/3. The area was alternatively covered by scanning along orthogonal RA and Dec directions, for an efficient removal of the scan noise. Each coverage took about 20 minutes and was repeated 15 to 40 times, depending on the target, frequency and observing conditions.

The spectral cubes were reduced using the proprietary Single-dish Spectral-polarimetry Software (SCUBE; [41]) that accounts for calibration, RFIs removal, baseline removal and map-making. Maps for total and polarised intensity were made with pixel size $15''$. Total intensity maps are shown in Fig. 2. Even though some mechanisms proposed for AME are expected to produce polarised emission [25], we detected none.

4 Preliminary results

Using the apertures shown in Fig. 2, we measured the global flux density for our targets at 18.6 GHz and 24.6 GHz. These new datapoints were analized together with the mid-radio continuum from KINGFISHER. For NGC 4254 and NGC 4736, the SRT observations are consistent with the synchrotron and free-free fits to data for $\nu \leq 10$ GHz. For NGC 3627 and NGC 5055, the 18.6 GHz flux density is still compatible with the fit in [9]; the 24.6 GHz flux density, instead, is in excess of the fit extrapolation and larger than that at 18.6 GHz (see Fig. 1 for NGC 3627).

With the current data, it is difficult to asses if the 24.6 GHz excess seen in NGC 3627 and NGC 5055 is due to an additional component, such as AME, or if it is just a statistical fluctuation. Observations at larger frequencies, like those that will be possible with the forthcoming Q-band receiver for SRT, are needed to confirm this finding. In fact, a model including synchrotron and free-free only can be fit to the entire $\nu \leq 24.6$ Ghz range, resulting in minor modifications with respect to the fit for $\nu \leq 10$ GHz. The new parameters for the two radio components are still consistent with the spread of properties found in the KINGFISHER sample. From the free-free component, estimates of the global SFR can be obtained, which are consistent with those from the UV-to-submm SED fitting, though affected by large uncertainties [42]. Again, more precise determinations will need to wait for data at higher frequencies, or better to the full coverage of the $15 \leq \nu/\mathrm{GHz} \leq 300$ gap, for a simultaneous analysis of all the emission components, dust included.

Currently, we can only put upper limits on the AME contribution at 30 GHz. We find that AME(30 GHz)/I(100 μm) is at least a factor 5 smaller that the values measured in

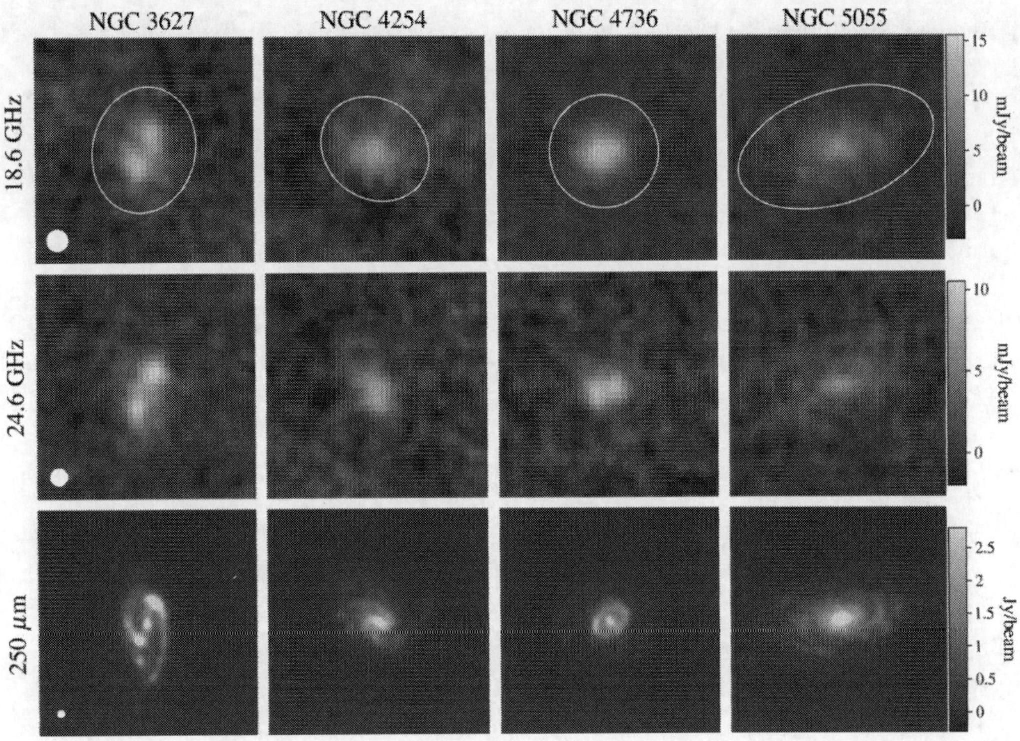

Figure 2. Total intensity maps for our targets (top and middle panels). For reference, 250 μm *Herschel*-SPIRE maps from DustPedia are shown in the bottom panels. We also show the areas used for photometry and the HPBW. Each panel is $10' \times 10'$.

M 31 and the MW (see the comparison between predicted AME and SRT observations in Fig. 1). This does not necessarily imply that the AME properties are different. In fact, the AME(30 GHz)/I(100 μm) ratio is an ill-defined characterisation of the AME emissivity, as it is expected to be lower in objects where the dust temperature is higher [43]. Indeed, the average dust temperatures of our galaxies, as well as those in three other targets with no AME detection [44] are higher than those of M 31 and the MW. When the AME emissivity is derived in terms of the column density of dust, available from the DustPedia analysis, the upper limits in our galaxies become consistent with the AME detections. Apparently, AME may not be detected in our targets and those in [44] because of their relatively stronger radio-continuum than in the Milky Way and Andromeda galaxy [42]. Future searches for AME might need to consider this in the selection of targets.

Acknowledgements

We acknowledge support from the INAF mainstream 2018 program "Gas-DustPedia: A definitive view of the ISM in the Local Universe" and from grant PRIN MIUR 2017-20173ML3WW_001.

References

[1] Bianchi, S., De Vis, P., Viaene, S., et al., A&A **620**, A112 (2018)

[2] Boquien, M., Burgarella, D., Roehlly, Y., et al., A&A **622**, A103 (2019)
[3] Galliano, F., MNRAS , **476**, 1445 (2018)
[4] Nersesian, A., Xilouris, E. M., Bianchi, S., et al., A&A **624**, A80 (2019)
[5] Aniano, G., Draine, B. T., Hunt, L. K., et al., ApJ **889**, 150 (2020)
[6] De Looze, I., Lamperti, I., Saintonge, A., et al., MNRAS **496** 3668 (2020)
[7] Galliano, F., Nersesian, A., Bianchi, S., et al., A&A **649**, A18 (2021)
[8] Davies, J. I., Baes, M., Bianchi, S., et al., PASP **129**, 044102 (2017)
[9] Tabatabaei, F. S., Schinnerer, E., Krause, M., et al., ApJ **836**, 185 (2017)
[10] Clark, C. J. R., Verstocken, S., Bianchi, S., et al., A&A **609**, A37 (2018)
[11] Battistelli, E. S., Fatigoni, S., Murgia, M., et al., ApJL **877**, L31 (2019)
[12] Planck Collaboration, A&A **594**, A25 (2016)
[13] Galliano, F., Galametz, M., Jones, A. P., ARA&A **56**, 673 (2018)
[14] Compiègne, M., Verstraete, L., Jones, A., et al., A&A **525**, A103 (2011)
[15] Demyk, K., Meny, C., Lu, X. H., et al., A&A **600**, A123 (2017)
[16] Bianchi, S., Casasola, V., Baes, M., A&A **631**, A102 (2019)
[17] Clark, C. J. R., De Vis, P., Baes, M., et al., MNRAS **489**, 5256 (2019)
[18] Katsioli, S., et al., these proceedings
[19] Ejlali, G., et al., these proceedings
[20] Murphy, E. J., Condon, J. J., Alberdi, A., et al., ASP Conference Series **517**, 421 (2018)
[21] Prandoni, I., Murgia, M., Tarchi, A., et al., A&A **608**, A40 (2017)
[22] D'Alessandro, G., et al., these proceedings
[23] Orfei, A., in RadioNet Workshop on Future Trends in Radio Astronomy Instrumentation, MPIfR, Bonn (2020)
[24] Orfei, A., Carbonaro, L., Cattani, A., et al., IEEE Antennas and Propagation Magazine **52**, 62 (2010)
[25] Dickinson, C., Ali-Haïmoud, Y., Barr, A., et al., New A Rev. **80**, 1 (2018)
[26] Planck collaboration, A&A **565**, A103 (2014)
[27] Draine, B. T., Lazarian, A., ApJ **494**, L19 (1998a)
[28] Draine, B. T., Lazarian, A., ApJ **508**, 157 (1998b)
[29] Hensley, B. S., Draine, B. T., ApJ, **836**, 179 (2017)
[30] Draine, B. T., Lazarian, A., ApJ **512**, 740 (1999)
[31] Jones, A. P., A&A **506**, 797 (2009)
[32] Nashimoto, M., Hattori, M., Génova-Santos, R., et al., PASJ **72**, 6 (2020)
[33] Ysard, N., Miville-Deschênes, M. A., Verstraete, L., A&A **509**, L1 (2010)
[34] Harper, S. E., Dickinson, C., Cleary, K., MNRAS **453**, 3375 (2015)
[35] Bell, A. C., Onaka, T., Galliano, F., et al., PASJ **71**, 123 (2019)
[36] Hensley, B. S., Draine, B. T., Meisner, A. M., ApJ **827**, 45 (2016)
[37] Murphy, E. J., Helou, G., Condon, J. J., et al., ApJL **709**, L108 (2010)
[38] Murphy, E. J., Linden, S. T., Dong, D., et al., ApJ **862**, 20 (2018)
[39] Murphy, E. J., Hensley, B. S., Linden, S. T., et al., ApJL **905**, L23 (2020)
[40] Melis, A., Concu, R., Trois, A., et al., Journal of Astronomical Instrumentation **7**, 1850004 (2018)
[41] Murgia, M., Govoni, F., Carretti, E., et al., MNRAS **461**, 3516 (2016)
[42] Bianchi, S., et al., in preparation
[43] Tibbs, C. T., Paladini, R., Dickinson, C., Advances in Astronomy **2012**,124931 (2012)
[44] Peel, M. W., Dickinson, C., Davies, R. D., et al., MNRAS **416**, L99 (2011)

Searching for high-z DSFGs with NIKA2 and NOEMA

L. Bing[12,*], *R.* Adam[3], *P.* Ade[4], *H.* Ajeddig[5], *P.* André[5], *E.* Artis[1], *H.* Aussel[5], *A.* Beelen[6], *A.* Benoît[7], *S.* Berta[8], *M.* Béthermin[12], *O.* Bourrion[1], *M.* Calvo[7], *A.* Catalano[1], *M.* De Petris[9], *F.-X.* Désert[2], *S.* Doyle[4], *E. F. C.* Driessen[8], *A.* Gomez[10], *J.* Goupy[7], *F.* Kéruzoré[1], *C.* Kramer[11], *B.* Ladjelate[11], *G.* Lagache[12], *S.* Leclercq[8], *J.-F.* Lestrade[13], *J.-F.* Macías-Pérez[1], *A.* Maury[5], *P.* Mauskopf[4,16], *F.* Mayet[1], *A.* Monfardini[7], *M.* Muñoz-Echeverría[1], *R.* Neri[8], *A.* Omont[14], *L.* Perotto[1], *G.* Pisano[4], *N.* Ponthieu[2], *V.* Revéret[5], *A. J.* Rigby[4], *A.* Ritacco[6,15], *C.* Romero[15], *H.* Roussel[14], *F.* Ruppin[15], *K.* Schuster[8], *S.* Shu[17], *A.* Sievers[11], *C.* Tucker[4], and *R.* Zylka[8]

[1]Univ. Grenoble Alpes, CNRS, Grenoble INP, LPSC-IN2P3, 53, avenue des Martyrs, 38000 Grenoble, France
[2]Univ. Grenoble Alpes, CNRS, IPAG, 38000 Grenoble, France
[3]LLR (Laboratoire Leprince-Ringuet), CNRS, École Polytechnique, Institut Polytechnique de Paris, Palaiseau, France
[4]School of Physics and Astronomy, Cardiff University, Queen's Buildings, The Parade, Cardiff, CF24 3AA, UK
[5]AIM, CEA, CNRS, Université Paris-Saclay, Université Paris Diderot, Sorbonne Paris Cité, 91191 Gif-sur-Yvette, France
[6]Institut d'Astrophysique Spatiale (IAS), CNRS, Université Paris Sud, Orsay, France
[7]Institut Néel, CNRS, Université Grenoble Alpes, France
[8]Institut de RadioAstronomie Millimétrique (IRAM), Grenoble, France
[9]Dipartimento di Fisica, Sapienza Università di Roma, Piazzale Aldo Moro 5, I-00185 Roma, Italy
[10]Centro de Astrobiología (CSIC-INTA), Torrejón de Ardoz, 28850 Madrid, Spain
[11]Instituto de Radioastronomía Milimétrica (IRAM), Granada, Spain
[12]Aix Marseille Univ, CNRS, CNES, LAM (Laboratoire d'Astrophysique de Marseille), Marseille, France
[13]LERMA, Observatoire de Paris, PSL Research University, CNRS, Sorbonne Université, UPMC, 75014 Paris, France
[14]Institut d'Astrophysique de Paris, CNRS (UMR7095), 98 bis boulevard Arago, 75014 Paris, France
[15]Department of Physics and Astronomy, University of Pennsylvania, 209 South 33rd Street, Philadelphia, PA, 19104, USA Laboratoire de Physique de l'École Normale Supérieure, ENS, PSL Research University, CNRS, Sorbonne Université, Université de Paris, 75005 Paris, France Kavli Institute for Astrophysics and Space Research, Massachusetts Institute of Technology, Cambridge, MA 02139, USA
[16]School of Earth and Space Exploration and Department of Physics, Arizona State University, Tempe, AZ 85287, USA
[17]Caltech, Pasadena, CA 91125, USA

Abstract. As the possible progenitors of passive galaxies at z=2-3, dusty star-forming galaxies (DSFGs) at z>4 provide a unique perspective to study the formation, assembly, and early quenching of massive galaxies in the early Universe. The extreme obscuration in optical-IR makes (sub)mm spectral scans the

*e-mail: longji.bing@lam.fr

most universal and unbiased way to confirm/exclude the high-z nature of candidate dusty star-forming galaxies. We present here the status of the NIKA2 Cosmological Legacy Survey (N2CLS), which is the deepest wide-area single-dish survey in the millimeter searching for high-z DSFGs. We also introduce a joint-analysis method to efficiently search for the spectroscopic redshift of high-z DSFGs with noisy spectra and photometric data and present its success in identifying the redshift of DSFGs found in NIKA2 science verification data.

1 Introduction

In the past decades, optical-IR deep-field observations have contributed to the studies on galaxy assembly in the early Universe [1]. The rest-frame UV-optical emission from young stars is widely used to constrain the cosmic star formation rate density (SFRD) [2], one of the key quantities to calibrate cosmological simulations and galaxy evolution models [3]. However, UV emission is also heavily affected by the extinction caused by dusty ISM. Dust extinction becomes increasingly prevalent in massive high-z galaxies, making the majority of them almost invisible even in near-IR. Thus, the SFRD and space density of massive galaxies might be underestimated in studies with only optical and near-IR data [4].

While most of the UV photons are absorbed, their energy heat the dusty ISM and is re-emitted mostly in far-IR. This makes dusty star-forming galaxies (DSFGs) bright in far-IR to the millimeter and constitutes the majority of sources in blind ALMA and single-dish observations. Complementary to optical and near-IR observations, far-IR to millimeter surveys provide additional constraints on total SFRD and the rise of massive galaxies at z<3 [5]. However, SFRDs measured at z>4 still differ from each other by more than a factor of ten [6–8]. This is mainly due to the difficulties in defining a large unbiased high-z DSFG sample and properly reconstructing the infrared luminosity function (IRLF). Current ALMA surveys [6, 9] uncover a few blindly detected $z > 4$ sources. However, their small area limits the sample size, hence the bright end of the IRLF is still poorly sampled.

To efficiently uncover a large sample of these rare sources, large area blind surveys are required. Nowadays, such surveys could only be achieved by large field-of-view continuum cameras on single-dish telescopes, like SCUBA-2 on Jame Clark Maxwell Telescope and NIKA2 on IRAM 30m [10–13]. However, the large beam size of these data and the high frequency of optical/radio dark sources make it difficult to identify their counterparts and measure their redshifts, infrared luminosity (L_{IR}), and other properties. NOEMA and ALMA interferometers provide the best resolution and sensitivity at mm wavelengths to measure the accurate positions and detect the spectral lines emitted by these sources.

In this paper, we first present the current status of the NIKA2 Cosmological Legacy Survey (N2CLS), an ongoing NIKA2 guaranteed time observation (GTO) large program searching for large samples of high-z DSFGs. We also introduce a new method to search for the redshift of DSFGs combining broadband photometry and mm spectral scan.

2 Basic Information and Status of N2CLS

N2CLS is a NIKA2 GTO large program observing 2 deep fields in the northern/equatorial sky: GOODS-N and COSMOS. The observations are designed to have good statistics on both bright and faint sources. The survey in GOODS-N covers a relatively narrow area of 160 arcmin^2 but is designed to approach the confusion limit of the IRAM 30m telescope. The survey in COSMOS covers a much larger area of ~1100 arcmin^2 with a shallower depth, to get a larger sample of bright sources and very high redshift DSFGs. These two fields

Table 1. Basic information of N2CLS and other single-dish (sub)mm surveys in GOODS-N and COSMOS. For each survey, the root mean square (RMS) noise is normalized to 1.2/2mm using the average IR SED of star-forming galaxies at z=3 [14]. For NIKA2, the RMS of each band is obtained with current data in hands.

Field	Survey	ν_{obs} (GHz)	Area (arcmin^2)	RMS (mJy/beam)
GOODS-N	AzTEC	273	245	1.06
	SCUBA2	353	~140	~0.16
	GISMO	150	38	0.14
	N2CLS	255 & 150	160	0.19 & 0.057
COSMOS	AzTEC-ASTE	273	2592	1.26
	MAMBO	255	400	1.00
	S2COSMOS	353	5760	~0.47
	GISMO	150	250	0.23
	N2CLS	255 & 150	1100	0.55 & 0.18

benefit from the multi-wavelength coverage by deep observations from the X-ray to radio. The sources detected by NIKA2 are cross-matched to existing catalogs across the whole electromagnetic spectrum to identify their counterparts, derive their redshift, and study their physical properties.

We have already completed 85.2h observations on GOODS-N and 82.9h on COSMOS, by March 2021. The data are reduced by PIIC, the data reduction pipeline supported by IRAM[15], as well as the NIKA2 collaboration internal pipeline. Table. 1 shows the current noise level in the central region of N2CLS fields compared to other (sub)mm surveys in the same fields. The data available so far in GOODS-N have already surpassed the depth of any other millimeter surveys performed with AzTEC [16], MAMBO [17] and GISMO [18], and they match with the deepest SCUBA2 survey at higher frequency [19]. 62/25 sources are detected with SNR>4 at 1.2/2mm. As for COSMOS, the survey has also been the deepest single-dish millimeter survey compared to AzTEC [20], MAMBO [21] and GISMO [22]. The normalized depth is still slightly shallower than the S2COSMOS survey [23] at 850μm and more observations in the coming season in winter 2021-2022 are expected to make the 2 surveys comparable in depth. 157/63 sources are detected in COSMOS with SNR>4 at 1.2/2mm.

The N2CLS survey aims to produce a full census of high-z DSFGs in these two fields. Combining the two fields, we expect to improve the current estimate on the millimeter number counts and provide new means to rule out or confirm state-of-the-art galaxy evolution models and simulations. In particular, combined with the wealth of ancillary data, N2CLS will put new solid constraints on the obscured SFRD at z>3. To reach that goal, we need first to obtain the redshift of N2CLS sources. The deep optical-IR data have been extensively used to obtain photometric redshift of sources in GOODS-N and COSMOS with novel spectral energy distribution (SED) fitting methods. However, the high dust attenuation makes some DSFGs completely dark in optical, making it difficult to get their redshift with these methods. The blind mm spectral scans will be the only solution to measure their spectroscopic redshift.

3 Searching for the Redshift of DSFGs Scanned by NOEMA

Before the start of N2CLS, we made several science verification observations to test the on-sky performance of the NIKA2 instrument, including one centered on the z=5.243 lensed DSFG J0918+5142 [26]. The NIKA2 observation covers ~175 arcmin^2 and reaches RMS of 0.45/0.11 mJy/beam at 1.2/2 mm. The surveyed area overlaps with the SPIRE footprint of Herschel Lensing Survey (HLS) [27] but is almost not covered by deep optical/near-IR observations. In this field, we detected 4 sources (HLS-2,3,4 and 22) bright in millimeters and with cold IR SEDs consistent with z=3-7 DSFGs. However, the lack of optical/near-IR data makes it difficult to obtain their accurate photometric redshift. To study their properties in detail, we made NOEMA band1 spectral scan observations from ~71GHz to ~102GHz on these sources to search for their redshift at first. Additional follow-up observations were made later in band1 and band2 based on the initial analysis on HLS-22 and HLS-2+3, respectively.

At least 2 lines detected with high significance are needed to get an unambiguous redshift solution, while we only got no more than 1 line detection on each source in our first observations. To design efficient follow-up observations, we develop a new method to search for the redshift of DSFGs jointly with spectral scans and broadband SED fitting, which could derive the redshift of DSFGs with high robustness from noisy mm spectra in our first observations.

The basic concept of this method is using the redshift probability distribution and IR luminosity from the SED fitting, in addition to mm spectra, to jointly constrain the redshift. With far-IR and (sub-)mm photometric data, we compute the best estimate of IR luminosity at any given redshift, $L_{IR}(z)$. With the $L_{IR}(z)$, we further predict the total luminosity of CO, [CI] and [CII] lines using the empirical L_{IR}-L_{line} relations [24] and then generate the model spectra at the given redshifts with our code. All spectral lines have Gaussian profiles of 500km/s line widths (FWHM) and fluxes scaled from the predicted luminosity at given redshifts. The line widths are chosen to be the average among bright sub-mm galaxies [25], which compromise the robustness and significance of the line identification under the interference of narrow noise spikes and a wide variety of line widths in different sources (see Fig. 3). Then, We compare the observed spectra with the predicted model at any given redshift to derive the goodness of match using the chi-square, noted as $\chi^2_{spec}(z)$. Combining with the chi-square of the SED fitting at any redshift (χ^2_{SED}), we derive the likelihood of source redshift with Eq. 1:

$$\mathcal{L}_{joint}(z) \propto \exp\{-[\chi^2_{SED}(z) + \chi^2_{spec}(z)]/2\} \quad (1)$$

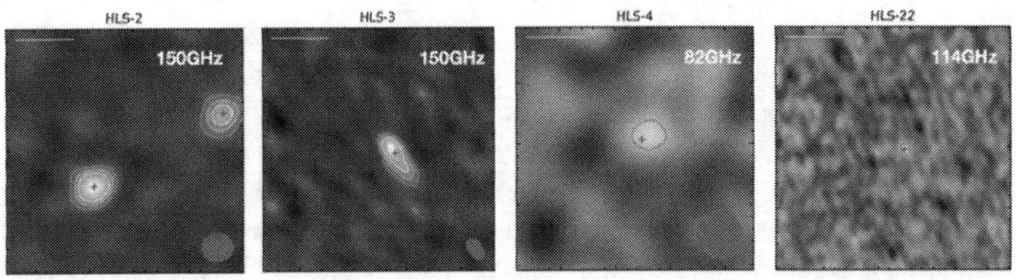

Figure 1. NOEMA continuum image of 4 sources in HLS field. Contour levels correspond to -3.5σ (purple), 3.5σ, 5σ and 6.5σ (yellow), respectively. The NOEMA beams are shown in the bottom-right.

Fig. 1. shows the cleaned continuum image from NOEMA at the highest frequency setup. We make blind detection in these continuum uv-tables and keep sources with SNR>4 by NOEMA (red markers in Fig. 1). in the following study. Observation on HLS-2 reveals 2 sources with nearly equal fluxes with a separation of ~10", while the rest 3 do not show evidence of multiplicity. The spectra of the sources are then extracted from the full uv table, with the position and shape fixed to be the same as the continuum sources.

We first applied the joint-analysis method to the spectra taken in 2017. We fit the SPIRE+NIKA2 SED with the SED templates of high-z star-forming galaxies from [14], generate the model spectra, and derive the joint log-likelihood following the described procedures. Fig. 2. shows the log-likelihood of the redshift and spectral lines of HLS-2-1, HLS-3, and HLS-22 from the joint analysis and the SED only, with the maximum log-likelihood normalized to 1. HLS-2-1 has one marginally detected line at ~92.3GHz, which matches with the discrete peaks in the normalized log-likelihood maximized at z=5.241. HLS-3 has no line detected in the band1 observation, and thus no significant peaks in log-likelihood. HLS-22 has one significant detection at ~85.7 GHz, which also matches with the discrete peaks maximized at z=3.036 and z=4.380 in the log-likelihood.

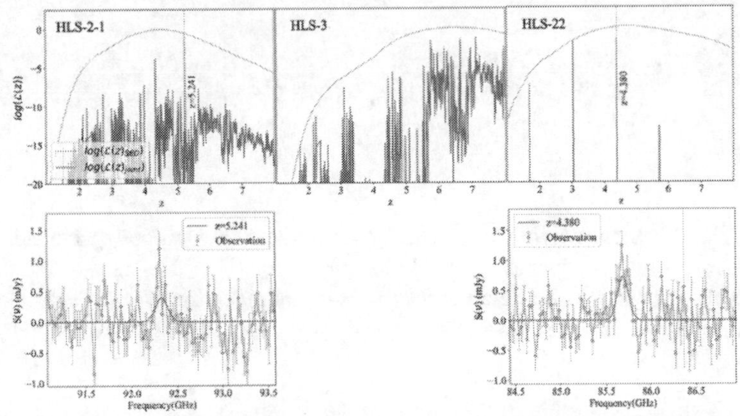

Figure 2. Results of joint-analysis method with the extracted source spectra of 3 HLS sources observed in 2017. First row: The normalized log-likelihood of 3 HLS sources derived by our joint-analysis method. Second row: The blindly extracted source spectra and the model at the best redshift solution of HLS-2-1 and HLS-22 generated by the joint-analysis method.

We present the results of the joint analysis with further follow-up data in Fig. 3. For HLS-2-1, we confirm the redshift (z=5.241) in the previous analysis with a new line detection at 147.7 GHz. With a new line detected at 114 GHz, the redshift of HLS-22 is also confirmed to be at z=3.036, matching with one of the prominent peaks in the previous joint log-likelihood. As for HLS-3, the new analysis reveals one prominent peak in log-likelihood at z=3.123 matching to the line detected at 139.8 GHz. This is not contradicting the previous results showing relatively high log-likelihood at this redshift under non-detection. The consistency of the results from our joint-analysis method indicates the promising potential of applying it to the search of the redshift of DSFGs observed with blind millimeter spectral scans.

Acknowledgements

We would like to thank the IRAM staff for their support during the campaigns. The NIKA2 dilution cryostat has been designed and built at the Institut Néel. In particular, we acknowledge the crucial contribution of the Cryogenics Group, and in particular Gregory Garde, Henri Rodenas, Jean Paul Leggeri, Philippe Camus. This work has been partially funded by the Foundation Nanoscience Grenoble and the LabEx FOCUS ANR-11-LABX-0013. This work is supported by the French National Research Agency under the contracts "MKIDS", "NIKA" and ANR-15-CE31-0017 and in the framework of the "Investissements d'avenir" program (ANR-15-IDEX-02). This work has benefited from the support of the European Research Council Advanced Grant ORISTARS under the European Union's Seventh Framework Programme (Grant Agreement no. 291294). F.R. acknowledges financial supports provided by NASA through SAO Award Number SV2-82023 issued by the Chandra X-Ray Observatory Center, which is operated by the Smithsonian Astrophysical Observatory for and on behalf of NASA under contract NAS8-03060.

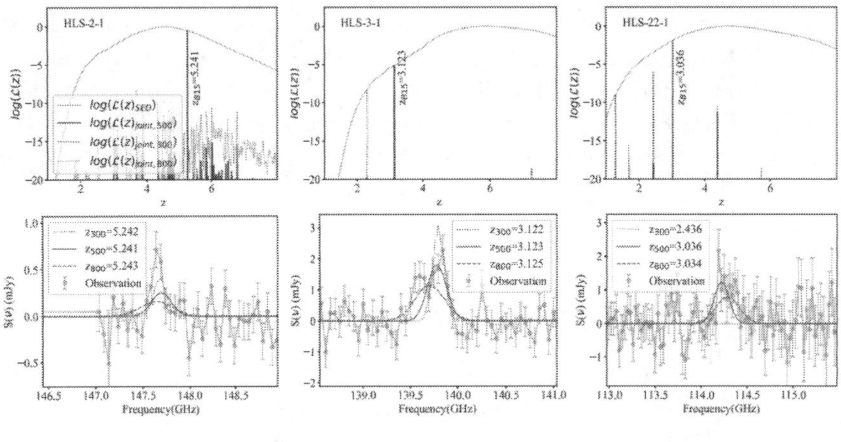

Figure 3. Similar to Fig. 2, the results of the joint-analysis with new observations added and the spectra plus the model at the redshift with the maximum log-likelihood. Here we also present the results under FWHM 300/800 km/s for comparison. Except for the 300km/s model on HLS-22 affected by noise spikes, the redshift solutions remain consistent with those using the default 500 km/s model.

References

[1] P. Madau & M. Dickinson, ARA&A **52**, 415 (2014).
[2] R. J. Bouwens, L. Bradley, A. Zitrin, *et al.* ApJ **795**, 126 (2014)
[3] A. Pillepich, V. Springel, D. Nelson, *et al.* MNRAS **473**, 4077 (2018)
[4] T. Wang, C. Schreiber, D. Elbaz, *et al.* Nature **572**, 211 (2018)
[5] J. S. Dunlop, R. J. McLure, A. D. Biggs, *et al.* MNRAS **466**, 861 (2017)
[6] C. Gruppioni, M. Béthermin, F. Loiacono, *et al*, Astron. Astrophys. **643**, A8 (2020)
[7] U. Dudzevičiūtė, I. Smail, A. M. Swinbank, *et al*, MNRAS. **494**, 3828 (2020)
[8] J. A. Zavala, C. M. Casey, S. M. Manning, *et al*, ApJ. **909**, 165 (2021)
[9] M. Franco, D. Elbaz, M. Béthermin, *et al*, Astron. Astrophys. **620**, A152 (2018)
[10] L. Perotto, N. Ponthieu, J.-F. Macías-Pérez, *et al.*, Astron. Astrophys. **637**, A71 (2020)
[11] R. Adam *et al.*, Astron. Astrophys. **609**, A115 (2018)
[12] M. Calvo *et al.*, Journal of Low Temperature Physics, **184**, 816 (2016)
[13] O. Bourrion *et al.*, Journal of Instrumentation, **11**, 11001 (2016)
[14] M. Béthermin, E. Daddi, G. Magdis, *et al*, Astron. Astrophys. **573**, A113 (2015)
[15] R. Zylka, Astrophysics Source Code Library. ascl:1303.011 (2013)
[16] T. A. Perera, E. L. Chapin, J. E Austermann, *et al*, MNRAS. **391**, 1227 (2008)
[17] , T. R. Greve, A. Pope, D. Scott, *et al*, MNRAS **389**, 1489 (2008)
[18] J. G. Staguhn, A. Kovács, R. G. Arendt, *et al*, ApJ. **790**, 77 (2014)
[19] L. L. Cowie, A. J. Barger, L.-Y. Hsu, *et al*, ApJ. **837**, 139 (2017)
[20] I. Aretxaga, G. W. Wilson, E. Aguilar, *et al*, MNRAS. **415**, 3181 (2011)
[21] F. Bertoldi, C. Carilli, M. Aravena, *et al*, ApJS. **172**, 132 (2007)
[22] B. Magnelli, A. Karim, J. Staguhn, et al. *et al*, ApJ. **877**, 45 (2019)
[23] J. M. Simpson, I. Smail, A. M. Swinbank, et al., *et al*, ApJ. **880**, 43 (2019)
[24] D. Liu, Y. Gao, K. Isaak, *et al*, ApJL. **810**, L14 (2015)
[25] J. S. Spilker, D. P. Marrone, J. E. Aguirre, *et al*, ApJ. **785**, 149 (2014)
[26] F. Combes, M. Rex, T. D. Rawle, *et al*, Astron. Astrophys. **538**, L4 (2012)
[27] E. Egami, M. Rex, T. D. Rawle, *et al*, Astron. Astrophys. **518**, L12 (2010)

On the dynamical and morphological state of the CHEX-MATE clusters

Maria Giulia Campitiello[1,2,*], *Stefano* Ettori[1,3], *Lorenzo* Lovisari[1,4], and the CHEX-MATE collaboration

[1]INAF, Osservatorio di Astrofisica e Scienza dello Spazio, via Piero Gobetti 93/3, 40129 Bologna, Italy
[2]Dipartimento di Fisica e Astronomia, Università di Bologna, Via Gobetti 92/3, 40121, Bologna, Italy
[3]INFN, Sezione di Bologna, viale Berti Pichat 6/2, 40127 Bologna, Italy
[4]Center for Astrophysics | Harvard & Smithsonian, 60 Garden Street, Cambridge, MA 02138, USA

Abstract. The CHEX-MATE sample was built to provide an overview of the statistical properties of the underlying cluster population and to set the stage for future X-ray missions. In this work, we perform a morphological analysis of the 118 clusters included in the sample with the aim to provide a classification of their dynamical state which will be useful for future studies of the collaboration.

1 Introduction

Clusters of galaxies are the largest virialised systems of the Universe and include between 100 - 1000 galaxies. However, the galactic and star component represents only the 15% of their barionic mass. The bulk of the baryons resides in the Intra Cluster Medium (ICM), a hot ionised plasma that emits in the X-ray band. By identifying specific features in the X-ray morphology of clusters is possible to derive their dynamical state, which is an important property to consider in both astrophysical and cosmological studies. Clusters show a large variety of dynamical states, but we can essentially identify two extreme classes of objects. On the one hand there are the relaxed systems, whose X-ray emission is characterised by a spherical symmetry and are assumed to be in hydrostatic equilibrium allowing a proper mass estimate [e.g., 1, 2], which is an important cosmological quantity. On the other hand, there are the disturbed clusters which show clear signs of mergers and which are useful for astrophysical studies related, for example, to turbulence or particle re-acceleration mechanisms [e.g., 3, 4]. Therefore, a classification of the dynamical state is crucial when dealing with large samples of clusters, since it allows to recognise the most proper objects to use in specific analysis. One of the most recent samples of clusters observed in the X-ray band is the CHEX-MATE sample [the Cluster HEritage project with XMM-Newton Mass Assembly and Thermodynamics at the Endpoint of structure formation; see 5], which is a 3 Ms Multi-Year XMM-Newton Heritage Programme to obtain X-ray observations of a minimally-biased, signal-to-noise-limited sample of clusters detected by Planck through the Sunyaev–Zeldovich effect (SZE). The goals of the project are to provide an accurate vision of the statistical properties of the underlying population, measure how the gas properties are shaped by collapse into the dark matter halo, uncover the provenance of non-gravitational heating, and resolve the major uncertainties in

*e-mail: giulia.campitiello@inaf.it

mass determination that limit the use of clusters for cosmological studies. In this work, we present the morphological analysis of the clusters in the CHEX-MATE sample with the aim to provide a classification of their dynamical state, which will be useful for future studies of the collaboration. Furthermore, we will retrace the steps taken over the last years to test and check the procedure elaborated until now in this field.

2 Dataset and methods

The CHEX-MATE[1] sample consists of 118 clusters detected by Planck at high signal-to-noise (S/N> 6.5) through their SZE signal, spanning a wide range of masses and redshifts. The sample is divided in two groups: the *Tier* 1, consisting of 61 objects located at low redshift ($0.05 < z < 0.2$) in the Northern sky (DEC > 0) with a total mass of 2×10^{14} $M_\odot < M_{500} < 9 \times 10^{14}$ $M_\odot$, providing an unbiased view of the population at the most recent time; the *Tier* 2, including the most massive systems to have formed thus far in the history of the Universe ($z < 0.6$ with $M_{500} > 7.25 \times 10^{14}$ $M_\odot$). Four clusters are in common between these two sub-samples. The XMM-Newton X-ray images (FWHM of the PSF ~ 6 arcsec) used in this analysis are filtered in the in the [0.7-1.2 keV] band, background-subtracted, and exposure corrected. All point-sources were masked to avoid any contamination of our results. Finally, the signal to noise ratio that characterised the images is S/N~150, for most of the objects, providing a uniform characterisation of the entire sample.

2.1 Visual classification

As first step of our study, we considered the oldest method used to classify the clusters X-ray morphology, i.e. the visual classification. In particular, 7 astronomers inspected the cluster images by eye, and assigned a grade from 0 to 3 to each object, where 0 are the most relaxed systems and 3 the most disturbed ones. The results of the 7 classifications were then averaged and clusters with rounded values equal to 0 were classified as relaxed (R, 18 clusters), clusters with rounded values equal to 3 as disturbed (D, 32 clusters), and all the other clusters as mixed (M, 68 clusters). The visual classification is a very subjective and time-consuming procedure and it is not suitable for large samples of objects or for future surveys. However, in our case, the involvement of 7 people allows to reduce biases related to the subjectivity of the method, and consequently we will use the results obtained as reference for the rest of the analysis.

2.2 Morphological parameters

The limits of the visual classification described above led to the definition of more robust indicators able to objectively quantify even small deviations from a perfectly regular and spherically-symmetric emission. Below we report the parameters considered in this work:

- the concentration, c, defined as the ratio of the surface brightness (SB) inside two concentric apertures [6]: $c = SB_{(r<0.15\,R_{500})}/SB_{(r<R_{500})}$;
- the centroid shift, w, [7] defined as the standard deviation of the projected separation between the X-ray peak and the centroid of the emission computed within N (=10 in our case) apertures of increasing radius: $w = \frac{1}{R_{500}}\sqrt{\frac{1}{N-1}\sum_i(\Delta_i - \bar{\Delta})^2}$ where Δ_i is the distance between the X-ray peak and the centroid of the i-th aperture, and R_{500} is the radius of the largest aperture.

[1] http://xmm-heritage.oas.inaf.it/

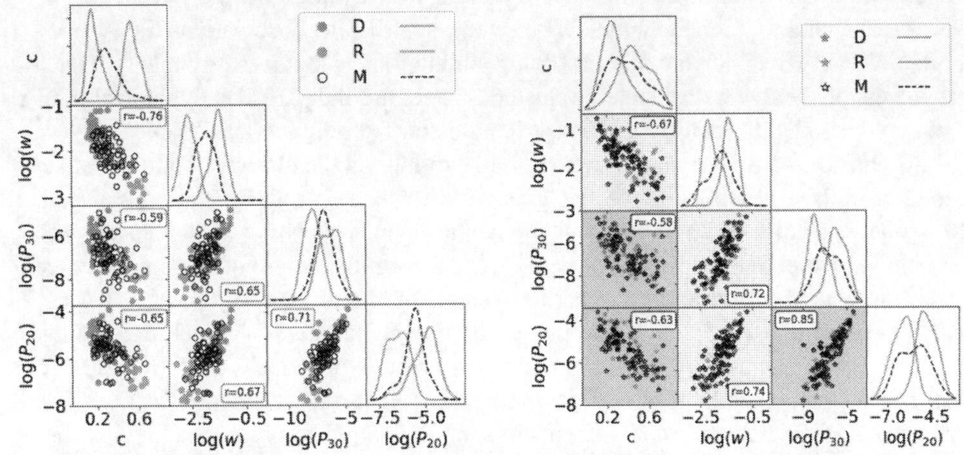

Figure 1. Correlations between the morphological parameters estimated from the observations (left) and simulations (right). Relaxed (R), disturbed (D), and mixed (M) clusters are shown in orange, cyan, and black, respectively. The background colours of the right plot represent the level of confidence between the observed and simulated Spearman coefficient, r (green stays for a level of confidence below 1 σ, yellow is for a level of confidence between 1 - 3 σ and red is for a level of confidence above 3 σ).

- the power ratios [8], which are a multipole decomposition of the X-ray surface brightness inside a certain aperture. The m-order (m >0) power ratio is defined as P_m/P_0, with: $P_0 = [a_0 \ln(R_{500})]^2$ and $P_m = (a_m^2 + b_m^2)/(2m^2 R_{500}^{2m})$, where a_0 is the total intensity within R_{500} and the moments a_m and b_m are calculated by: $a_m(R) = \int_{R<R_{500}} S(x) R^m \cos(m\phi) d^2x$ and $b_m(R) = \int_{R<R_{500}} S(x) R^m \sin(m\phi) d^2x$, where $S(x)$ is the SB at the position x=(R, ϕ). In this work we are going to consider the ratios P_2/P_0 and P_3/P_0 (hereafter P_{20} and P_{30}, respectively), which are respectively a measure of the ellipticity of clusters and of the presence of asymmetries or substructures.

3 Morphological analysis

3.1 The observed sample

We computed the parameters described above for each cluster and we present the results in Fig. 1 (left panel) where we also highlight the clusters dynamical state. We investigated the presence of correlations between the parameters by computing the Spearman coefficient. We found good correlations ($r > 0.5$) between all parameters with the best (anti-) correlation observed for the couple $c - w$ (r = -0.76). By using the visual classification of the dynamical state we can see that all parameters are able to separate the relaxed and disturbed populations, with c and w showing the smaller overlap between the distributions.

3.2 The simulated sample

In order to obtain an objective view of the ability of the morphological parameters to separate the relaxed and disturbed populations, we repeat our analysis using a simulated sample provided by The Three Hundred collaboration [e.g., 9]. This sample considered in the analysis is composed of 878 clusters spanning a wide range of redshift ($0 < z < 0.6$) and masses ($M_{500} > 4.6 \times 10^{14}$ h^{-1} $M_{\odot}$). The lack of low-mass simulated objects may induce discrepancies between the observed and simulated sample, which will be better investigated in future analysis.

In order to reproduce the quality of the *XMM-Newton* observations, we smoothed the simulated images with a Gaussian function of $\sigma = 6$ arcsec, which represents the FWHM of the *XMM-Newton* PSF, and we binned them using the same scale of the observations (i.e., 1 pixel = 2.5 arcsec). The great advantage in considering simulations is that it is possible to have a priori knowledge of the dynamical state of clusters. There are indeed some dynamical indicators that could be used to identify the relaxed and disturbed populations. In our analysis, the indicators considered are the mass fraction of all sub-halo in the cluster, f_s, and the offset of the centre of mass, Δ_r, defined respectively as: $F_S = \Sigma_i M_i / M_{500}$ and $\Delta_r = |r_{cm} - r_c| / R_{500}$, where M_i is the mass of the sub-halo, r_{cm} is the centre-of-mass position of the cluster, and r_c is the position of the highest density peak (i.e., the theoretical centre of the cluster). We defined as most relaxed (disturbed) clusters, the systems with both $\Delta_r < 0.1$ and $f_s < 0.1$ ($\Delta_r > 0.1$ and $f_s > 0.1$, [9, 10]). Having determined the dynamical state of the clusters of the simulated sample, we estimated the morphological parameters presented in Section 2.2. To compare these new results with the ones obtained for the CHEX-MATE sample, we build 10^4 sub-samples of 118 simulated clusters. Each sub-sample is build to reproduce the distribution in redshift and mass of the CHEX-MATE sample. In particular, we randomly extracted from each redshift snapshot of the simulations a number of objects equal to the number of CHEX-MATE systems located at that redshift. Hereinafter the results shown for the simulated sample are the media of the results obtained for each of the 10^4 sub-samples. As first step we compute the level of correlation between the parameter pairs, by determining the Spearman coefficient, and we compare it with the coefficient obtained from simulations. The level of concordance between observations and simulations is reported in Fig. 1 (right panel): in green are represented the pair of parameters for which the level of agreement is lower than 1σ, in yellow the pairs for where the level of confidence is included between 1 and 3σ, and in red the pairs with a level of confidence higher than 3σ. From this comparison it arises that all the pairs show a level of confidence lower than 3 σ, with the exception of the $P_{20} - P_{30}$ whose trend does not reproduce the one arising from the observations.

4 ROC curves

To explore the ability of the morphological parameters in detecting the relaxed and disturbed systems we build the so called ROC curves. ROC curves are essentially a comparison between two classifications: the first one, hereafter *real classification*, is based on the dynamical state obtained using the dynamical indicators described in Section 3.2; the second one, hereafter *morphological classification*, is obtained by defining for each morphological parameter a threshold, V_T, above (or below) which, clusters are classified as disturbed (or as relaxed). By comparing the results provided by these two classifications it is possible to define as TP or TN the objects for which the two classifications show agreement, and as FP or FN the objects classified as relaxed from the morphological classification instead as disturbed and vice versa. The ROC curves is then build by varying V_T, and computing for each value of V_T the true and the false positive rate (respectively TPR and FPR), defined as: TPR= TP/(TP+FN) and FPR=FP/(FP+TN). By computing the area under the curve (AUC) obtained it is possible to quantify the probability of the considered parameter to correctly identify the dynamical state of clusters: in the perfect case (i.e., 100% probability of correct detection) the curves obtained is a step function with an AUC = 1. On the contrary, if the probability of the parameter to correctly detect the relaxed and disturbed population is 50%, the ROC curve obtained is an identity line with AUC = 0.5. By testing this method on our parameters we found that the most powerful parameter is w, showing an AUC = 0.89±0.05, followed by the power ratios P_{30} and P_{20}, showing respectively AUC = 0.79±0.07 and AUC = 0.78±0.07. For c, we found

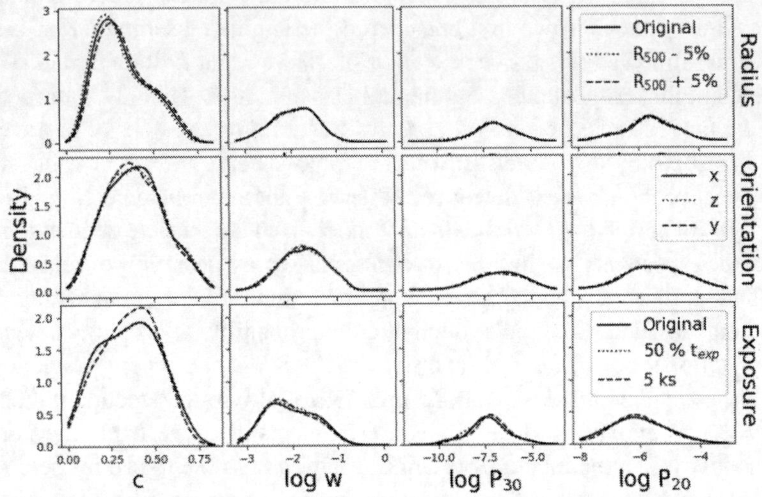

Figure 2. Results of the test reported in Section 5. Top: comparison between the morphological parameters computed in regions with different radii. Center: comparison between the morphological parameters estimated for different orientations of the same cluster. Bottom: comparison between the morphological parameters estimated from original images, images with halved exposure times and images with $t_{exp} \sim 5$.

AUC = 0.74±0.08 suggesting a lower ability to disentangle between the two populations of clusters.

5 Study of the systematics

In order to investigate the robustness of the morphological parameters and their dependence on the quality of the images used, we realised the following tests:

- decrease of the exposure time: we considered a sub-sample of 20 CHEX-MATE clusters and we repeat the morphological analysis on both images with halved exposure time and images with exposure time equal to 5 ks;
- variation the orientation of clusters: we considered the simulated sample which provides for each object three images related to three different orientations (x, y and z). We then compare the values of the morphological parameters estimated along these three different orientations;
- change of the radius: we estimate the morphological parameters using two circular regions of of radius $r = 1.05 * R_{500}$ and $0.95 * R_{500}$. We then compare these two estimations with the original ones (obtained using $r = R_{500}$).

The results of these tests are reported in Fig. 2. The very good agreement between the results indicates that these parameters robustly determine the dynamical state of the clusters.

6 Combining the morphological parameters

One of aims of this analysis is to obtain a single quantity which is able to define the grade of relaxation of a system. Therefore, we defined a new parameter M as the combination of the other four parameters [10–12]: $M = \frac{1}{N}\sum A_{Par} \times \frac{Par - m_{Par}}{\sigma_{Par}}$, where N is the number of parameters used, σ and m represent respectively the standard deviation and the median of the considered parameter and A_{Par} is equal to -1 for the concentration and 1 for the other parameters. To test the power of this parameter, we first considered the simulated sample. For each of the 10^4 sub-samples we followed this procedure: first of all we identify the number of relaxed, f_r, and disturbed, f_d, clusters as defined by the real classification. Then, we compute M and check whether the first f_r (or f_d) objects with the lowest (or highest) values of M are classified as relaxed (or disturbed) by the real classification (so no specific threshold values of M are adopted). If so, we talk about correct detection, otherwise about wrong detection. To estimate the probability of M to correctly or wrongly detect the relaxed objects, we defined the ratio C_R and W_R which are respectively the number of correct or wrong detection of relaxed systems divided for f_r (and similarly C_D and W_D which are the number of correct or wrong detection of disturbed systems divided for f_d). We found that the probability to correctly detect relaxed and disturbed systems is $C_R = 61 \pm 6$ % and $C_D = 54 \pm 8$ %. These percentage are not too high, but could be explained by observing the probability of wrong detection, which is very low, with $W_R = 9 \pm 4$ and $W_D = 10 \pm 5$. These latter result suggest that the non-correct detection are mostly due to the mixed population. As it is possible to see in the corner plot of Figure 1 (right panel) this class of objects is indeed uniformly distributed and could thus alter our results. We then apply the same procedure on the CHEX-MATE sample using as reference the visual classification. We found that the probability of correct detection are C_R = 78 % and C_D = 73 %, while the probability of wrong detection are zero for both the relaxed and disturbed populations.

7 Conclusions

We realised a morphological analysis of the 118 clusters of the CHEX-MATE sample. In particular, we investigate the link between the distribution of the X-ray emission and the dynamical state by means of four parameters: c, w, P_{20}, and P_{30}. We found that they all show good correlations and are powerful in identifying the most disturbed and most relaxed systems of a sample. Furthermore, we verify that these four indicators are robust and are not strongly influenced by the quality of the images used. We thus combined them in a single quantity, M, that allows to build a continuous classification of the grade of relaxation of the systems of the CHEX-MATE sample and we verify that the extreme objects of this classification are effectively relaxed and disturbed systems.

References

[1] Ettori S. et al., 2013, Space Sci. rev, 177, 119
[2] Pratt G. et al., 2019, Space Sci. rev, 215, 25
[3] Cassano R. et al., 2010, Apj, 721, L82
[4] Mann A. W. et al., 2012, MNRAS, 420, 2120
[5] CHEX-MATE Collaboration, 2021, A&A, 650, 104
[6] Santos J. S. et al., 2008, A&A, 483, 35
[7] Mohr J. J. et a., 1995, ApJ, 447,8
[8] Buote D. A. et al., 1995, ApJ, 452,522
[9] Cui W. et al., 2018, MNRAS, 480, 2898
[10] De Luca F. et al., 2021, MNRAS, 504, 5383
[11] Rasia E. et al., 2013, The Astronomical review, 8, 40
[12] Cialone G. et al., 2018, MNRAS, 477, 139

Morphological analysis of SZ and X-ray maps of galaxy clusters with Zernike polynomials

Valentina Capalbo[1,*], *Marco* De Petris[1], *Federico* De Luca[2], *Weiguang* Cui[3], *Gustavo* Yepes[4,5], *Alexander* Knebe[4,5,6], *Elena* Rasia[7,8], *Florian* Ruppin[9], and *Antonio* Ferragamo[1]

[1]Dipartimento di Fisica, Sapienza Università di Roma, Piazzale Aldo Moro 5, I-00185 Roma, Italy
[2]Dipartimento di Fisica, Università di Roma 'Tor Vergata', Via della Ricerca Scientifica 1, I-00133 Roma, Italy
[3]Institute for Astronomy, University of Edinburgh, Royal Observatory, Edinburgh EH9 3HJ, UK
[4]Departamento de Física Teórica, Módulo 8, Facultad de Ciencias, Universidad Autónoma de Madrid, E-28049 Madrid, Spain
[5]Centro de Investigación Avanzada en Física Fundamental (CIAFF), Facultad de Ciencias, Universidad Autónoma de Madrid, E-28049 Madrid, Spain
[6]International Centre for Radio Astronomy Research, University of Western Australia, 35 Stirling Highway, Crawley, Western Australia 6009, Australia
[7]National Institute for Astrophysics, Astronomical Observatory of Trieste (INAF-OATs), Via Tiepolo 11, I-34131 Trieste, Italy
[8]Institute for Fundamental Physics of the Universe (IFPU), Via Beirut 2, I-34014 Trieste, Italy
[9]Kavli Institute for Astrophysics and Space Research, Massachusetts Institute of Technology, Cambridge, MA 02139, USA

Abstract. Several methods are used to evaluate, from observational data, the dynamical state of galaxy clusters. Among them, the morphological analysis of cluster images is well suited for this purpose. We report a new approach to the morphology, which consists in analytically modelling the images with a set of orthogonal functions, the Zernike polynomials (ZPs). We validated the method on mock high-resolution Compton parameter maps of synthetic galaxy clusters from The Three Hundred project. To classify the maps for their morphology we defined a single parameter, C, by combining the contribution of some ZPs in the modelling. We verify that C is linearly correlated with a combination of common morphological parameters and also with a proper 3D dynamical-state indicator available for the synthetic clusters we used. We also show the early results of the Zernike modelling applied on Compton parameter maps of local clusters ($z < 0.1$) observed by the *Planck* satellite. At last, we report the preliminary results of this kind of morphological analysis on mock X-ray maps of The Three Hundred clusters.

1 Introduction

Classify galaxy clusters based on their dynamical state is crucial to correctly infer other physical properties of those systems. For example, their mass can be estimated by exploiting the

*e-mail: valentina.capalbo@uniroma1.it

hypotheses of hydrodynamical and thermal equilibrium with an assumption of spherical distributions for both dark matter and baryonic components. However, in many cases they prove to be too simplified approximations. In fact, it is well known that galaxy clusters are dynamically active systems and their physical state does not always reflect a condition of equilibrium. Several studies are then focused on defining valuable methods to infer the real state of the clusters from observational data. Among the others, the analysis of the morphological appearance of multiwavelength images is a common approach used to classify clusters in different dynamical classes. Several morphological parameters can be defined based on characteristic features in the images and also used in some combinations (see e.g. [1], [2], [3] and references therein).

Here, we report a new approach to the morphological analysis, developed in [4], which consists in modelling the cluster images with a set of functions, the Zernike polynomials (ZPs). This new method was validated, at first, on mock projections maps realized by exploiting the Sunyaev-Zel'dovich (SZ) effect [5], i.e. Compton parameter maps. We defined a single parameter to quantify the morphological differences between the cluster maps and we verified that it is well correlated with some common morphological parameters largely used in literature and also with a proper dynamical-state classification available from 3D data for the simulated clusters we used. We also present the preliminary results of this method applied on the real Compton parameter maps of galaxy clusters from the *Planck*-SZ catalogue and on mock X-ray maps.

2 Analytical modelling with Zernike polynomials

The ZPs are a complete basis of orthogonal functions defined on a unit disk. Completeness and orthogonality make them well adapted to model functions or images in circular domains. ZPs are defined as follows [6]:

$$Z_n^m(\rho,\theta) = N_n^m R_n^m(\rho)\cos(m\theta) \ , \ Z_n^{-m}(\rho,\theta) = N_n^m R_n^m(\rho)\sin(m\theta) \tag{1}$$

where ρ is the normalized radial distance ($0 \le \rho \le 1$), θ is the azimuthal angle ($0 \le \theta \le 2\pi$), n and m are, respectively, the polynomial order and the angular frequency such that $m \le n$ and $n - m =$ even, $N_n^m = \sqrt{2(n+1)/(1+\delta_{m0})}$ is a normalization factor in which δ_{m0} is the Kronecker delta and $R_n^m(\rho) = \sum_{s=0}^{(n-m)/2} \frac{(-1)^s(n-s)!}{s!\left(\frac{n+m}{2}-s\right)!\left(\frac{n-m}{2}-s\right)!}\rho^{n-2s}$ is the radial term. We emphasize that the goal of our work was not to reconstruct the cluster maps in details, but rather to reveal the main features able to identify different dynamical classes. For example, very regular patterns, mostly circular, can be related to relaxed systems, while asymmetries and substructures are signals of a disturbed dynamical state. In [4] we analyzed mock Compton parameter maps at high resolution (see Sect. 3) and within a circular aperture of radius equal to R_{500}, therefore we used ZPs up to the eight order n (i.e. 45 terms, see the ordering scheme in [6]) in order to obtain a spatial resolution of $\sim 0.5R_{500}$ in the modelling (see also Appendix A in [4]). We divided the ZPs in two classes, based on their 2D projections: terms with $m = 0$, which show a circular symmetry and terms with $m \neq 0$, which fit azimuthal inhomogeneities. We are currently using the same number of ZPs to model mock X-ray maps as well. Each map is then expressed as a combinations of ZPs:

$$S_{y,X} = \sum_{n=0}^{8}\sum_{m=0}^{n} c_{nm} Z_{nm} \tag{2}$$

where $S_{y,X}$ indicates the spatial distribution of the Compton parameter y or the X-ray surface brightness and c_{nm} are the expansion coefficients of the single polynomials. In the preliminary analysis on the *Planck* Compton parameter maps we decided to reduce the number of ZPs in the fit since the resolution in the maps is significantly lower. Internal regions ($< R_{500}$) are poorly resolved, therefore we limit our modelling to a spatial resolution of $\sim R_{500}$, by using ZPs up to the fifth order (i.e. 21 terms).

3 Data sets

In this Section we describe three different types of maps used in our analyses: the mock Compton parameter and X-ray maps for synthetic galaxy clusters in the The Three Hundred project [7], a large catalogue with 324 massive clusters ($M_{200} > 6 \times 10^{14} h^{-1} M_\odot$ at $z = 0$) generated through hydrodynamical simulations; the real Compton parameter maps of galaxy clusters from the *Planck* observations.

- **Mock y-maps** Mock maps of the thermal component of the SZ effect [5] were realized for all The Three Hundred clusters. The maps are 2D projections of the Compton parameter y (i.e. y-maps) which characterizes the effect. They were mimicked with the PyMSZ code[1] [7], with a spatial resolution of 10 kpc per pixel corresponding to angular resolutions of 5.25″, 1.14″ and 0.59″ respectively at $z = 0$, 0.45 and 1.03, the reference redshifts analysed in [4]. We normalized each map to the maximum and performed the Zernike fitting within a circular aperture of radius R_{500} centred on the y-centroid. Note that the centroid was computed by fixing at first the circular aperture on the centre of the maps, which corresponds to the highest density peak.
- **Mock X-ray maps** Mock X-ray maps were realized for The Three Hundred clusters by using the PyXSIM code[2] [8]. The maps are in terms of photon number counts, with 10 ks exposure time and with a fixed spatial resolution of 10 kpc per pixel (same angular resolution as for the y-maps). They are related to the spectral band 0.2-15 keV, convolved with the response file of the Wide Field Imager instrument for the *Athena* satellite [9]. We analysed them within a circular aperture of radius R_{500} centred on the X-ray centroid, computed as in the mock y-maps. However, in this case we normalized the maps to the mean inside a region of radius $0.05 R_{500}$, to smooth the contribution from saturated pixels.
- ***Planck* SZ-selected clusters** In the preliminary analysis of ZPs applied to real y-maps we selected a sample of clusters at $z < 0.1$ from the PSZ2 catalogue [10]. We used the public available full sky y maps [11] realized with two component separation algorithms: MILCA [12] and NILC [13]. These are maps in HEALPIX format with pixel size of 1.7′ and with a final angular resolution of 10′. Therefore, we only selected clusters with $R_{500} \leq 10'$ (i.e. resolved clusters). We extracted gnomonic projections from the HEALPIX y-maps, centred on the cluster coordinates and with a side-length of $2R_{500}$ for each cluster. We also used the available point source mask [11] to select maps with a low residual contamination from point sources, based on the spatial resolution of the Zernike modelling described in Sect. 2. Our final sample was composed by 135 galaxy clusters.

To estimate the capability of the Zernike morphological analysis in recognizing different dynamical classes in the *Planck* cluster sample, we also used *Planck*-like maps realized for The Three Hundred clusters described above. For those clusters, in fact, we can evaluate *a priori* the dynamical state from 3D data. The mock *Planck* maps were redone to mimic real *Planck* maps in these following steps: at first the maps were convolved with a 10′ beam and gridded in pixels of 1.7′; a full-sky noise map was realized by using the *Planck* noise power spectrum and the public available full-sky maps of the standard deviation of the Compton

[1] https://github.com/weiguangcui/pymsz
[2] https://github.com/jzuhone/pyxsim

parameter; a patch was then extracted in this noise map at a random position (considering only the positions of detected clusters) to generate the final noise added in the y-maps. Note that the mock maps do not include point source contamination. We analysed them within a circular aperture of radius R_{500}, as for the real *Planck* maps. In particular, we used the mock maps realized at four snapshots of redshift ($z = 0.02, 0.04, 0.07, 0.09$) to cover the redshift range of the *Planck* clusters.

4 Results

At first, we report the application of the Zernike fitting on cluster y-maps. We summarize the results in [4] for mock y-maps and then the preliminary analysis on real maps for clusters in the PSZ2 catalogue. At last, we briefly discuss the ongoing analysis on X-ray maps.

4.1 Zernike analysis for y-maps

The synthetic clusters we analysed in [4] were previously classified in [3] for their dynamical state by using a combination of some 3D dynamical indicators, i.e. the χ parameter (see also [14]). In addition, a morphological analysis of their y-maps were also performed in [3] by applying a combined morphological parameter, M (see also [2] for details on the parameters combined in M). Those results represented our reference to check the efficiency of the Zernike modelling. When fitting the mock y-maps as in eq. 2, we verified that ZPs with $m \neq 0$ have c_{nm} values negligible in case of regular (mostly circular) distributions in the maps, while they increase when dealing with complex patterns involving e.g. asymmetries or substructures. On the contrary, ZPs with $m = 0$ have c_{nm} values almost invariant when fitting different morphologies. This behaviour is clarified in Fig. 1, in which we show the results of the Zernike fitting for a relaxed and a disturbed cluster (as classified with χ and M). Therefore, we defined a single parameter, C, as follows:

$$C = \sum_{n,m\neq 0} |c_{nm}|^{1/2} \tag{3}$$

We verified that C is able to distinguish different morphologies, showing a linear correlation of $\sim 78\%$ at $z = 0$ with M, that is quite stable at higher z. The correlation with χ is lower, $\sim 62\%$ at $z = 0$, and it is mild descreasing along z. We concluded that the Zernike fitting is a valid method to evaluate the cluster morphology and infer the cluster dynamical state. In particular, the parameter we extracted shows similar performances with respect to a more complex combination of several parameters.

For the current analysis of the *Planck* y-maps we are also using *Planck*-like maps generated for The Three Hundred clusters in order to check the efficiency of the Zernike fitting in drawing a dynamical state evaluation for the *Planck* cluster sample. In Fig. 2 (left panel) we show the cluster distribution in the mass-redshift plane, for both real and synthetic clusters. In each redshift bin (marked by thin vertical lines in the figure) we perform random extractions of the same number of synthetic clusters as for the corresponding *Planck* sub-sample, applying an upper-mass limit for the synthetic clusters defined by considering the value of the *Planck* masses in each bin (see the horizontal dashed lines in the figure). In this way we simulate several mock *Planck* samples and for each one we compute the C parameter on all the y-maps. In the right panel in Fig. 2 we show the distributions of C for the mock samples extracted (grey lines) and for the *Planck* clusters (red and blue lines). The distributions in the two cases are quite different. However, we note that the *Planck* clusters have, on average, lower masses with respect to the synthetic clusters (see left panel in Fig. 2). This is particularly evident at low redshift ($z \lesssim 0.05$), where the number of *Planck* clusters we selected

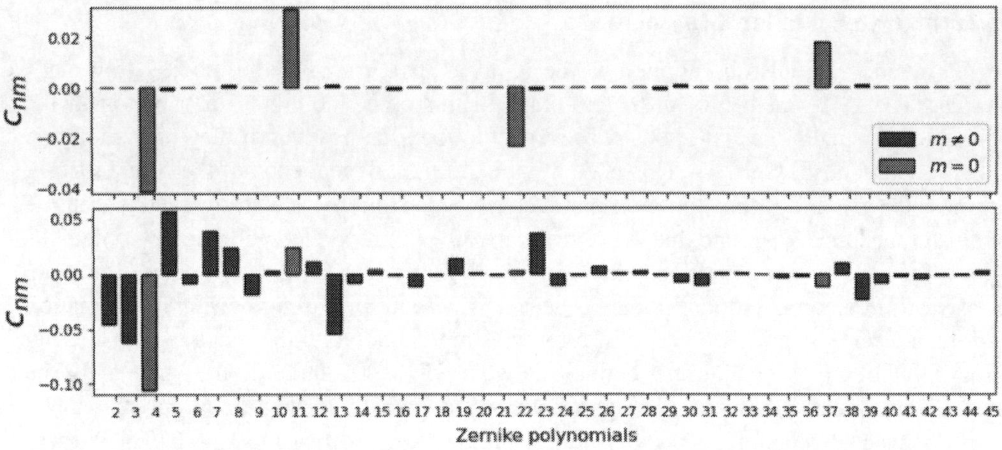

Figure 1. Zernike coefficients c_{nm} computed by fitting y-maps of a relaxed (top) and a disturbed (bottom) cluster. Red bars refer to ZPs with $m = 0$, while blue bars are for $m \neq 0$. The ZPs are ordered by following the Noll's scheme [6]. Note that the first term (Z_0^0) is neglected because the corresponding c_{00} is simply equal to the mean value of the signal inside the circular aperture in the map.

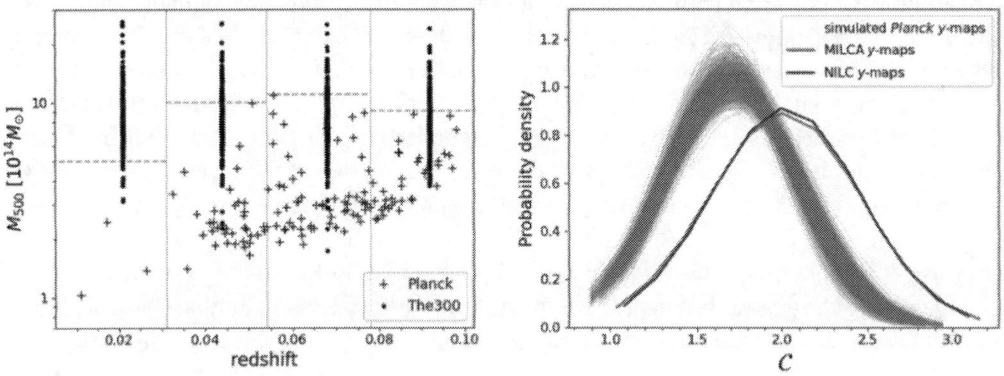

Figure 2. Left: mass-redshift distribution of 135 *Planck*-SZ clusters and of the synthetic clusters generated in The300 project at four redshifts. The vertical thin lines mark the redshift bins we used to compare the Zernike analysis on *Planck* and mock (*Planck*-like) y-maps. The horizontal dashed lines indicate the upper-mass limits defined to randomly extract the synthetic clusters in each bin. Right: probability density for the distributions of C computed on mock y-maps of the synthetic clusters (grey) and on MILCA (red) and NILC (blue) y-maps of the *Planck* clusters.

is very low. This discrepancy could have an impact on the distributions we compared in the right panel in Fig. 2. Therefore, we plan to improve the analysis by considering a new sample of synthetic clusters which should be better representative of the *Planck* mass distribution. However, in this first application we also performed a check to verify the correlation between C computed on the low-angular resolution (10′) *Planck*-like maps and the χ indicator defining the dynamical state of the synthetic clusters. From the random extractions described above we computed a mean (linear) correlation of $\sim 46\%$, which we can consider as the efficiency of the Zernike fitting in evaluating the cluster dynamical state from *Planck*-like maps.

4.2 Zernike analysis for X-ray maps

The ZPs method can also be applied to the X-ray maps. We used the mock X-ray maps from The Three Hundred project to test that. In this case, the signal distribution in the maps varies on smaller spatial scales with respect to the y-maps (even if they have the same angular resolution, see Sect. 3). This is due to the different dependence of the X-ray surface brightness ($\propto n_e^2$) and of the Compton parameter ($\propto n_e$) from the electron density n_e in the intracluster medium. We found that the X-ray maps are poor modelled with the low-order ZPs we used for the y-maps. However, increasing the number of ZPs in the fit should be more expansive from a computational point of view, and without providing new key information about the cluster dynamical state. Therefore, we simply smoothed the maps by considering the logarithm of the photon number counts and we used 45 ZPs in the fit, the same as for the y-maps. Note that also these X-ray maps were analysed in [3] by using the M parameter. We obtained a good correlation at $z = 0$ between C and both M ($\sim$ 80%) and χ ($\sim$ 65%). We also estimated a correlation of $\sim$ 84% between C computed on the X-ray maps and on the y-maps. We plan to extend the X-ray analysis at higher redshifts and to compare in more details the results with the study of the y-maps in [4], in order to check if and how the morphological classification of the clusters changes in the two surveys.

5 Conclusions

The modelling of projection maps of galaxy clusters with ZPs results in a valuable approach to study their morphology and infer their dynamical state. By using a set of mock y and X-ray maps we verified that it is possible to extract a single parameter from the Zernike fitting which is well correlated with combinations of more common morphological parameters or dynamical-state indicators. The advantage of this approach is its flexibility, i.e. the chance to simply tune the accuracy of the modelling by changing the number of ZPs used, by exploiting their completeness and orthogonality. This is particularly useful when dealing with real maps at low resolution and/or with a residual noise contamination. The method is well adapted to study clusters in large samples even if, in general, such a morphological analyses are limited by the angular resolution in the maps. We are currently using ZPs to study y-maps of local ($z < 0.1$) *Planck* SZ-selected clusters. The final results will be presented in an upcoming work.

References

[1] E. Rasia *et al.*, Astron. Rev., **8**, 40 (2013)
[2] G. Cialone *et al.*, MNRAS, **477**, 139 (2018)
[3] F. De Luca *et al.*, MNRAS, **504**, 5383 (2021)
[4] V. Capalbo *et al.*, MNRAS, **503**, 6155 (2021)
[5] R. A. Sunyaev, I. B. Zeldovich, ARA&A, **18**, 537 (1980)
[6] R. J. Noll, J. Opt. Soc. Am., **66**, 207 (1976)
[7] W. Cui *et al.*, MNRAS, **480**, 2898 (2018)
[8] J. A. ZuHone, E. J. Hallman, ASCL, record ascl:1608.002 (2016)
[9] N. Meidinger *et al.*, SPIE, **106991F** (2018)
[10] Planck Collaboration XXVII, A&A, **594**, A27 (2016)
[11] Planck Collaboration XXII, A&A, **594**, A22 (2016)
[12] G. Hurier *et al.*, A&A, **558**, A118 (2013)
[13] M. Remazeilles *et al.*, MNRAS, **410**, 2481 (2011)
[14] R. Haggar, *et al.*, MNRAS, **492**, 6074 (2020)

Many-probes multi-object spatially-resolved analyses of galaxy clusters in the big data era

Fabio Castagna[1,2,*], *Stefano* Andreon[2], *Alberto* Trombetta[1], and *Marco* Landoni[3]

[1]Department of Pure and Applied Sciences, Computer Science Division, Insubria University, 21100 Varese, Italy
[2]INAF–Osservatorio Astronomico di Brera, via Brera 28, 20121 Milano, Italy
[3]INAF-Osservatorio Astronomico di Brera, via Emilio Bianchi 46, 23807 Merate, Italy

Abstract. The thermal Sunyaev-Zeldovich (SZ) effect and the X-ray emission offer separate and highly complementary probes of the thermodynamics of the intracluster medium, particularly on their radial dependence. We already released `JoXSZ`, the first publicly available code designed to jointly fit SZ and X-ray data coming from various instruments to derive the thermodynamic radial profiles of galaxy clusters, including mass. `JoXSZ` follows a fully Bayesian forward-modelling approach, adopts flexible parametrization for the thermodynamic profiles and includes many useful options that users can customize according to their needs. We are including shear measurement in our joint analysis, and moving from single-cluster to multi-cluster analyses, allowing to quantify the heterogeneity of thermodynamic properties within the cluster population. At the same time, we are creating a suitable framework that efficiently stores and optimally processes huge volumes of data being released by the current and new generation surveys.

1 Introduction

Galaxy clusters are the largest and most massive gravitationally bound objects in the Universe, and thus they offer a unique tracer of cosmic evolution [1]. The thermodynamic properties of a galaxy cluster can be probed through observations in the optical, X-ray, or microwave bands via the Sunyaev-Zeldovich (SZ) effect [2, 3], that we are combining into a single joint analysis.

We have already released `JoXSZ` [4, source code available on GitHub[1]], the first publicly available code designed to jointly fit SZ and X-ray data to derive the thermodynamic profiles of galaxy clusters.

We are currently working to include shear data as a third complementary component in the fit and to move from single-cluster to multi-cluster analyses, allowing to gather thermodynamic measurements on populations of clusters. Since it will involve large amount of observations coming from current and next generation facilities, we are creating a suitable framework that efficiently stores and optimally processes huge volumes of data.

*e-mail: fabio.castagna@inaf.it
[1]https://github.com/fcastagna/joxsz

2 `JoXSZ`

`JoXSZ` is a Python code that combines SZ and X-ray data into a single analysis following a fully Bayesian forward-modelling approach. `JoXSZ` accounts for beam smearing and data analysis transfer function, for the SZ calibration uncertainty and X-ray and SZ background level systematics. It adopts extremely flexible parametrization for the thermodynamic profiles and it employs a consistent temperature across the various parts of the code, allowing for differences between X-ray and SZ gas mass weighted temperatures when required by the user, and calculates the correct Poisson-Gauss expression for the joint likelihood.

`JoXSZ` is built upon the SZ data fitting pipeline described in `PreProFit` [5, source code available on GitHub[2]] and an updated version of the X-ray data cube fitter `MBProj2`, originally developed by [6]. `JoXSZ` merges and extends these two processes into a unique joint and consistent model based on a Markov chain Monte Carlo (MCMC) fitting algorithm.

2.1 Inputs and Outputs

For each cluster, `JoXSZ` requires the user to specify the SZ surface brightness radial profile, and the X-ray data cube in the form of multiple energy band radial profiles. The SZ analysis requires the point spread function (PSF), the SZ transfer function (a correction for data affected by filtering), and the conversion factor from Compton parameter to the observed data unit with temperature dependence (see [5] for more details). In X-ray, by using a data cube we exploit the information about temperature contained in the X-ray spectra. The components required for processing X-ray data, *i.e.* the response matrix file (RMF) and the ancillary response file (ARF), must be provided by the user (see [4] for more details).

Through MCMC computation, `JoXSZ` samples the joint and marginal posterior distribution of parameters and derives the radial profiles of main thermodynamic properties, *i.e.* pressure, electron density, temperature, entropy. Other thermodynamic profiles are computed as usual [6], including mass profile and gas fraction under the assumption of hydrostatic equilibrium.

2.2 Program flow

Figure 1 depicts the block diagram describing the program flow of `JoXSZ`. In the SZ pipeline, outlined in Fig. 2, `JoXSZ` parametrizes the pressure profile, then projects it onto a two-dimensional map using forward Abel transform, convolves the map with the instrumental beam and the transfer function, and finally derives the surface brightness profile through appropriate conversion factors. If the beam file already includes the transfer function, the user has to specify it and `JoXSZ` will correctly handle it. Regarding the X-ray analysis, shown in Fig. 3, `JoXSZ` makes full use of the spatial-spectral X-ray data cube, profiting of the better angular resolution of X-ray data, when available, to measure thermodynamic profiles at small radii unprobed by lower resolution SZ data.

`JoXSZ` is compatible with SZ data coming from different sources: NIKA, MUSTANG and SPT, whose provenence should be indicated because teams releasing the data adopts different standards. After specification, `JoXSZ` will process the data accordingly. `JoXSZ` has been already used for data coming from Chandra, NIKA and MUSTANG [4, 7], and is being used for data from XMM, XRT, SPT, and Planck [8]. `JoXSZ` is also able to read data in different formats, and to automatically handle measurement units throughout the code using `astropy.units` [9] upon correct specification from the user of the measurement units of input data.

[2] https://github.com/fcastagna/preprofit

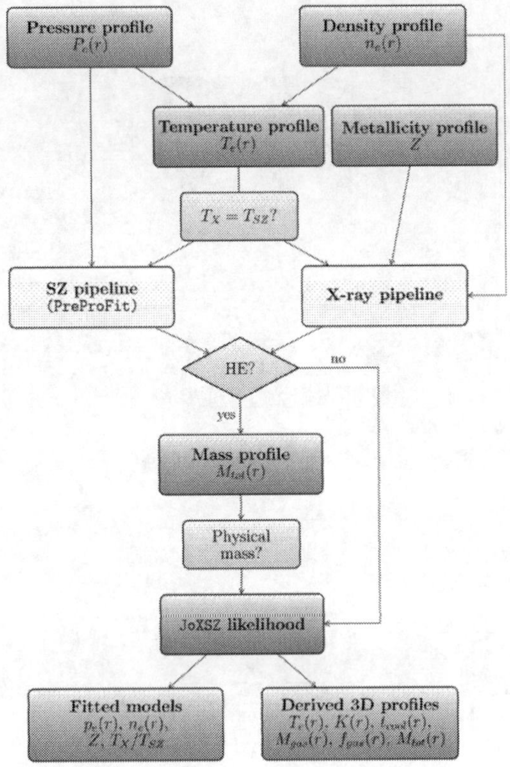

Figure 1. Block diagram showing the program flow. Radial profiles are in red. Options are in green. Analysis pipelines are in yellow. Data enter in the blue box.

2.3 Thermodynamic profiles modelization

Users are allowed to formulate the pressure profile through either parametric model or binned profile: a generalized Navarro, Frenk & White (gNFW) model [10], a cubic spline model or a power-law interpolation model. Optionally, users can apply a prior constraint on the outer slope of the pressure profile. The multi-parameter (up to 10) model introduced by Vikhlinin [11] is used to parametrize the electronic density profile, and given the large number of possible parameters it should be flexible enough to satisfy the needs of most users.

The temperature profile is simultaneously influenced by the pressure constraints from SZ observations and the X-ray data cube, *i.e.* the X-ray spectra at various distances from the cluster center. `JoXSZ` allows users either to consider a unique temperature profile $T_{SZ} = T_X$, or to make a distinction between the gas mass weighted temperature T_{SZ} and the X-ray temperature T_X, introducing the multiplicative parameter $\log(T_X/T_{SZ})$.

`JoXSZ` assumes a flat metallicity profile, whose value can be fitted, or fixed, at user request.

2.4 Additional functionalities

Additional functionalities are implemented in the code, each user can choose whether to adopt them or not. By default, `JoXSZ` includes a calibration parameter accounting for uncertainties in the SZ measurements and a X-ray background scaling parameter that accounts for

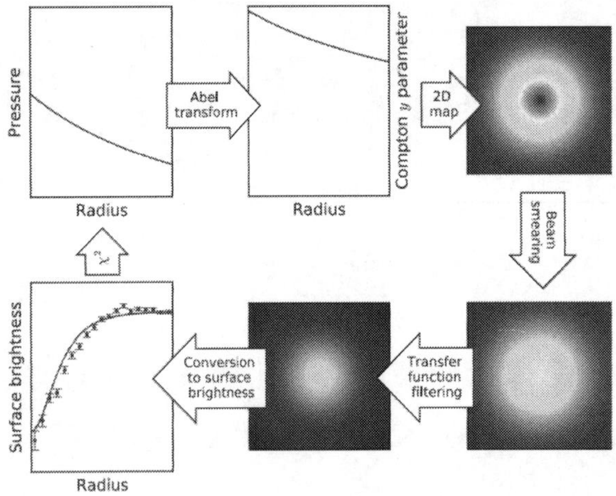

Figure 2. Graphical representation of the SZ single-cluster analysis pipeline within JoXSZ. The multi-object version loops over the clusters.

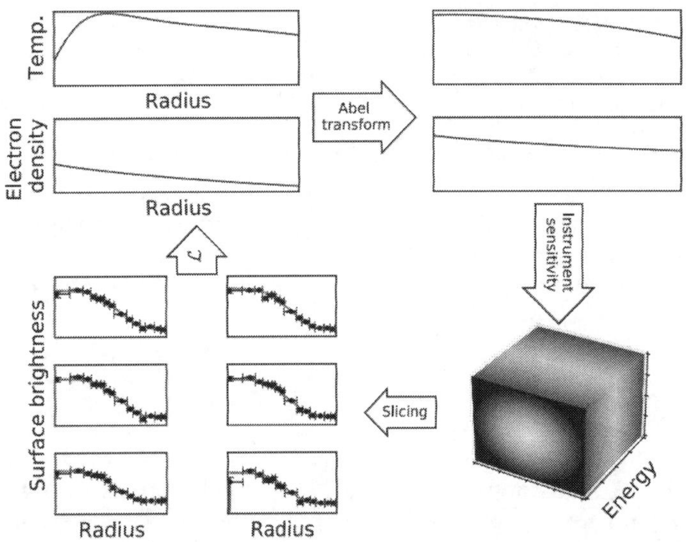

Figure 3. Graphical representation of the X-ray single-cluster analysis pipeline within JoXSZ. The multi-object version loops over the clusters.

differences in background level between the cluster and control field directions. In the SZ analysis, a pedestal parameter for the surface brightness can be added in the case of maps with non-zero level. Another option is to apply a prior constraint on the integrated Compton parameter. JoXSZ can be easily used for feasibility studies adopting simulated data, e.g. Gaussian approximations for the PSF and the transfer function.

If dealing with a large filtering image, JoXSZ gives the opportunity to neglect extremely small scales to increase the execution speed, but at the user's own risk.

To further save CPU time, users may ask to start the exploration of the parameter space not far from the likelihood maximum by specifying a (guessed) value for the cluster's characteristic radius, from which pressure parameters are derived assuming the universal profile defined by Arnaud et al. [12] as first approximation. This option is also available for a binned pressure model: JoXSZ automatically converts the parameters from the universal profile into the required binning values.

3 Current and future developments

To fully take advantage of the multi-probe potential of current and incoming surveys, such as Vera Rubin Observatory, Euclid, eROSITA, Roman Space Telescope, ACT, and SPT, we are currently adding shear measurements to our joint SZ and X-ray analysis. At the same time, we are paving the way to extend the analysis from single clusters at a time to multiple clusters at the same time, meaning that parameter estimation will be simultaneously conducted at both individual level and population of clusters level through a Bayesian hierarchical model. Such a tool will be of great interest for the astrophysical community, giving the opportunity to answer more complex questions [13], e.g. testing General Relativity [14] or assessing the heterogeneity of thermodynamic properties within the cluster population, such as the scatter in pressure or mass profiles.

Since the code of `JoXSZ` includes highly time-consuming operations, we are currently working on optimizing the computation, both reducing redundant operations and exploiting parallel computing techniques. In view of the extension from single to multi-object analyses, we need to develop typical features of Big Data Analytics, in order to transform the huge avalanche of data into accurate information. To do so, we are constructing an adequate infrastructure which is based on state-of-the-art providers for data storage (Amazon Web Services, e.g. see Landoni et al. [15]) and data processing (Apache Spark [16]). Frameworks like Apache Spark allow us to define very specific functionalities on datasets of large dimensions, and to define computation procedures that can be easily scaled on parallel and distributed architectures. Furthermore, they are able to integrate data analysis techniques based on deep learning, such as convolution neural networks, recently used to estimate cluster masses through X-ray data.

References

[1] G.M. Voit, Reviews of Modern Physics **77**, 207 (2005), astro-ph/0410173
[2] R.A. Sunyaev, Y.B. Zeldovich, Ap&SS **7**, 3 (1970)
[3] R.A. Sunyaev, Y.B. Zeldovich, Comments on Astrophysics and Space Physics **4**, 173 (1972)
[4] F. Castagna, S. Andreon, A&A **639**, A73 (2020), 2005.12315
[5] F. Castagna, S. Andreon, A&A **632**, A22 (2019)
[6] J.S. Sanders, A.C. Fabian, H.R. Russell, S.A. Walker, Monthly Notices of the Royal Astronomical Society **474**, 1065 (2018), 1705.09299
[7] S. Andreon, C. Romero, F. Castagna, A. Ragagnin, M. Devlin, S. Dicker, B. Mason, T. Mroczkowski, C. Sarazin, J. Sievers et al., MNRAS **505**, 5896 (2021), 2106.11327
[8] S. Andreon, et al., in prep. (2021)
[9] Astropy Collaboration, T.P. Robitaille, E.J. Tollerud, P. Greenfield, M. Droettboom, E. Bray, T. Aldcroft, M. Davis, A. Ginsburg, A.M. Price-Whelan et al., A&A **558**, A33 (2013), 1307.6212
[10] D. Nagai, A.V. Kravtsov, A. Vikhlinin, ApJ **668**, 1 (2007), astro-ph/0703661
[11] A. Vikhlinin, A. Kravtsov, W. Forman, C. Jones, M. Markevitch, S.S. Murray, L. Van Speybroeck, The Astrophysical Journal **640**, 691 (2006), astro-ph/0507092
[12] M. Arnaud, G.W. Pratt, R. Piffaretti, H. Böhringer, J.H. Croston, E. Pointecouteau, A&A **517**, A92 (2010), 0910.1234
[13] E.D. Feigelson, R.S. de Souza, E.E.O. Ishida, G. Jogesh Babu, Annual Review of Statistics and Its Application **8**, 493 (2021), 2005.13025

[14] V.F. Cardone, P. Karmakar, M. De Petris, R. Maoli, Phys. Rev. D **103**, 064065 (2021), 2008.00580

[15] M. Landoni, P. Romano, S. Vercellone, J. Knödlseder, A. Bianco, F. Tavecchio, A. Corina, ApJS **240**, 32 (2019), 1901.00410

[16] M. Zaharia, M. Chowdhury, M.J. Franklin, S. Shenker, I. Stoica et al., HotCloud **10**, 95 (2010)

CONCERTO at APEX: Installation and first phase of on-sky commissioning

A. Catalano[4,6,*], P. Ade[1], M. Aravena[2], E. Barria[3,4], A. Beelen[5], A. Benoit[3,4], M. Béthermin[5], J. Bounmy[6,4], O. Bourrion[6,4], G. Bres[3,4], C. De Breuck[7], M. Calvo[3,4], F.-X. Désert[8,4], C.A Durán[9], G. Duvauchelle[5], L. Eraud[6], A. Fasano[3,4], T. Fenouillet[5], J. Garcia[5], G. Garde[3,4], J. Goupy[3,4], C. Groppi[10], C. Hoarau[6,4], W. Hu[5], G. Lagache[5], J.-C. Lambert[5], J.-P. Leggeri[3,4], F. Levy-Bertrand[3,4], J. Macías-Pérez[6,4], H. Mani[10], J. Marpaud[6,4], M. Marton[6], P. Mauskopf[10], A. Monfardini[3,4], G. Pisano[1], N. Ponthieu[8,4], L. Prieur[5], G. Raffin[6], S. Roni[6], S. Roudier[6], D. Tourres[6,4], C. Tucker[1], and L. Vivargent[6]

[1]Astronomy Instrumentation Group, University of Cardiff, The Parade, CF24 3AA, United Kindgom
[2]Núcleo de Astronomía, Facultad de Ingeniería y Ciencias, Universidad Diego Portales, Av. Ejército 441, Santiago, Chile
[3]Univ. Grenoble Alpes, CNRS, Grenoble INP, Institut Néel, 38000 Grenoble, France
[4]Groupement d'Interet Scientifique KID, 38000 Grenoble and 38400 Saint Martin d'Héres, France
[5]Aix Marseille Université, CNRS, LAM (Laboratoire d'Astrophysique de Marseille), F-13388 Marseille, France
[6]Univ. Grenoble Alpes, CNRS, LPSC/IN2P3, 38000 Grenoble, France
[7]European Southern Observatory, Karl Schwarzschild Straße 2, 85748 Garching, Germany
[8]Univ. Grenoble Alpes, CNRS, IPAG, 38400 Saint Martin d'Héres, France
[9]European Southern Observatory, Alonso de Cordova 3107, Vitacura, Santiago, Chile
[10]School of Earth and Space Exploration and Department of Physics, Arizona State University, Tempe, AZ 85287, USA

Abstract. CONCERTO (CarbON CII line in post-rEionisation and ReionisaTiOn) is a large field-of-view (FoV) spectro-imager that has been installed on the Cassegrain Cabin of Atacama Pathfinder EXperiment (APEX) telescope in April 2021. CONCERTO hosts 2 focal planes and a total number of 4000 Kinetic Inductance Detectors (KID), with an instantaneous FoV of 18.6 arcminutes in the range of 130-310 GHz. The spectral resolution can be easily tuned down to 1 GHz depending on the scientific target. The scientific program of CONCERTO has many objectives, with two main programs focused on mapping the fluctuations of the [CII] line intensity in the reionisation and post-reionisation epoch (4.5<z<8.5), and on studying galaxy clusters via the thermal and kinetic Sunyaev-Zel'dovich (SZ) effect. CONCERTO will also measure the dust and molecular gas contents of local and intermediate-redshift galaxies, it will study the Galactic star-forming clouds and finally it will observe the CO intensity fluctuations arising from 0.3<z<2 galaxies.
The design of the instrument, installation at APEX and current status of the commissioning phase and science verification will be presented. Also we describe the deployment and first on-sky tests performed between April and June 2021.

*e-mail: catalano@lpsc.in2p3.fr

1 Introduction

Starting from January 2019, the Grenoble collaboration (LPSC[1], Institut Néel and IPAG [2]) together with the LAM[3], funded by an Advanced Grant ERC has worked to design, fabricate, install, commission, and observe with the CONCERTO [1] instrument. CONCERTO is installed at the 12-metre APEX telescope and now it is open to the APEX community, the scientific goals will be manifold. The CONCERTO collaboration will focus on two cosmological goals:

- The measurement of variations of the [CII] line emission at redshift z>5.2. The atomic [CII] line is one of the most valuable tracers of star formation. At high redshifts it is observed in the sub-millimetre and millimetre atmospheric windows [2, 3]. We will use the [CII] line emission as a tracer of cosmic density structure, and make the first constraints on the power spectrum of dusty star-forming matter. Our experiment will also observe the CO intensity fluctuations arising from galaxies at redshifts between 0.2 and 2, mapping the spatial distribution and abundance of molecular gas over a broad range of cosmic time.
- Precise measurements of the shape of the SZ electromagnetic spectrum for cluster of galaxies for redshifts between 0.2 and 0.8 to disentangle the kinetic and relativistic corrections from the main thermal contribution [3].

Since April 2021 CONCERTO is installed at the 12 metre APEX telescope, located at a 5105 m altitude on the Llano de Chajnantor in Northern Chile [4]. The commissioning phase has been performed until the end of June. Starting from July 2021 the instrument has started its regular scientific program. In the next sections we will present a description of the instrument, its installation at the telescope and some technical and scientific commissioning results.

2 The CONCERTO Instrument

The instrument is based on several sub-systems: an optical box which selects the optical inputs to be injected in a second box that contains a Martin-Puplett interferometer. The signal is then focused on two arrays of about 4000 KID detectors, cooled at 100 mK through a closed-cycle dilution cryostat (see Fig. 1).

The design and the control of the systematic errors of such a configuration drives the major requirement for the whole instrument. This is because the MPI must be able to perform continuously up to 9 cm path interferograms, which gives a spectral resolution of about 1 GHz within the band, at a frequency of about 4 Hz in order to maintain a low sky noise level. The fast KID time response permits us to use this method without any loss of information. The readout sampling frequency is constrained by this requirement; data are sampled at 4 kHz producing roughly 128 MByte per second.

The optical system has been arranged to have a FoV of 18.6 arcmin. Since an MPI measures the difference between two optical inputs [5], we split the beam after passing through the telescope into two fully polarised beams before reaching the MPI. This can be done with the use an initial wire-grid polariser. After interfering in the MPI, the first half of the beam is focused onto the detectors, creating an image of 18.6 arcmin on the sky. The second half of the beam is first de-focused from a focused image to a pupil (corresponding to the image of the primary mirror of the telescope) and then collected to the detectors. This makes an initial hardware subtraction of the constant common mode from the atmosphere to simplify

[1] *Laboratoire de Physique Subatomique et Cosmologie*
[2] *Institut de planétologie et d'astrophysique de Grenoble*
[3] *Laboratoire d'Astrophysique de Marseille*

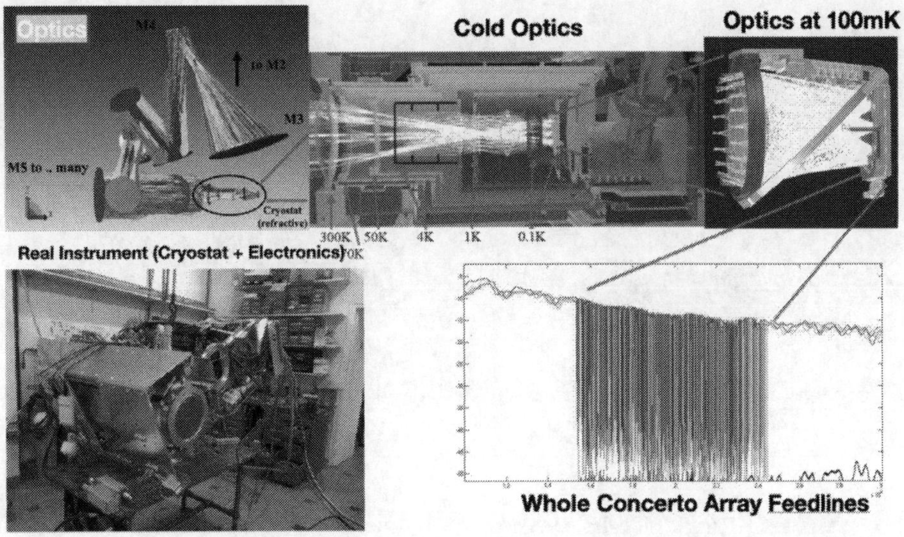

Figure 1. Top: CONCERTO optical design (using Zemax software with details of the cold optics inside the cryostat and the 100 mK stage cold pupil, polariser and the two arrays). Bottom left: picture of the fabricated CONCERTO cryostat and warm electronics in the laboratory at Neel in Grenoble. Bottom right: baseline sweeps for one array (6 feedlines) taken with Vector Network Analyser (VNA).

the off-line analysis. In addition, instead of the sky itself, the second input can be taken from a cold reference, which is a cold black-body at a given tunable temperature.

The 18.6 arcmin focal plane is fully covered by two arrays of single polarisation aluminium LEKID containing 2152 pixels each and separated by a 45 degrees polariser. Each array has a different optical band 195-310 GHz for the High frequency array (HF) and 130-270 GHz for low frequency array (LF).

We recall in Table 1 the main characteristics of CONCERTO.

Telescope primary mirror diameter [m]	12
Field-of-view diameter [arcmin]	18.6
Beam Widths [arcsec]	30 (HF) 35 (LF)
Absolute spectral resolution [GHz]	≥ 1
Relative spectral resolution R [#]	1–300
Frequency range HF | LF [GHz]	195–310 | 130–270
Pixels on Sky HF | LF [#]	2,152 | 2,152
Instrument geometrical throughput [sr m^2]	2.5×10^{-3}
Single Pixel geometrical throughput [sr m^2]	1.16×10^{-6}
Data rate [MBytes/sec]	128

Table 1. Main characteristics of CONCERTO at APEX.

3 Installation and first phase of on-sky commissioning

After installation of the whole instrument, the cryostat has been first cooled down. After reaching a base temperature of 70 mK CONCERTO has been switch on for the first time the

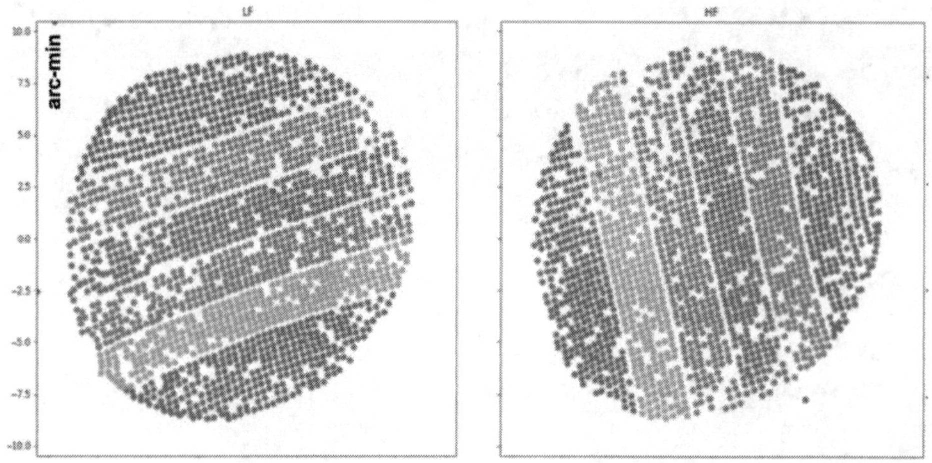

Figure 2. LF (left) and HF (right) focal plane geometries. Around 90% of the designed pixels exhibit a beam and are healthy.

12th of April 2021. The first phase of the technical commissioning started right after. During this phase, the sub-systems operations and typical electrical characteristic of the electronics and detectors (feedlines connections, quality factors of the detectors, feedline cosmetics, etc..) were checked. The optical alignment has been achieved using an optical laser and a very first light on Jupiter was obtained.

After these first checks we started to point the telescope in order to start the scientific commissioning. We list, among others, the following achievements:

- About 90% of the detectors are valid and able to detect bright sources individually on the sky.
- A fraction of about 70% of the designed KIDs exhibits a beam with an eccentricity lower than 0.7.
- A preliminary Noise Equivalent Temperature (NET) has been evaluated and equal to 2.2 mK/$\sqrt{Hz}$, which is well in agreement with the expected value. Note that we expect to have a better sensitivity per beam at low frequency where we have more than one KID per beam, and at intermediate frequencies where the two arrays overlap in frequency.

The presented results must be consolidated with detailed data analysis, which is still in progress. Before the end of the scientific commissioning few selected astronomical targets were observed to demonstrate the capabilities of the instrument and optimize the observing strategies. In particular in this paper we present in Figure 3 photometric maps (integrated emission on the full CONCERTO band) of the star forming region NGC6334 (the Cat Paw Nebula) and of the Crab nebula supernova remnant.

4 Conclusions, Future work and perspectives

The CONCERTO instrument has been successfully installed on the APEX telescope in April 2021. Operations were started few days after the installation and scientific commissioning was performed in less than two months.

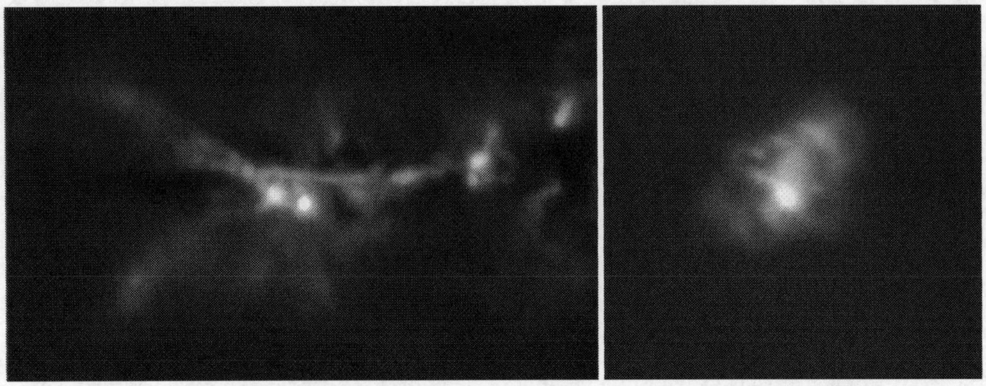

Figure 3. First results of the CONCERTO scientific commissioning. Two observations of continuum emission are shown as observed by the two bands of CONCERTO (1). Left plot: star forming region NGC6334 (Cat Paw Nebula) mapped in a field of 37×25 arcmin^2 for 16 minutes of integration. Right plot: Crab nebula observed with 2.5 minutes integration time mapped in a field of 10×10 arcmin^2. In both cases we represent composite images where the LF array is in blue, and the HF array in red (ESO/CONCERTO collaboration: https://www.eso.org/public/unitedkingdom/announcements/ann21010/?lang).

Here we presented preliminary results of the scientific commissioning showing estimates of the instrument performance well in agreement with expectations.

The regular scientific observations have already started in July 2021 thanks to the fact that the instrument has a complete remote operations design. Indeed, after installation it became impossible to travel to Chile, therefore operations of the instrument are all conducted remotely from France.

Spectra and further results of the commissioning will be presented in a more extensive paper in the coming months. Scientific commissioning data are available in the ESO data base at https://www.eso.org/sci/publications/announcements/sciann17437.html.

Acknowledgements. This project has received funding from the European Research Council (ERC) under the European Union's Horizon 2020 research and innovation programme (grant agreement No 788212) also from the Excellence Initiative of Aix-Marseille University-AMidex, a French "Investissements d'Avenir" programme and the labex FOCUS. We thank the electronics, mechanics, cryogenics and administrative groups at Institut Néel, LPSC and LAM. For the outstanding technological and human support we thank the whole APEX staff.

References

[1] A. Monfardini and G. Lagache, Nature Astronomy 5, 970 (2021)
[2] Kovetz, E. D., Viero, M. P., Lidz, A., et al. 2017, arXiv e-prints, arXiv:1709.09066
[3] CONCERTO collaboration, 2020, A&A 642, 60
[4] APEX Telescope https://www.eso.org/public/teles-instr/apex/
[5] D. H. Martin and E. Puplett, Polarised interferometric spectrometry for the millimeter and submillimeter spectrum, Infrared Physics, 10, 105-109 (1970)

The THREEHUNDRED project: The effect of baryon processes at galaxy cluster scale

Weiguang Cui[1,*]

[1]Institute for Astronomy, University of Edinburgh, Royal Observatory, Edinburgh EH9 3HJ, UK

Abstract. The role of baryon models in hydrodynamic simulations is still unclear. Future surveys that use cluster statistics to precisely constrain cosmological models require a better understanding of the baryonic effects. Using the hydro-simulated galaxy clusters from different baryon models (Gadget-MUSIC, Gadget-X and Gizmo-SIMBA) from the THREEHUNDRED project, we can address this question into more details. We find that the galaxy cluster mass change due to different baryon models is at a few per cent level. However, the mass changes can be positive or negative depending on the baryon models. Such a small mass change leaves a weak influence (slightly larger compared to the mass changes) on both the cumulative halo numbers and the differential halo mass function (HMF) above the mass completeness. Similarly to to the halo mass change, the halo mass or the HMF can be increased or decreased with respect to the dark-matter-only (DMO) run depending on the baryon models.

1 Introduction

More than a decade ago, theoretical studies on structure and galaxy formations were relying N-body DMO simulations, see the well-known Millennium simulations [1] and the Multi-Dark simulations [2], for example. As we can only directly observe the galaxies in the sky and simulations only contains gravitational bound objects – dark matter halos, there is a missing connection between observation – galaxies, and theoretical prediction – dark matter halos. Thus, numerous techniques, such as hydrodynamic simulations, semi-analytic models, empirical forward modelling, subhalo abundance modelling and halo occupation models (from more physical models to more empirical models), are developed to bridge this connection [see 3, for a recent review]. Ideally, we would like to model the baryon process as physical as we could. However, limited by the current computation power, unknown baryon processes or incomplete baryon models implemented in the simulation codes [see 4, for the current development of baryon models in hydrodynamic simulations] as well as the dependence on the simulation resolution[5], we are still facing difficulties to fully model galaxy formation at the non-linear scales. Nevertheless, the hydrodynamic simulations including baryons evolved simultaneously with dark matter under the same gravity potential is the only way to understand the difference between these DMO simulations and the effects of baryons on them. Numerous works have studied the effects from different aspects [see 6, for a review]. However, no clear agreement has been reached, especially at the non-linear scale.

*e-mail: weiguang.cui@ed.ac.uk

Galaxy clusters will be able to provide precise values of σ_8 and Ω_m with errors at around 0.1 percent from the next-generation space telescopes [see 7, for example], such as EUCLID, CSST. Therefore, theoretically understanding the cluster masses, especially their changes according to different baryon models, is essential to accurately constrain cosmology parameters [see 8, for a detailed discussion]. In this short article, we will address this question, focusing on the galaxy cluster scale with clusters simulated with different codes from the THREEHUNDREDproject[9]. The outline of this paper is following: the THREEHUNDREDproject is introduced at Sect. 2; the effects of baryons on the halo mass and HMF are presented in Sect. 3; we finally draw our conclusions and discuss our results in Sect. 4.

2 The simulated galaxy clusters from the THREEHUNDRED project

The THREEHUNDRED project [9] is a re-simulation of 324 most massive galaxy clusters ($M_{vir} > 8 \times 10^{14}\,h^{-1}M_{\odot}$)[1] from the MultiDark simulation (MDPL2, [2], also referred to as the DMO run in this paper) which utilises the cosmological parameters from the Planck mission [10], and has a periodic cube of comoving length 1 h^{-1} Gpc containing 3840^3 DM particles with a mass of $1.5 \times 10^9\,h^{-1}M_{\odot}$. Each cluster lies in the highest resolution region of a comoving radius of 15 h^{-1}Mpc (over 5 times R_{200}) for re-simulations with different baryonic models: Gadget-MUSIC[11], Gadget-X[12, 13], Gizmo-SIMBA([14] and Cui et al. 2021 in prep. for more details). **We note here that the three simulation codes are all based on Gadget-3, but they are different from the hydrodynamics (from classic SPH and Wendland SPH to meshless finite mass) to the detailed baryon models (see Cui et al. 2021 in prep.).**

These re-simulation regions are generated with the parallel GINNUNGAGAP code (https://github.com/ginnungagapgroup/ginnungagap: the highest resolution Lagrangian regions share the same mass resolution as the original MDPL2 simulation with gas particles ($M_{gas} = 2.36 \times 10^8\,h^{-1}M_{\odot}$) split from DM particles. The outside regions are degraded in multiple layers (with a shell thickness of $\sim 4\,h^{-1}$Mpc) with lower mass resolution particles (mass increased by eight times for each layer) that eventually provide the same tidal fields yet at a much lower computational costs. The halo catalogue used in this paper is generated with AHF[15] and we mainly focusing on halo masses within two interesting overdensities: 200 and 500, at $z = 0$, in this paper.

Benefiting from the large volume of these re-simulated cluster regions, detailed relations between the central cluster and the filaments connecting to it have been studied [16–18]. Furthermore, the cluster backsplash galaxies [19, 20] and shock radius [21] have also been well addressed. The advanced baryon models in hydrodynamic simulations allow us to perform a detailed investigation on the cluster properties, such as profiles [22, 23], substructure and its baryonic content [24–27], dynamical state and morphology [28, 29], cluster (non-)thermalization [30, 31], the fundamental plane [32], and the cluster mass biases [33, for hydrostatic-equilibrium assumption], [34, for sigma-mass relation] and [35, for weak-lensing]. Lastly, comparing to the void/field region runs in this project allows us to study the effect of environment [36]; including the self-interacting dark matter run allow us to constrain the dark matter cross-section[37]; It further help us to examine the chameleon gravity[38].

3 The effects on halo mass and HMF

The effect of baryons on HMF is simply a consequence of the halo mass change. Therefore, we need to first investigate the halo mass changes due to the effects of baryon models. This

[1] The halo mass is defined as the mass enclosed inside an overdensity of δ times the critical density of the universe: $\delta =\sim 98$ for virial mass and $M_{200,500}$ is with $\delta = 200, 500$ respectively.

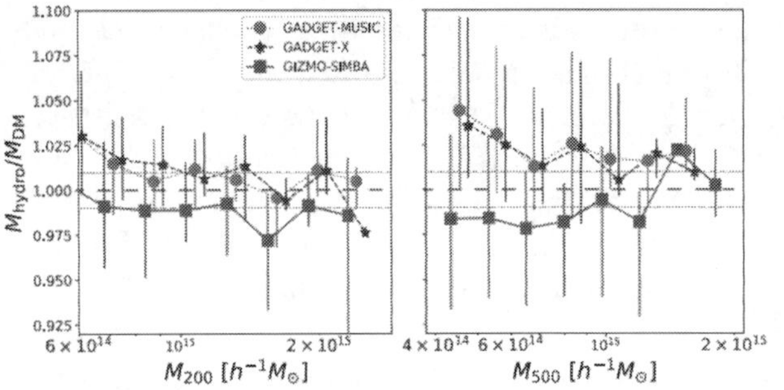

Figure 1. The halo mass differences at M_{200} (left panel) and M_{500} (right panel) with respect to the DMO run. X-axis shows the halo masses from the DMO run. As indicated in the legend, different symbols with colours show different simulation runs. The errorbars show the $16^{th} - 84^{th}$ percentiles. Only the mass-complete sample is presented here. The horizontal dashed line marks the unit value of 1 with the two dotted lines show the 1 percent difference in mass.

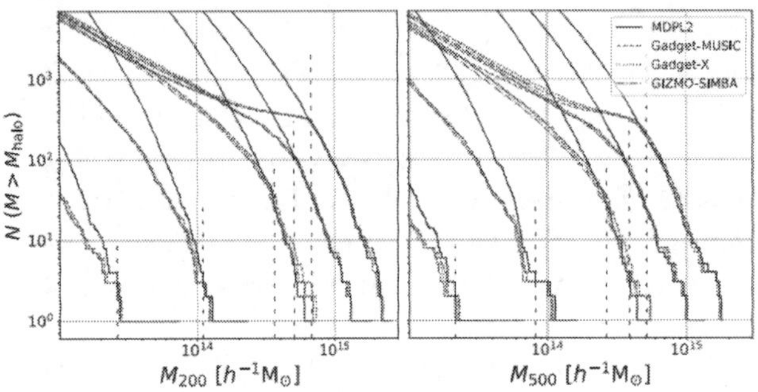

Figure 2. The cumulative HMF from MDPL2 (solid black lines), GADGET-MUSIC(red dashed lines), GADGET-X(blue dotted lines) and GIZMO-SIMBA(green dot-dashed lines). Left panel shows the results of M_{200} with M_{500} is presented on the right panel. In each panel, the results from $z = 4.0, 2.3, 1.0, 0.5$, and 0.0 are presented from left to right, respectively. Vertical dashed lines show the mass-completeness limitation at each redshift.

Table 1. The mass-complete samples at different redshifts. The second row shows the mass completeness thresholds for M_{200} with the numbers of clusters above that threshold in MDPL2, GADGET-MUSIC, GADGET-X and GIZMO-SIMBA, respectively. Similar results at M_{500} are at bottom.

Redshift	z=0.0	z=0.5	z=1.0	z=2.3	z=4.0
M_{200} $[10^{14}\,h^{-1}\mathrm{M}_{\odot}]$	6.71	5.02	3.62	1.10	0.26
Number of clusters	284/300/308/282	112/104/110/108	26/38/27/27	3/3/3/3	3/3/2/3
M_{500} $[10^{14}\,h^{-1}\mathrm{M}_{\odot}]$	5.16	3.90	2.65	0.82	0.21
Number of clusters	184/208/206/179	73/76/85/73	28/36/30/28	3/3/3/3	1/2/1/2

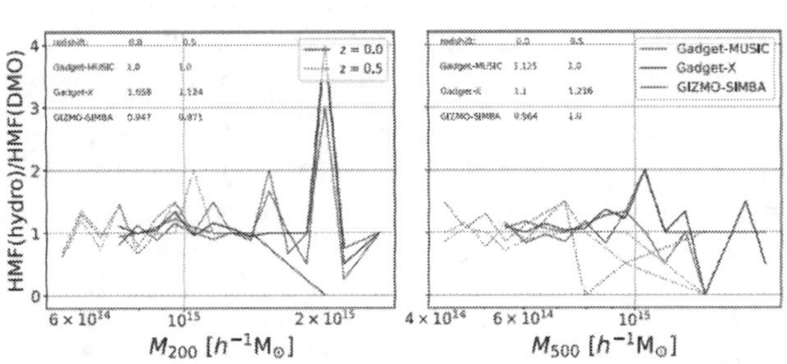

Figure 3. The differential HMF differences with respected to the one from MDPL2 above the mass completeness. The same mass bin with size of 0.05 in logarithm space is applied. Left panel shows the results of M_{200} with M_{500} is presented on the right panel. In each panel, only the results from $z = 0.5$ (dotted lines), and 0.0 (solid lines) are presented due to the statistic limits. The median changes for all bins with ≥ 8 clusters are shown in the top left corner of each panel.

requires cross-identifying the same galaxy cluster between different runs. For each DMO cluster, we simply use the cluster centre position ($\Delta_r < 200\ h^{-1}$kpc) and its mass ($\Delta_M/M_{200} < 0.1$) to find its counterpart.

Figure 1 shows the relative mass differences between the hydro-simulated clusters and their DMO counterparts. As there are many clusters, we simply bin them in the DMO cluster mass and show the median values with errorbars. At M_{200}, both Gadget-MUSIC and Gadget-X tend to have slightly (at about 1 percent) higher mass than Gizmo-SIMBA which seems about 1 percent lower than the DMO halo mass. Gadget-X shows weak dependence on halo mass, while Gizmo-SIMBA does not within this mass-complete range. At M_{500}, all the three hydro runs tends to deviate from the DMO halo mass larger towards opposite directions, besides the highest halo mass end. It also worth to note that the errorbars are slightly larger compared to the ones in the left panel.

At *lower* halo mass $M_{halo} \lesssim 10^{14}\, h^{-1}\mathrm{M}_\odot$, the hydro-simulated halo tends to gain mass without AGN feedback [see 39, for example]. However, it is well-known that it is very hard to cease the star formation in galaxies without AGN feedback which will result in unrealistic galaxy properties compared to observation. Including AGN feedback, the hydro-simulated halo tends to lose mass compared to the DMO run [see 40, 41, for example]. At more *massive* end, this requires a very large volume simulation or zoomed-in simulations like the THREEHUNDRED (this work) to provide statistical information. It seems that the *different* baryon models leave a weak influence (~ a few percents) on changing these massive cluster masses.

We show the cumulative HMFs in figure 2 for these different runs. The mass completeness of our cluster sample is shown by vertical dashed lines, which is determined by the crossing point between the Gizmo-SIMBA and MDPL2 lines. More details about the mass-complete sample can be found in table 1. Note that these values are slightly different to [9] which used the crossing point between Gadget-X and MDPL2. It is clear that the difference between these cumulative HMFs is very small for our mass-complete sample.

Lastly, in figure 3, the relative differences of these differential HMFs with respect to the DMO run seem to wiggle around 1 with no clear discrepancy within this limited mass range. Calculating the median raitos for all the mass bins with more than 8 clusters, we notice a few to about 10 percent changes depending on the simulations with lower values from Gizmo-SIMBA and higher values for Gadget-X. Similar to the halo mass changes, different baryon models can also change the HMF towards the same opposite directions.

4 Conclusion and discussion

Use the hydro-simulated galaxy clusters from the THREEHUNDRED project, which was performed using different simulation codes, we investigate the effects of baryon models on the halo mass and HMF by comparing to their DMO counterparts. Limited to our mass-complete sample, we can only statistically present the results at the massive cluster mass scale. We find that the cluster mass changes due to different baryon models is at few percent level. There is a slightly larger deviation for $\Delta_{M_{500}}$ than $\Delta_{M_{200}}$. The cluster mass can be increased or decreased with respect to the DMO run and that is model dependent. Such a small change in mass leaves weak influence on the HMF (although slightly higher fractions compared to mass changes). However, different models can alter the changes into different directions of increase or decease, the same as the mass changes.

The effect of baryons on halo mass is tightly connected with the halo mass definition. In this paper, we adopt the commonly used halo masses M_{200} and M_{500} which are based on the critical density of the universe. Within a fixed radius (not very large), we can expect that baryons condensed in the centre of the cluster will result in a high concentration, thus an increase in mass. While this may result a steep slope and the 200, 500 $\times\rho_{crit}$ can be reached at a shorter distance. Hence, maybe a lower $M_{200,500}$. Therefore, it is also not surprising to see that there is less mass change for M_{200} than M_{500} as shown in figure 1 [see also 39, 40]. We note here that the FoF (Friend-of-Friend) halo definition seems to have a similar problem [see 40, for example]. Furthermore, the effect of baryons may change the centre of the halo [42]. Although this may not happen very frequently, the change due to this mis-centring problem is generally believed to bias towards a lower halo mass.

Acknowledgement

The author would like to thank the amazing organisers of the NIKA2 conference and the THREEHUNDRED project for contribution and collaboration. WC is supported by the STFC AGP Grant ST/V000594/1 and the science research grants from the China Manned Space Project with NO. CMS-CSST-2021-A01 and CMS-CSST-2021-B01.

References

[1] V. Springel, S.D.M. White, A. Jenkins, C.S. Frenk, N. Yoshida, L. Gao, J. Navarro, R. Thacker, D. Croton, J. Helly et al., Nature**435**, 629 (2005), astro-ph/0504097
[2] A. Klypin, G. Yepes, S. Gottlöber, F. Prada, S. Heß, MNRAS**457**, 4340 (2016), 1411.4001
[3] R.H. Wechsler, J.L. Tinker, ARA&A**56**, 435 (2018), 1804.03097
[4] R.S. Somerville, R. Davé, ARA&A**53**, 51 (2015), 1412.2712
[5] A.D. Ludlow, J. Schaye, R. Bower, MNRAS**488**, 3663 (2019), 1812.05777
[6] W. Cui, Y. Zhang, *The Impact of Baryons on the Large-Scale Structure of the Universe*, in *Trends in Modern Cosmology* (2017), 1708.02644
[7] B. Sartoris, A. Biviano, C. Fedeli, J.G. Bartlett, S. Borgani, M. Costanzi, C. Giocoli, L. Moscardini, J. Weller, B. Ascaso et al., MNRAS**459**, 1764 (2016), 1505.02165
[8] S.N.B. Debackere, J. Schaye, H. Hoekstra, MNRAS**505**, 593 (2021), 2101.07800
[9] W. Cui, A. Knebe, G. Yepes, F. Pearce, C. Power, R. Dave, A. Arth, S. Borgani, K. Dolag, P. Elahi et al., MNRAS**480**, 2898 (2018), 1809.04622
[10] Planck Collaboration, P.A.R. Ade, N. Aghanim, M. Arnaud, M. Ashdown, J. Aumont, C. Baccigalupi, A.J. Banday, R.B. Barreiro, J.G. Bartlett et al., A&A**594**, A13 (2016), 1502.01589
[11] F. Sembolini, G. Yepes, M. De Petris, S. Gottlöber, L. Lamagna, B. Comis, MNRAS**429**, 323 (2013), 1207.4438
[12] E. Rasia, S. Borgani, G. Murante, S. Planelles, A.M. Beck, V. Biffi, C. Ragone-Figueroa, G.L. Granato, L.K. Steinborn, K. Dolag, ApJ**813**, L17 (2015), 1509.04247
[13] A.M. Beck, G. Murante, A. Arth, R.S. Remus, A.F. Teklu, J.M.F. Donnert, S. Planelles, M.C. Beck, P. Förster, M. Imgrund et al., MNRAS**455**, 2110 (2016), 1502.07358
[14] R. Davé, D. Anglés-Alcázar, D. Narayanan, Q. Li, M.H. Rafieferantsoa, S. Appleby, MNRAS**486**, 2827 (2019), 1901.10203
[15] S.R. Knollmann, A. Knebe, ApJS**182**, 608 (2009), 0904.3662
[16] A. Rost, U. Kuchner, C. Welker, F. Pearce, F. Stasyszyn, M. Gray, W. Cui, R. Dave, A. Knebe, G. Yepes et al., MNRAS**502**, 714 (2021), 2012.02850
[17] U. Kuchner, A. Aragón-Salamanca, F.R. Pearce, M.E. Gray, A. Rost, C. Mu, C. Welker, W. Cui, R. Haggar, C. Laigle et al., MNRAS**494**, 5473 (2020), 2004.08408
[18] U. Kuchner, A. Aragón-Salamanca, A. Rost, F.R. Pearce, M.E. Gray, W. Cui, A. Knebe, E. Rasia, G. Yepes, MNRAS**503**, 2065 (2021), 2102.12500

[19] R. Haggar, M.E. Gray, F.R. Pearce, A. Knebe, W. Cui, R. Mostoghiu, G. Yepes, MNRAS**492**, 6074 (2020), 2001.11518

[20] A. Knebe, M. Gámez-Marín, F.R. Pearce, W. Cui, K. Hoffmann, M. De Petris, C. Power, R. Haggar, R. Mostoghiu, MNRAS**495**, 3002 (2020), 2005.09896

[21] E.J. Baxter, S. Adhikari, J. Vega-Ferrero, W. Cui, C. Chang, B. Jain, A. Knebe, MNRAS**508**, 1777 (2021), 2101.04179

[22] R. Mostoghiu, A. Knebe, W. Cui, F.R. Pearce, G. Yepes, C. Power, R. Dave, A. Arth, MNRAS**483**, 3390 (2019), 1812.04009

[23] Q. Li, W. Cui, X. Yang, E. Rasia, R. Dave, M. De Petris, A. Knebe, J.A. Peacock, F. Pearce, G. Yepes, MNRAS**495**, 2930 (2020), 2005.06135

[24] J. Arthur, F.R. Pearce, M.E. Gray, A. Knebe, W. Cui, P.J. Elahi, C. Power, G. Yepes, A. Arth, M. De Petris et al., MNRAS**484**, 3968 (2019), 1901.05969

[25] R. Haggar, F.R. Pearce, M.E. Gray, A. Knebe, G. Yepes, MNRAS**502**, 1191 (2021), 2101.03178

[26] R. Mostoghiu, A. Knebe, F.R. Pearce, C. Power, C.D.P. Lagos, W. Cui, S. Borgani, K. Dolag, G. Murante, G. Yepes, A&A**652**, A10 (2021), 2105.01676

[27] R. Mostoghiu, J. Arthur, F.R. Pearce, M. Gray, A. Knebe, W. Cui, C. Welker, S.A. Cora, G. Murante, K. Dolag et al., MNRAS**501**, 5029 (2021), 2101.01734

[28] V. Capalbo, M. De Petris, F. De Luca, W. Cui, G. Yepes, A. Knebe, E. Rasia, MNRAS**503**, 6155 (2021), 2009.04565

[29] F. De Luca, M. De Petris, G. Yepes, W. Cui, A. Knebe, E. Rasia, MNRAS**504**, 5383 (2021), 2011.09002

[30] J. Sayers, M. Sereno, S. Ettori, E. Rasia, W. Cui, S. Golwala, K. Umetsu, G. Yepes, MNRAS**505**, 4338 (2021), 2102.06324

[31] M. Sereno, L. Lovisari, W. Cui, G. Schellenberger, MNRAS**507**, 5214 (2021), 2109.06199

[32] L.A. Díaz-García, K. Umetsu, E. Rasia, W. Cui, M. Meneghetti, arXiv e-prints arXiv:2108.13498 (2021), 2108.13498

[33] S. Ansarifard, E. Rasia, V. Biffi, S. Borgani, W. Cui, M. De Petris, K. Dolag, S. Ettori, S.M.S. Movahed, G. Murante et al., A&A**634**, A113 (2020), 1911.07878

[34] D. Anbajagane, H. Aung, A.E. Evrard, A. Farahi, D. Nagai, D.J. Barnes, W. Cui, K. Dolag, I.G. McCarthy, E. Rasia et al., arXiv e-prints arXiv:2110.01683 (2021), 2110.01683

[35] R. Herbonnet, A. Crawford, C. Avestruz, E. Rasia, C. Giocoli, M. Meneghetti, A. von der Linden, W. Cui, G. Yepes, arXiv e-prints arXiv:2109.01673 (2021), 2109.01673

[36] Y. Wang, F. Pearce, A. Knebe, G. Yepes, W. Cui, C. Power, A. Arth, S. Gottlöber, M. De Petris, S. Brown et al., ApJ**868**, 130 (2018), 1809.05244

[37] J. Vega-Ferrero, J.M. Dana, J.M. Diego, G. Yepes, W. Cui, M. Meneghetti, MNRAS**500**, 247 (2021), 2006.08596

[38] A. Tamosiunas, C. Briddon, C. Burrage, W. Cui, A. Moss, arXiv e-prints arXiv:2108.10364 (2021), 2108.10364

[39] W. Cui, S. Borgani, K. Dolag, G. Murante, L. Tornatore, MNRAS**423**, 2279 (2012), 1111.3066

[40] W. Cui, S. Borgani, G. Murante, MNRAS**441**, 1769 (2014), 1402.1493

[41] T. Castro, S. Borgani, K. Dolag, V. Marra, M. Quartin, A. Saro, E. Sefusatti, MNRAS**500**, 2316 (2021), 2009.01775

[42] W. Cui, C. Power, V. Biffi, S. Borgani, G. Murante, D. Fabjan, A. Knebe, G.F. Lewis, G.B. Poole, MNRAS**456**, 2566 (2016), 1512.01253

https://doi.org/10.1051/epjconf/202225700012

Millimetric Sardinia radio Telescope Receiver based on Array of Lumped elements kids

G. D'Alessandro[1,2,*], *E.* Barbarava[1], *E.S.* Battistelli[1,2], *P.* de Bernardis[1,2], *F.* Cacciotti[1], *V.* Capalbo[1], *E.* Carretti[3], *F.* Columbro[1,2], *A.* Coppolecchia[1,2], *A.* Cruciani[2], *M.* De Petris[1,2], *F.* Govoni[4], *G.* Isopi[1], *L.* Lamagna[1,2], *P.* Marongiu[4], *S.* Masi[1,2], *L.* Mele[1,2], *M.* Molinari[4], *M.* Murgia[4], *A.* Navarrini[4], *A.* Orlati[3], *A.* Paiella[1,2], *G.* Pettinari[5], *F.* Piacentini[1,2], *T.* Pisanu[4], *S.* Poppi[4], *G.* Presta[1,2], and *F.* Radiconi[1,2]

[1]*Sapienza* Università di Roma, Roma, Italy
[2]INFN sezione di Roma, 00185 Roma, Italy
[3]INAF - Istituto di Radioastronomia
[4]INAF - Osservatorio Astronomico di Cagliari
[5]CNR-IFN - Istituto di Fotonica e Nanotecnologie, Roma, Italy

Abstract. MISTRAL is a millimetric camera working in the W-band (78–103 GHz) which will take data from the Sardinia Radio Telescope, the Italian 64-m radio telescope located 50 km form Cagliari, at 600 m above the sea level, in Sardinia. It is being built as a facility instrument by the Sapienza University for INAF, that manages the radio telescope, under a PON contract. It will consist of a compact cryostat hosting the re–imaging optics, cooled at 4 K, and a 408–pixel array of photon–noise limited lumped element kinetic inductance detectors fabricated at CNR-IFN and cooled at a base temperature lower than 300 mK. MISTRAL will be able to investigate a long list of scientific targets spanning from extragalactic astrophysics to solar system science, with high angular resolution ($\sim$ 12 arcsec), including Sunyaev Zel'dovich effect measurements and the study of the Cosmic Web.

1 Introduction

Sardinia Radio telescope (SRT) is a multipurpose telescope aimed to measure the electromagnetic radiation from 300 MHz to 116 GHz. After the scientific commissioning, in 2015, 4 receivers have been installed and successfully operated at frequency lower than 26.5 GHz. New instruments are going to be installed in the next future, among which, MISTRAL will measure in the W-band.

High angular resolution W-band measurements open at a number of possibilities of scientific cases, including galaxies spectral energy distribution measurement and their continuum observation [1], AGN and radio galaxies [2], detection of circumstellar disks [3], dense core in giant molecular gas investigation [4], synchrotron non thermal jet and hot spot [5], intra-clusters filaments, cosmic web, and other S-Z related observation [6, 7]; and even more, by correlating with other experiment.

*e-mail: giuseppe.dalessandro@roma1.infn.it

Because of the large diameter telescope needed to perform high angular resolution measurements only few cameras around the world are devoted to this, such as MUSTANG2 [8] at the Green Bank Telescope.

MISTRAL aims to perform high angular resolution measurement with an array of lumped element kids detectors coupled, thanks to a re-imaging optic system, with SRT. MISTRAL is in an advanced phase of manufacturing and all the parts of the experiment are going to be assembled together at the beginning of 2022. MISTRAL will be installed in March 2022 and commissioning activities will start in spring 2022.

In section 2 we describe the Sardinia Radio Telescope enlightening its optical system and the typical observational conditions. Section 3 describes the instrument and its main parts.

2 Sardinia Radio Telescope

SRT [9] is a multipurpose instrument operated in either single dish or Very Long Baseline Interferometer mode. It is located 50km north of Cagliari in Sardinia (Italy), Lat. 39.4930N - Long. 9.2451E, 600 m above sea level. Construction was completed in August 2012. The technical commissioning phase to validate scientific performances was managed by National Institute for Astrophysics and concluded in 2014. The Early Science Program observations started in 2016, and regular observations in 2018.

The primary antenna (M1) is fully steerable, 64 m in diameter. It is composed of 1008 aluminum elements controlled by electromechanical actuators. An f/0.33 primary focus occurs near the sub-reflector (M2), 7.9 m in diameter, composed of 49 aluminum elements. Its position can be changed for focus adjustment. M1 and M2 are shaped to minimize spillover and the standing waves between the feed and the sub-reflector. The Gregorian focus, f/2.34 occurs around 20 meters below M2 in the Gregorian room, the violet box in Figure 1. MISTRAL will be placed in this room by using the Gregorian focus of SRT.

SRT has been designed to perform measurements from 305 MHz to 116 GHz [10]. Four receivers have been successfully installed: L-band (305 MHz - 410 MHz), P-band(1.3 GHz - 1.8 GHz), C-band (5.7 GHz - 7.7 GHz) and K-band (18 GHz - 26.5 GHz). In the next future, new receivers will increase the frequency coverage [10]: Q-band (33 GHz - 50 GHz), polarization-sensitive W-band (16 beams), K/Q/W band (3 beams) polarization-sensitive, and MISTRAL[1].

Table 1. SRT optical and mechanical main features

M1 diameter	64 m
M2 diameter	7.9 m
Primary focus	f/0.33
Gregorian focus	f/2.34
Pointing accuracy	2-13 arcsec
Range in elevation	5-90 deg
Range in azimuth	180 ± 270 deg

Among other activities, a metrological activity to measure with high precision deformations of the SRT due to gravitational and thermal effects is undergoing [10]; by means of the

[1]The Enhancement of the Sardinia Radio Telescope (SRT) for the study of the Universe at high radio frequencies is financially supported by the National Operative Program (Programma Operativo Nazionale - PON) of the Italian Ministry of University and Research Research and Innovation 2014-2020, Notice D.D. 424 of 28-02-2018 for the granting of funding aimed at strengthening research infrastructures, in implementation of the Action II.1 – Project Proposal PIR01_00010

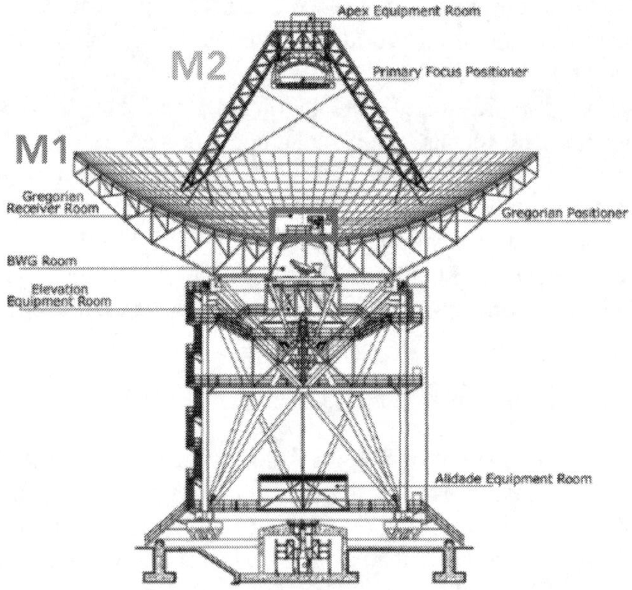

Figure 1. SRT schematic. The primary mirror (M1) consists of 1008 individual aluminum panels. Each panel has an area ranging from 2.4 to 5.3 m^2. It is built using aluminum sheets glued with epoxy resin. The sub-reflector (M2) consists of 49 individual aluminum panels. Near M2 occurs a primary f/0.33 focus. MISTRAL will be placed in the Gregorian room (violet box) where the f/2.34 Gregorian focus occurs.

active surface it will be possible to correct the deformations of the main mirror allowing the surface roughness needed for observations up to 116 GHz. Also, the metrological system will allow high precision pointing measuring and correcting the subreflector displacement from the optical axis together with the deformation of the whole structure.

2.1 Site condition

Estimation of sky opacity, based on recorded atmospheric data, forecasts by Navarrini et al. (http://hdl.handle.net/20.500.12386/28787) report $\tau < 0.15$ (50th percentile) at 93 GHz during the winter nights. The average PWV in the same conditions is mainly 8 mm. 50 years of radiosonde profiles taken at Cagliari airport (30 km far, at sea level) and scaled for SRT site shows PWV < 11 mm (50th percentile) and opacity < 0.2 (50th percentile) at 100 GHz [11]. What mainly concern W-band measurement a dynamic scheduling will be applied to optimize the observational time and weather conditions.

3 MISTRAL instrument

MISTRAL is a W-band camera for SRT. The focal plane is composed of 408 Lumped element kinetic inductance detectors (LEKID), fabricated at CNR-IFN, arranged on a 97 mm circular focal plane. The focal plane is optically coupled with SRT thanks to a re-imaging optics system composed of two silicon lenses coated with an anti-reflection system. The focal plane is cooled down at 270 mK with a cryostat provided by QMC instruments from Cardiff (UK).

3.1 Cryostat

The cryostat is provided by QMC instruments[2]. It is composed of two radiation shields at 40 K and 4 K cooled down by a pulse tube cryocooler. Another shield, cooled at 1 K by a He^4 fridge, surrounds the focal plane assembly. The detectors reach 270 mK thanks to the use of two He^3 refrigerators. The external shell, made on Aluminum, is 700 mm diameter and 1680 mm high, see Figure 2. It is closed with a 190 mm UHMW-PE window [12]

The pulse tube is a Sumitomo RP-182B2S-F100H with 1.5 W at 4.2 K and 36 W at 48 K, and uses a remote valve in order to decrease the vibrations on the focal plane. The pulse tube head is mechanically disconnected to the shell thanks to a vibrations dumper. The pulse tube is water-cooled and we plan to use 100 m Helium lines[13].

The sub-kelvin stages are cooled down with a Chase Twin GL10 fridge: two He^3 reaching 251 mK with loading of 20 μW for the focal plane, a He^3 reaching 332 mK with 30 μW for the focal plane support and a He^4 reaching 840 mK with 150 μW for the focal plane shield.

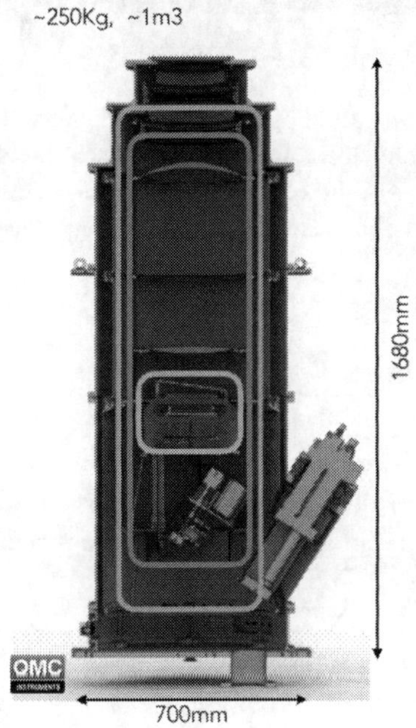

Figure 2. MISTRAL cryostat receiver provided by QMC. A pulse tube refrigerator cooled down two radiation shields, at 40 K (orange box) and 4 K (green one). The fridges cooled down the 1 K shield (light blue box), the focal plane support and the focal plane (dark blue box). The focal plane is cooled down at ~ 270 mK

3.1.1 Magnetic shield

The experiment will move with the telescope during the observations. A magnetic shield surrounds the detectors, fridges, and relevant read-out parts to mitigate the Earth's magnetic field effects. The shield (1 mm thick) is made of Cryoperm 10 with $\mu_r > 70000$.

[2]http://www.terahertz.co.uk/qmc-instruments-ltd

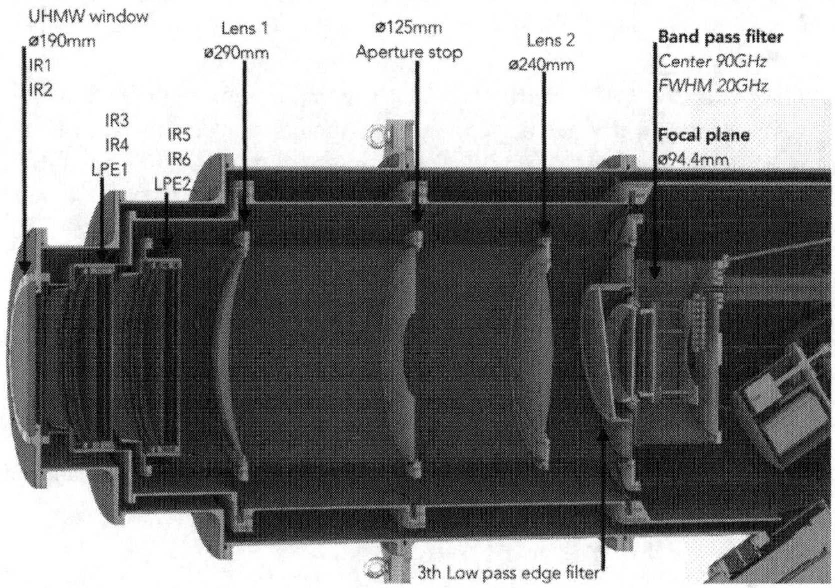

Figure 3. MISTRAL optics chain. A UHMW window closes the vacuum shell with a clearance of 190 mm. The 40 K and 4 K shields are closed by a series of stack filters: for each shield, two thermal filters and a low pass edge filter. At 4 K the re-imaging optical system has been placed. The last low pass edge filter closes the 1 K shield. The last filter is a band-pass filter which defines the MISTRAL absorption band.

3.2 Optics

MISTRAL uses a re-imaging optical system, see Figure 3, to couple the Gregorian focus with the array focal plane. The re-imaging optics is composed of two Silicon lenses (L1 and L2) and a cold stop which is the aperture stop of the system. L1 is a convex-concave aspherical lens 290 mm in diameter. The cold stop is a 125 mm circular aperture coated with an absorber material (i.e. Eccosorb AN72). L2 is a convex-convex aspherical lens. The relay optics has been optimised to operate at 4 K assuming a Silicon refractive index of 3.4. Each lens surface will be covered with an antireflection coating based on one layer of PTFE; the efficiency in transmission is always > 90% over the MISTRAL frequency band. The field of view of the system is 4 arcmin by assuming an array focal plane 94.4 mm in diameter. The focal plane scale ratio is 2.54 arcsec/mm. The Strehl ratio on-axis is 0.97 with an FWHM of 12.2 arcsec; at the edge of the focal plane, the Strehl ratio decrease at 0.91, and the FWHM became 12.7 arcsec. The overall efficiency of MISTRAL optics chain is 90 % on-axis and decrease around 60 % near the focal plane edge.

3.3 Focal Plane

MISTRAL focal plane will be populated with 408 LEKIDs arranged on a triangular scheme with 4.2 mm side. Each absorber is a front-illuminated, 3rd order Hilbert, 3 mm × 3 mm resonator, see Figure 4. The wafer is a 4 inches Silicon wafer 235 μm thick, while the LEKIDs are made of superconducting Ti-Al bilayer (10 nm + 30 nm), as in [14, 15]. Due to the re-imaging optics two close-by pixels separation is 10.6 arcsec. The band-pass filter, placed in front of the wafer, closes the holder. It is centered at 90 GHz with ~ 25 GHz of bandwidth. The readout system is based on an FPGA, successfully used for OLIMPO [16].

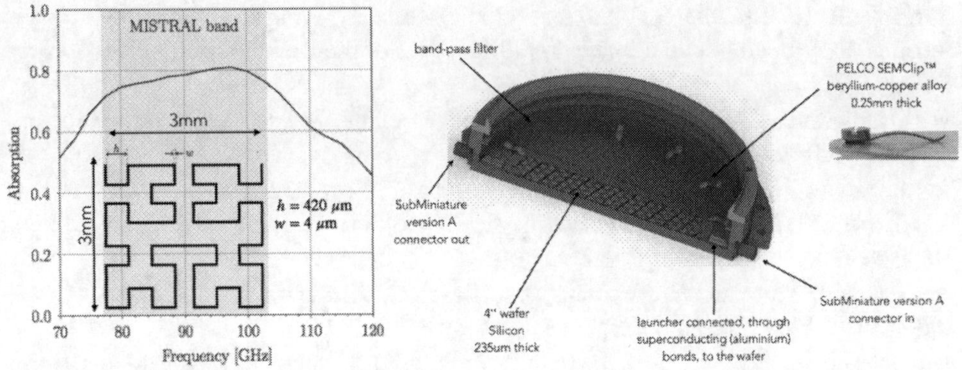

Figure 4. *Left*: HFSS simulations have been performed to optimize the pixel absorption. *Right*: The 4 inches silicon wafer is anchored inside an aluminum holder thanks to beryllium-copper spring. The holder is closed with the band-pass filter and the continuity is guaranteed with SMA connectors.

3.3.1 Sensitivity estimation

To predict the performance on the sky, the measured noise equivalent power (NEP) has been compared with photon noise due to warm optics and atmospheric emission, Figure 4 in [17]. The latter, should be considered in its emission part and turbulence part which will act as correlated noise. A preliminary noise equivalent power about 3.5×10^{-16} W/ $\sqrt{\text{Hz}}$ has been measured at 270 mK by using a test-bed cryostat. By using experimental data, taken at SRT in the K-band, the atmospheric fluctuation noise, using modelling software such as *am*[3], increases the atmospheric noise estimation a factor ≈ 10 with respect to the photon noise level even in the best 5% percentile resulting in a noise equivalent flux density of ≈ 3 mJy/beam.

4 Conclusions

SRT is a multipurpose observatory designed to measure a wide range of radio wavelengths: from 300 MHz to 116 GHz. At SRT, the sky opacity in winter is < 0.15 (50th percentile) at 93 GHz MISTRAL will be coupled with SRT with a re-imaging optical system. The minimum spatial resolution (FWHM) is 12.2 arcsec with a field of view of 4 arcmin. The 408 LEKIDs array has been optimized for the best 90 GHz absorption and for the background at SRT.

The scientific commissioning activities will start in summer 2022.

References

[1] W.F. Wall, I. Puerari, R. Tilanus, F.P. Israel, J.E. Austermann, I. Aretxaga, G. Wilson, M. Yun, K.S. Scott, T.A. Perera et al., Monthly Notices of the Royal Astronomical Society **459**, 1440 (2016), 1603.07736

[2] A. Humphrey, M. Zeballos, I. Aretxaga, D.H. Hughes, M.S. Yun, R. Cybulski, G.W. Wilson, J. Austermann, H. Ezawa, R. Kawabe et al., Monthly Notices of the Royal Astronomical Society **418**, 74 (2011), 1107.3120

[3] M.S. Petersen, R.A. Gutermuth, E. Nagel, G.W. Wilson, J. Lane, Monthly Notices of the Royal Astronomical Society **488**, 1462 (2019)

[3]https://www.cfa.harvard.edu/~spaine/am/

[4] A. Sokol, R. Gutermuth, R. Pokhrel, A. Gómez-Ruiz, G. Wilson, S. Offner, M. Heyer, A. Luna, F. Schloerb, D. Sánchez, Monthly Notices of the Royal Astronomical Society **483**, 407 (2019)

[5] M.J. Hardcastle, L.W. Looney, Monthly Notices of the Royal Astronomical Society **388**, 176 (2008), 0804.3369

[6] T. Mroczkowski, D. Nagai, K. Basu, J. Chluba, J. Sayers, R. Adam, E. Churazov, A. Crites, L. Di Mascolo, D. Eckert et al., Space Science Reviews **215**, 17 (2019), 1811.02310

[7] A. de Graaff, Y.C. Cai, C. Heymans, J.A. Peacock, Astronomy & Astrophysics **624**, A48 (2019), 1709.10378

[8] S.R. Dicker, P.A.R. Ade, J. Aguirre, J.A. Brevik, H.M. Cho, R. Datta, M.J. Devlin, B. Dober, D. Egan, J. Ford et al., Journal of Low Temperature Physics **176**, 808 (2014)

[9] P. Bolli, A. Orlati, L. Stringhetti, A. Orfei, S. Righini, R. Ambrosini, M. Bartolini, C. Bortolotti, F. Buffa, M. Buttu et al., Journal of Astronomical Instrumentation **04**, 1550008 (2015), https://doi.org/10.1142/S2251171715500087

[10] F. Govoni et al., 2021 XXXIVth General Assembly and Scientific Symposium of the International Union of Radio Science **-**, 1 (2021)

[11] F.T. Nasir, F. Buffa, G.L. Deiana, Experimental Astronomy **29**, 207 (2011)

[12] G. D'Alessandro, A. Paiella, A. Coppolecchia, M. Castellano, I. Colantoni, P. de Bernardis, L. Lamagna, S. Masi, Infrared Physics & Technology **90**, 59 (2018)

[13] A. Coppolecchia, E. Battistelli, Masi et al., Pulse tube cooler with > 100m flexible lines optimized for operation of cryogenic detector arrays at large radiotelescopes (2021)

[14] Catalano, A., Goupy, J., le Sueur, H., Benoit, A., Bourrion, O., Calvo, M., D´addabbo, A., Dumoulin, L., Levy-Bertrand, F., Macías-Pérez, J. et al., Astronomy & Astrophysics **580**, A15 (2015)

[15] A. Paiella, A. Coppolecchia, M.G. Castellano, I. Colantoni, A. Cruciani, A. D'Addabbo, P. de Bernardis, S. Masi, G. Presta, Journal of Low Temperature Physics **184**, 97 (2016)

[16] A. Paiella, E.S. Battistelli, M.G. Castellano, I. Colantoni, F. Columbro, A. Coppolecchia, G. D'Alessandro, P. de Bernardis, S. Gordon, L. Lamagna et al., Journal of Physics: Conference Series **1182**, 012005 (2019)

[17] A. Coppolecchia, A. Paiella, L. Lamagna, G. Presta, E.S. Battistelli, P. de Bernardis, M.G. Castellano, F. Columbro, S. Masi, L. Mele et al., Journal of Low Temperature Physics **199**, 130 (2020)

Mass Estimation of Planck Galaxy Clusters using Deep Learning

Daniel de Andres[1,*], *Weiguang* Cui[2], *Florian* Ruppin[3], *Marco* De Petris[4], *Gustavo* Yepes[1], *Ichraf* Lahouli[5], *Gianmarco* Aversano[5], *Romain* Dupuis[5], and *Mahmoud* Jarraya[5]

[1]Departamento de Física Teórica M-8 and CIAFF, Universidad Autónoma de Madrid, Cantoblanco 28049, Madrid, Spain
[2]Institute for Astronomy, University of Edinburgh, Blackford Hill, Edinburgh, EH9 3HJ, UK
[3]Kavli Institute for Astrophysics and Space Research, Massachusetts Institute of Technology, Cambridge, MA 02139, USA
[4]Dipartimento di Fisica, Sapienza Universitá di Roma, Piazzale Aldo Moro, 5-00185 Roma, Italy
[5]EURANOVA, Mont-Saint-Guibert, Belgium

Abstract. Galaxy cluster masses can be inferred indirectly using measurements from X-ray band, Sunyaev-Zeldovich (SZ) effect signal or optical observations. Unfortunately, all of them are affected by some bias. Alternatively, we provide an independent estimation of the cluster masses from the Planck PSZ2 catalog of galaxy clusters using a machine-learning method. We train a Convolutional Neural Network (CNN) model with the mock SZ observations from The Three Hundred (the300) hydrodynamic simulations to infer the cluster masses from the real maps of the Planck clusters. The advantage of the CNN is that no assumption on a priory symmetry in the cluster's gas distribution or no additional hypothesis about the cluster physical state are made. We compare the cluster masses from the CNN model with those derived by Planck and conclude that the presence of a mass bias is compatible with the simulation results.

1 Introduction

Galaxy clusters are the biggest gravitational bound objects in the Universe and their number density as a function of their mass and redshift is very sensitive to the matter content of the Universe and the expansion history. Therefore, they constitute one of the best cosmological probes to constrain different cosmological parameters [1]. However, the total mass of a cluster is not a direct measurable quantity. It is usually inferred from several observables under some physical assumptions. An accurate determination of the total mass of galaxy clusters is in fact a key problem in physical cosmology.

Among the different cluster mass definitions, we are considering M_{500} defined as the total mass contained in a sphere of radius R_{500}, within which the density is 500 times the critical density:

$$M_{500} = \frac{4\pi}{3} R_{500}^3 500 \rho_c(z), \tag{1}$$

here z is the redshift and $\rho_c(z) \equiv 3H(z)^2/8\pi G$ is the critical density.

*e-mail: daniel.deandres@uam.es

By observing the millimeter wavelength sky one can measure the spectral distortions of the cosmic microwave background (CMB) photon due to the inverse-Compton scattering by the free electrons of the intra cluster medium (ICM), i.e. the Sunyaev-Zeldovich (SZ) effect [2]. Through the SZ effect, thousands of clusters have been observed, and discovered, by the South Pole Telescope (SPT; [3]), the Atacama cosmology telescope (ATC; [4]), and the Planck telescope [5]. The intensity of the thermal component of the SZ effect (tSZ) over the line of sight is quantified by the Compton parameter y, defined as

$$y = \frac{\sigma_T k_B}{m_e c^2} \int n_e T_e dl \,, \tag{2}$$

where σ_T is the Thomson cross section, k_B is the Boltzmann constant, c the speed of light, m_e the electron rest-mass, n_e the electron number density, T_e the gas temperature.

One procedure to derive the total mass of the cluster from the y-Compton parameter map is using a scaling relation $Y_{500} - M_{500}$, where Y_{500} denotes the integrated value of the y parameter within R_{500}.

Nevertheless, in the Planck catalog, the calibration of this scaling relation is carried out assuming hydrostatic equilibrium. Therefore the estimated mass , M_{est}, is expected to be biased. This bias parameter b, is usually defined as $1 - b = M_{est}/M_{true}$, where M_{true} is the total mass of the clusters. The determination of cosmological parameters, such as the matter density Ω_m and the present amplitude of density fluctuations σ_8 is sensitive to the particular value of b. Particularly, the bias value b needed to find no tension with CMB observations is $(1-b) = 0.58 \pm 0.04$ [6], or the latest result $(1-b) = 0.62 \pm 0.05$ [7]. However, the bias value is about $1 - b \simeq 0.8$ from numerical simulations, which is clearly in a disagreement with CMB observations (e.g. [8, 9]). In addition, recent analyses of weak lensing observations led to a value of the bias $1 - b = 0.84 \pm 0.04$ [10], which is in agreement with simulations.

In recent development, machine learning techniques have been used in simulations to provide an unbiased estimation of the cluster total mass by training the algorithms using mock observation in X-ray frequencies [11], simulated SZ observations [12], and in a multi-channel approach combining X-ray, SZ and stellar mock maps [13].

In this work, we aim at predicting the total mass of real galaxy clusters selected from the Planck Compton-y parameter maps by training a Convolutional Neural Network (CNN) on simulated maps extracted from a large dataset of hydrodynamic simulations provided by The Three Hundred collaboration [14].

2 Data set

The mock Compton-y parameter maps were computed from the results of the The Three Hundred hydrodynamic simulations. They consist on the resimulations of spherical regions of radius $15h^{-1}Mpc$ centered around the most massive 324 clusters found at $z = 0$ MULTIDARK dark-matter-only simulation (MDPL2; [15]) with cosmological parameters from the Planck collaboration [16]. Particularly, in this work we have used the results from the Gadget-X [17, 18] runs. The cluster size halos are identified with the AHF algorithm [19]. Then, we have chosen halos with masses greater than $10^{14}\, h^{-1}M_\odot$ at redshift $z < 1$. Moreover, the tSZ signal corresponding to these clusters is therefore computed using equation 2 with the PYMSZ public software [20]. The images are generated to have a resolution of 1920×1920 pixels with a fixed angular resolution of 5" so that all images will cover at least R_{200} for all clusters.

A Gaussian smoothing is then applied to these maps in order to mimic the impact of the Planck Beam (FWHM = 10 arcmin) and we further add a similar statistical noise to mimic the real Planck observations. In summary, the simulated data set consists on 7016 selected clusters with 27 random rotations each, amounting to a total of 191,862 maps. They have been particularly selected in order to cover the same mass and redshift range of the selected Planck cluster sample. The real Planck cluster catalog consists of 1,096 observed tSZ maps selected from the Second Planck Catalogue of Sunyaev-Zeldovich Sources (PSZ2; [21]) with known redshifts. In figure 1, we show the Probability Distribution Function (PDF) corresponding to the distribution in redshift and mass of the simulated Planck clusters and real Planck clusters used in this work. An example of the y-maps for the simulated clusters is presented in figure 2.

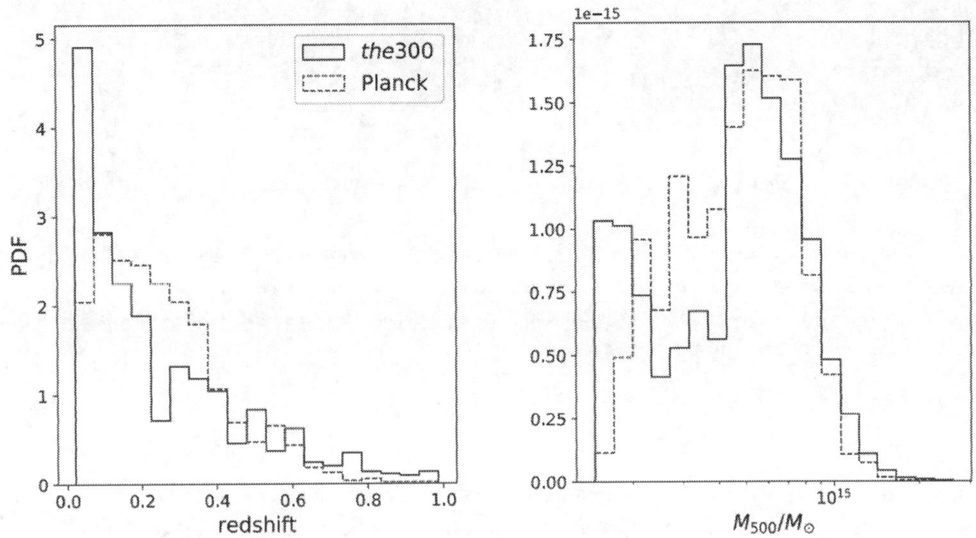

Figure 1. PDF of the cluster samples (The Three Hundred simulation in solid black line and Planck PSZ2 clusters in blue dashed line) corresponding to the distribution in redshift (left panel) and mass (right panel).

3 Algorithm

In this work, we train a deep learning model (e.g. [22]) using the mock images of our simulated cluster sample. Our particular model is based on a CNN model, that can find a mapping from a two dimensional image I_{ij} to a scalar quantity, i.e. M_{500}. Our CNN architecture is a simplified version of the architecture proposed by Simonyan & Zizzerman commonly known as VGGNet [23]. This architecture has already been used to infer cluster masses from simulated observations in [11] and [13]. Furthermore, we have checked that more dense and complicated VGGNet architectures reproduce similar results based on our data set. To train our algorithm, we consider the following loss function

$$\mathcal{L} = \frac{1}{N} \sum_{i=1}^{N} (\log M^{i}_{\text{true}} - \log M^{i}_{\text{pred}})^2 , \tag{3}$$

where M^{i}_{pred} is the predicted CNN mass for cluster i and N is the total number of cluster images in the training set. We then split our data set in 80% training, 10% validation, and 10%

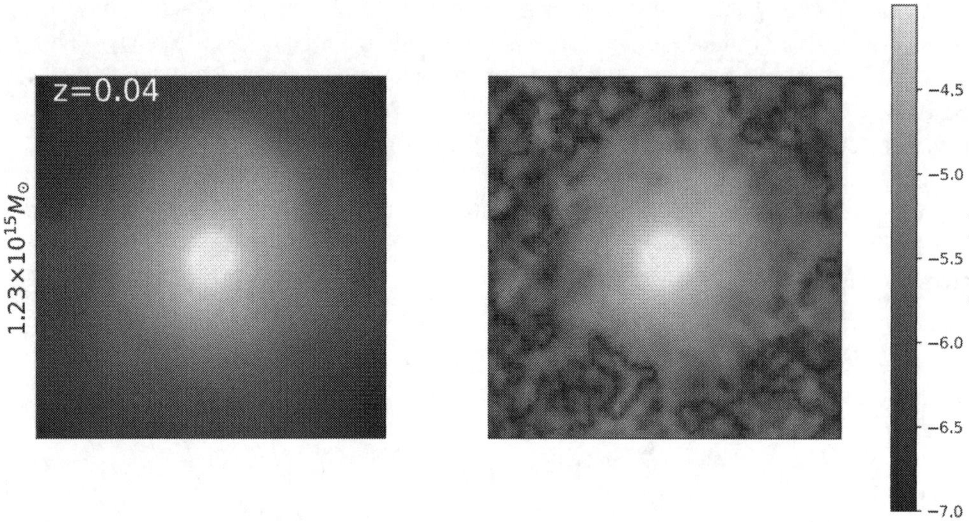

Figure 2. Example of simulated tSZ maps. The left panel corresponds to the clean map and the right panel represents the same image with Planck instrumental beam size and noise. The value of this cluster mass is $M_{500} = 1.23\times10^{15}$ M$_\odot$ and the corresponding redshift is z=0.04. In addition, the color bar shows the logarithmic value of the Compton-y parameter.

test. The validation data set is used to ensure that the model produces an unbiased estimation of the mass. The test set is used exclusively to asses the final accuracy of the model. We have observed that mixing all the redshifts together yields to a very poor performance. Therefore, we have trained 4 CNNs by diving the data set in 4 different redshift bins: $z \leq 0.1$; $0.1 < z \leq 0.2$; $0.2 < z \leq 0.4$ and $0.4 > z$.

Only open source python libraries have been used in this project. Particularly, Keras [24] with Tensorflow [25] GPU acceleration.

4 Results

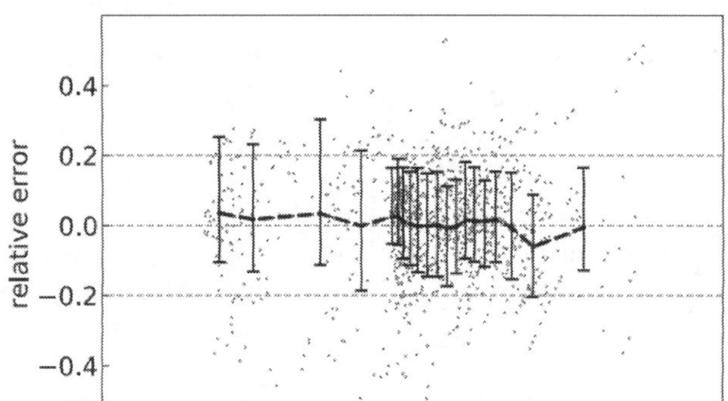

Figure 3. The value of the relative error $(M_{\rm pred} - M_{\rm true})/M_{\rm pred}$ as a function of the predicted mass $M_{\rm pred}$ for The Three Hundred mock data. The dashed black line represents the median for each mass bin while the error bars represents the 68% confident interval. Furthermore, red dashed lines represent the 0 relative error line and ±20% error. Only 1000 points (in red) drawn randomly from the original distribution are displayed. The error bars are computed using the full test set ($\simeq$ 20000 maps).

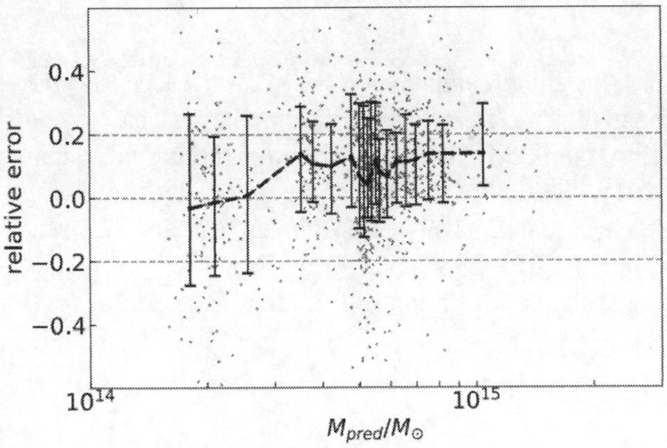

Figure 4. The value of the relative error $(M_{\rm pred} - M_{\rm Planck})/M_{\rm pred}$ as a function of the predicted mass $M_{\rm pred}$ for the **real Planck data set.** The error bars and lines are the same as figure 3.

After training the CNN models with Planck-like simulated Compton-y parameter maps, we predict the masses of the PSZ2 cluster catalog. In order to study the performance of these predictions, we compute the relative error defined as the relative difference of the Planck $M_{\rm Planck}$ mass and the predicted CNN mass $M_{\rm pred}$: relative error = $\frac{M_{CNN}-M_{Planck}}{M_{CNN}}$.

In figure 3, we show the relative error $(M_{\rm pred} - M_{\rm true})/M_{\rm pred}$ as a function of the predicted mass $M_{\rm pred}$ using simulated data. According to our results depicted in the figure 3, the relative error with respect to the 3D dynamical simulated mass is always smaller than 5% for this considered mass range and the scatter (standard deviation σ) is 17%. A similar result by applying the same CNN to real Planck maps is shown in figure 4. However, note that the CNN mass is an unbiased estimator of the cluster mass as shown in figure 3, while the Planck cluster mass show systematic difference to the CNN mass. The bias parameter b can then be computed as the mean of relative difference between $M_{\rm pred}$ and $M_{\rm Planck}$: $b \simeq (M_{\rm pred} - M_{\rm Planck})/M_{\rm pred}$. We would like to highlight that the relative error is biased $1 - b = 0.86$ for massive clusters $M_{\rm pred} > 4 \times 10^{14}$ $M_{\odot}$. However, in the low mass range the CNN predictions seem unbiased with respect to the masses estimated by Planck.

A possible explanation for that is by taking into consideration that Planck collaboration [26] calibrated the scaling relation from a mass-proxy relation [27] with a slope of 1.79, whereas the Gadget-X clusters of The Three Hundred simulations is in agreement with a self-similar scaling relation [14] with a slope of 5/3.

5 Conclusion

In this work, we have estimated the mass of clusters by exploring Compton-y parameter maps provided by Planck. In order to do so, we have created a catalog of almost 200,000 mock Compton-y maps mimicking the noise levels of Planck observations. Then, a CNN is trained with the mock images aiming at predicting the mass of real Planck observations.

Our results show the presence of a mass bias b dependent on cluster mass. The masses estimated by Planck are unbiased with respect to our predicted masses for low mass clusters, while the bias for massive clusters is $1 - b = 0.86^{+0.12}_{-0.7}$.

Otherwise, in hydrodynamic simulations, the value of this bias does not depend on the cluster mass [9] and its value is of order 10-20%. This is in agreement with the fact that the parameters of The Three Hundred $Y_{500} - M_{500}$ scaling relation are compatible with the self-similar relation.

Acknowledgments

We would like to thank the Red Española de Supercomputación for granting access to the Marenostrum supercomputer where most of The Three Hundred simulations have been performed. GY and DA would like to thank MICIN/Feder (Spain) for partial financial support under project grant PGC2018-094975-C21. WC is supported by the STFC AGP Grant ST/V000594/1 and the science research grants from the China Manned Space Project with NO. CMS-CSST-2021-A01 and CMS-CSST-2021-B01. MDP acknowledges support from Sapienza Università di Roma thanks to Progetti di Ricerca Medi 2020, prot. RM120172B32D5BE2.

References

[1] Planck Collaboration et al., A&A, 641, A6 (2020)
[2] Sunyaev R. A., Zeldovich Y. B., Comments on Astro. & Sp. Sci. Physics, 4, 173 (1972)
[3] Bleem L. E., et al., ApJS, 247, 25 (2020)
[4] Hilton M., et al., ApJS, 235, 20 (2020)
[5] Planck Collaboration et al., A&A, 594, A27 (2016)
[6] Planck Collaboration et al., A&A, 594, A24 (2016)
[7] Salvati, et al., A&A, 626, A27 (2019)
[8] Makiya R., Hikage C., Komatsu E., PASJ, psz147 (2020)
[9] Gianfagna G., et al., MNRAS, 502, 5115 (2021)
[10] Herbonnet R., et al., MNRAS, 497, 4684 (2020)
[11] Ntampaka M., et al., ApJ, 876, 82 (2019)
[12] Gupta N., Reichardt C. L., 2020b, The Astrophysical Journal, 900, 110
[13] Yan Z., etal., MNRAS, 499, 3445 (2020)
[14] Cui W., et al., MNRAS, 480, 2898 (2018)
[15] Klypin A., Yepes G., Gottlöber S., Prada F., Heß S., MNRAS, 457, 4340 (2016)
[16] Planck Collaboration et al., A&A, 594, A13 (2016)
[17] Murante G.,etal., MNRAS, 405, 1491 (2010)
[18] Rasia E., et al., ApJ, 813, L17 (2015)
[19] Knollmann S. R., Knebe A., ApJS, 182, 608 (2009)
[20] Cui W., et al., MNRAS, 480, 2898 (2018)
[21] Planck Collaboration XXVII. A&A, 594, A27 (2016)
[22] Goodfellow I. et al, *Deep Learning*. MIT Press (2016)
[23] Simonyan K., Zisserman A., arXiv preprint arXiv:1409.1556 (2014)
[24] Chollet F., et al., Keras, https://keras.io (2015)
[25] Abadi M., et al., arXiv preprint arXiv:1603.04467 (2016)
[26] Planck Collaboration et al., A&A, 571, A20 (2014)
[27] Kravtsov A. V., Vikhlinin A., Nagai D., ApJ, 650, 128 (2006)

CMB at small scales: Cosmology from tSZ power spectrum

Marian Douspis[1,*], *Laura* Salvati[1], *Adélie* Gorce[1,2], and *Nabila* Aghanim[1]

[1]Institut d'Astrophysique Spatiale, CNRS, Université Paris Saclay, Orsay, France
[2]Department of Physics and McGill Space Institute, McGill University, Montreal, QC, Canada H3A 2T8

Abstract. Small scale CMB angular power spectrum contains not only primordial CMB information but also many contaminants coming from secondary anisotropies. Most of the latter depend on the cosmological model but are often marginalised over. We propose a new analysis of the SPT data focusing on the cosmological contribution of the thermal Sunyaev Zel'dovich (tSZ) effect. We model the tSZ angular spectrum with the halo model and train a random forest algorithm to speed up its computation. We show that using the cosmological information of the tSZ on top of the primordial CMB one contained in SPT data bring more constraints on cosmological parameters. We also combine for the first time Planck tSZ angular power spectrum with SPT ones to put further constraints. This proof of concept study shows how much a proper modelling of the foregrounds in the cosmological analyses is needed.

1 Introduction

The thermal SZ effect [1] is a good tracer of the hot gas in the Universe, and is often use to detect or characterise clusters of galaxy. It is also a secondary anisotropy which contaminates CMB data and analyses. Frequency maps observed in the millimetre range contain the sum of many signals coming form the last scattering surface (primordial CMB) and from all gravitational and electromagnetic effects along the line of sight (one of which is the tSZ). Using an adapted component separation method, it is possible to retrieve the tSZ effect from several frequency map and so create a map tracing the hot gas [2, 3]. For example, using for the first time data from two different experiments, Planck and ACT, [4] were able to reconstruct an optimal tSZ map probing both large scales (from Planck) and small scales (from ACT). From the Planck tSZ map, the angular power spectrum can be estimated and further used to put constraints on the cosmological parameters (eg. [5]).

If no component separation is applied, each frequency map and thus each corresponding angular power spectrum, contains a contribution from tSZ. It is consequently necessary to be able to model all the contributions (including the tSZ angular power spectrum) to perform cosmological analyses. Most current analyses of CMB observations at small scales are done following the same approach [see, e.g., 6–9]. A theoretical CMB power spectrum is added to templates of each non-CMB signal to reproduce the observed power spectrum. As mentioned above, apart from the tSZ, the remaining non-CMB components are kinetic SZ, thermal dust emission from dusty star-forming galaxies (DSFG) – both the Poisson and spatially clustered

*e-mail: marian.douspis@ias.u-psud.fr

component, the radio galaxy emission, the Galactic cirrus signal and the cosmic infrared background (CIB). The cosmological parameters enter only the CMB power spectrum, whilst the amplitudes of all templates are set free and marginalised over:

$$C_\ell^{\rm obs} \equiv C_\ell^{\rm CMB}(\Theta) + \sum_i D_{3000,i} \times T_i, \tag{1}$$

where Θ is the set of free cosmological parameters and $D_{3000,i}$ is the free amplitude of a template T_i for foreground component i, normalised to one at $\ell = 3000$. All but CMB signals are con- sidered nuisance quantities and the cosmological information is retrieved only from the CMB part. There are two disadvantages to this approach. First, no cosmological information is retrieved from the non-CMB contribution while tSZ, kSZ, CIB contain such information. Second, the analysis is most often not coherent: the template of contribution i is obtained using a given cosmological parameter set which is different from the one used for computing the template j.

Our goal is thus to propose an approach which alleviate these two imperfections, by doing a coherent analysis retrieving the maximum of the cosmological information from all contributions (primordial and secondary). In this first study, we focus on using the tSZ as an example of foreground containing cosmological information, and leave the study of kSZ signal [10] or a complete analysis of the combination of Planck and SPT [11] to companion papers. Our approach consists in substituting templates of non CMB signals with models depending on cosmological parameters. We will thus assume:

$$C_\ell^{\rm obs} \equiv C_\ell^{\rm CMB}(\Theta) + C_\ell^{\rm tSZ}(\Theta, \Sigma) + \sum_j D_{3000,j} \times T_j \tag{2}$$

where Σ is the set of scaling relation parameters and j runs over all the remaining non-CMB components.

We apply our approach to the SPT power spectra from [8] where the tSZ dominates the CMB signal. A full analysis is presented in [12] but we will focus on the approximated modelling of the tSZ angular spectrum in the next section, the data we use in section 3 and the main results in section 4.

2 tSZ power spectrum

The derivation of the tSZ power spectrum based on the halo model is described in [5, 12] and citations therein. It implies the computation and integration of the Fourier transform of the Compton-y parameter over a large range of masses, redshifts and k-modes. As the multipole range increases towards higher ℓ with high resolution instruments like SPT, the time to compute the C_ℓ^{tSZ}'s increases and slows down the sampling.

Therefore, we turn to machine learning techniques to approximate the full "true" C_ℓ 's computed with the halo model. We choose Random Forest (RF) algorithm for its simplicity, rapidity and ability to learn non linear relation between inputs and outputs [e.g. 13–15]. We train the RF with 15000 models for which we give as input 5 cosmological parameters and 3 scaling relation parameters, and as output the values of the C_ℓ^{tSZ}'s computed with the halo model at 10 given ℓ's. Then, we interpolate these 10 values to obtain the atSZ angular power spectrum at all ℓ's needed by the SPT likelihood. Figure 1 shows the comparison of 10 predicted C_ℓ's versus the test ones. The score is 96.2% and the precision lower than 2%, low enough for today's observations.

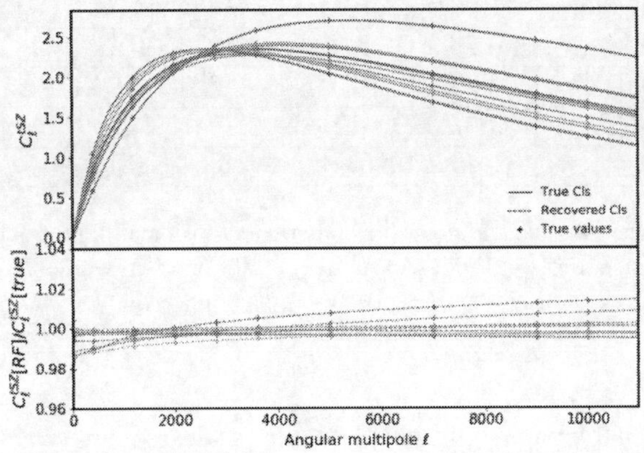

Figure 1. Comparison of the 'true' tSZ C_ℓ's values at given ℓ (crosses, and solid lines) with the RF-inferred C_ℓ's after interpolation (dashed lines) for a sample of models (units are arbitrary before renormalisation in y-units at 143 GHz). The lower part of the plot shows the ratio of RF C_ℓ's over 'true' C_ℓ's.

3 Data and method

We consider the observed signal on small scales (high multipoles ℓ) and, therefore, make use of the SPT temperature-only data and likelihood introduced in [8] and made publicly available by the authors. We refer the interested reader to this work for details on the dataset, but highlight the fact that the data considered spans the range $2000 \leq \ell \leq 11000$ at frequencies 95, 150 and 220 GHz. The likelihood makes use of auto- and cross-spectra and marginalises over calibration and beam parameters. In the baseline analysis of Reichardt et al., the total signal is modelled by CMB, tSZ, kSZ, galactic cirrus contamination, radio and infrared galaxies spectra and tSZ-CIB cross-spectra. In this work, we modified only the tSZ spectra and consequently the tSZ-CIB cross spectra contributions, computed in the SPT likelihood as the sampled correlation co- efficient times a function of the tSZ Cl's (as defined in Zahn et al. 2012). Implementing new CIB and kSZ modelling will be the focus of upcoming works.

We have modified and used the publicly available Monte Carlo Markov Chain CosmoMC code [16] in order to compute the tSZ spectrum with our RF implementation. We consider five cosmological parameters : the baryon and cold dark matter densities, $\Omega_b h^2$ and $\Omega_c h^2$; the ratio of the sound horizon to the angular diameter distance at decoupling θ; the scalar spectral index, n_s; the overall normalisation of the spectrum, A_s; the reionisation optical depth τ is assumed fixed. For the scaling relations, we sample three parameters :the mass bias $(1-b)$, the mass slope α and the normalisation Y_*. We assume the self-similar evolution for halos, and fix the redshift evolution to $2/3$. We focus here our results on the parameters most sensitive to the tSZ effect: Ω_m, σ_8, H_0 , and the mass bias parameter $(1-b)$ while others are marginalised over. We assume Gaussian priors on n_s and $\Omega_b h^2$ as given by Planck.

4 Results

We first apply our approach on SPT data alone and study the consequences of substituting the tSZ template with cosmology-dependent spectra. Figure 2 compares the constraints in the two configurations. We can see that using SPT data alone with the tSZ template (in black) allows for a large range of parameter values, including a narrow degeneracy between H_0 and Ω_m. These constraints are driven by the CMB spectrum dominating other signals on the

	SPT (tSZ template)	SPT (tSZ RF)	SPT + Planck (RF) [U]	SPT + Planck (RF) [G]
Ω_m	0.40 ± 0.11	0.38 ± 0.06	0.35 ± 0.05	0.31 ± 0.03
σ_8	0.83 ± 0.07	0.825 ± 0.048	0.774 ± 0.048	0.744 ± 0.027
H_0	63.2 ± 8.12	64.2 ± 3.4	64.4 ± 2.8	65.9 ± 2.35
$(1-b)$	-	0.49 ± 0.16	0.62 ± 0.17	0.76 ± 0.09

Table 1. Constraints on cosmological and mass bias parameters for the different dataset combinations. We report the 68% confidence levels. [U] goes for uniform prior on (1-b) while [G] goes for the CCCP Gaussian prior, $(1-b) = 0.79 \pm 0.09$, applied on (1-b)

range $2000 < \ell < 4000$ but being negligible at smaller scales. As expected, substituting the template by our RF tSZ power spectrum allows to exploit the full cosmological information (in green), and the constraints are largely tightened with a factor roughly two of improvement.

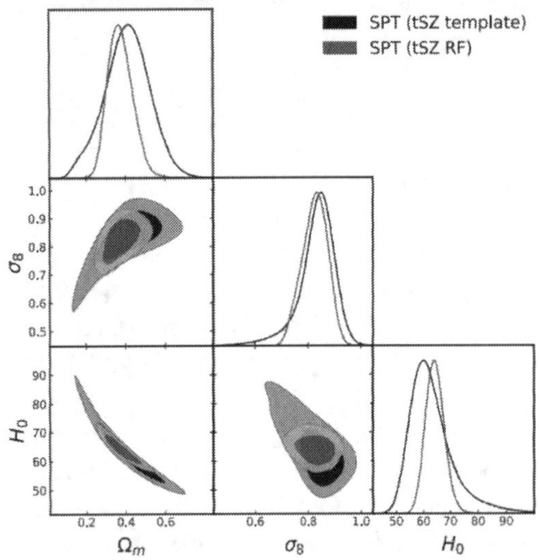

Figure 2. Constraints on cosmological parameters obtained with SPT high-ℓ data using the tSZ template as done in [8, in black] and using the tSZ power spectra derived from the halo model (in green).

We then add the Planck tSZ angular power spectrum [3] in the likelihood to add contraints coming from large scale ($100 < \ell < 1000$). We only consider the case using the RF modelling of the tSZ spectrum. Adding Planck tSZ data does not improve drastically the constraints (Fig. 3), but move the best parameters within 1-σ. Ω_m and σ_8 are shifted to lower values, while the mass bias parameter is shifted high, towards a better agreement with the combination of Planck CMB and cluster number counts [5]. The constraints are still not good enough anyway to determine the mass bias. We thus study how the constraints vary when adding a Gaussian prior on the latter (from CCCP [17]). The degeneracies are broken, and low values of Ω_m and σ_8 are preferred. All results are summarise in Table 1.

5 Conclusions

This work is the first attempt to bring all full cosmological information from secondary anisotropies and foregrounds in CMB analysis at small scales. In this study we focused on tSZ effect only and SPT data. We have shown that substituting the tSZ template by a cosmology dependent spectrum brings more constraints on cosmological parameters. Adding the Planck measurement of tSZ spectrum at large scales does not improve much the constraints but eliminates too large or too small values of the parameters (eg. $\Omega_m = 0.5$). Such a

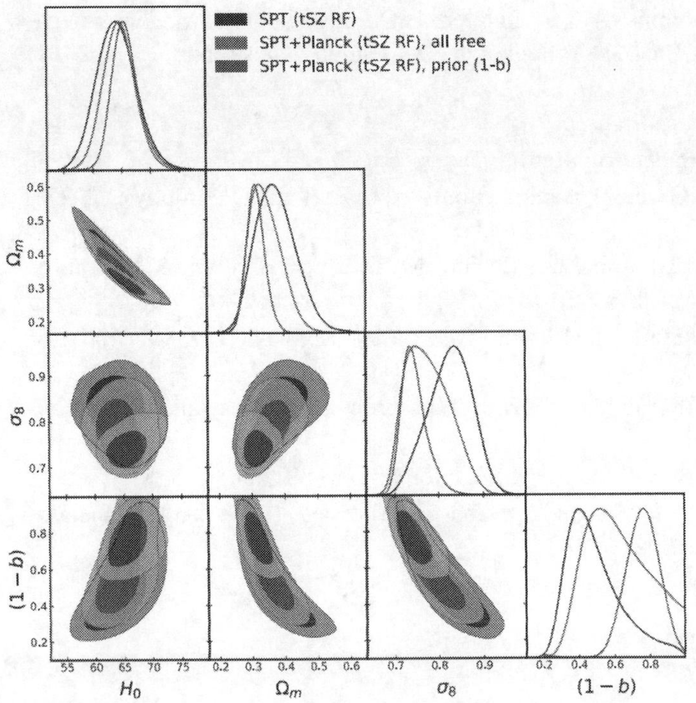

Figure 3. Constraints on cosmological and scaling relation parameters using different data sets and priors: in blue, SPT high-ℓ data alone, and in red, with *Planck* tSZ power spectra added. Results when CCCP prior on the mass bias is added are shown in green.

combination prefers low values of σ_8 0.77 compared to *Planck* CMB analysis. We are now proceeding in implementing kSZ and CIB cosmology dependent spectra in the likelihood (Gorce et al, Douspis et al., in prep.). On the methodological aspect, we built a modelling of the tSZ spectrum using machine learning technique (Random Forest) and implemented it in the COSMOMC framework. We gained a factor more than 100 in the computation of the spectra and a factor of 10 in running MCMC chains. Such approach (fast and coherent), maybe of great interest in the futur not only for constraining cosmological parameters but also astrophysical ones describing the properties of the large scale structures.

References

[1] R.A. Sunyaev, I.B. Zeldovich, ARA&A **18**, 537 (1980)

[2] Planck Collaboration, P.A.R. Ade, N. Aghanim, C. Armitage-Caplan, M. Arnaud, M. Ashdown, F. Atrio-Barandela, J. Aumont, C. Baccigalupi, A.J. Banday et al., Astron. & Astrophys. **571**, A21 (2014), 1303.5081

[3] Planck Collaboration, N. Aghanim, M. Arnaud, M. Ashdown, J. Aumont, C. Baccigalupi, A.J. Banday, R.B. Barreiro, J.G. Bartlett, N. Bartolo et al., Astron. & Astrophys. **594**, A22 (2016), 1502.01596

[4] N. Aghanim et al., Astron. Astrophys. **632**, A47 (2019), 1902.00350

[5] L. Salvati, M. Douspis, N. Aghanim, Astron. & Astrophys. **614**, A13 (2018), 1708.00697

[6] E.M. George et al., Astrophys. J. **799**, 177 (2015), 1408.3161

[7] F. Couchot, S. Henrot-Versillé, O. Perdereau, S. Plaszczynski, B. Rouillé d'Orfeuil, M. Spinelli, M. Tristram, Astron. & Astrophys. **597**, A126 (2017), 1510.07600

[8] C.L. Reichardt, S. Patil, P.A.R. Ade, A.J. Anderson, J.E. Austermann, J.S. Avva, E. Baxter, J.A. Beall, A.N. Bender, B.A. Benson et al., Astrophys. J. **908**, 199 (2021), 2002.06197

[9] D. Han, N. Sehgal, A. MacInnis, A. van Engelen, B.D. Sherwin, M.S. Madhavacheril, S. Aiola, N. Battaglia, J.A. Beall, D.T. Becker et al., Cosmology Astropart. Phys. **2021**, 031 (2021), 2007.14405

[10] M. Douspis, L. Salvati, Gorce (in prep)

[11] A. Gorce, M. Douspis, L. Salvati (in prep)

[12] M. Douspis, L. Salvati, A. Gorce, Aghanim, submitted to Astron. & Astrophys. (2021), 2109.03272

[13] V. Bonjean, N. Aghanim, P. Salomé, A. Beelen, M. Douspis, E. Soubrié, Astron. & Astrophys. **622**, A137 (2019), 1901.01932

[14] F. Hernandez Vivanco, R. Smith, E. Thrane, P.D. Lasky, MNRAS **499**, 5972 (2020), 2008.05627

[15] A. Kennedy, G. Nash, N.J. Rattenbury, A.W. Kempa-Liehr, Astronomy and Computing **35**, 100460 (2021)

[16] A. Lewis, S. Bridle, Phys. Rev. **D66**, 103511 (2002), astro-ph/0205436

[17] H. Hoekstra, R. Herbonnet, A. Muzzin, A. Babul, A. Mahdavi, M. Viola, M. Cacciato, MNRAS **449**, 685 (2015), 1502.01883

 https://doi.org/10.1051/epjconf/202225700015

Characterizing the bulk and turbulent gas motions in galaxy clusters

Simon Dupourqué[1,*], *Etienne* Pointecouteau[1], *Nicolas* Clerc[1], and *Dominique* Eckert[2]

[1]IRAP, Université de Toulouse, CNRS, UPS, CNES, Toulouse, France
[2]Department of Astronomy, University of Geneva, Ch. d'Ecogia 16, CH-1290 Versoix, Switzerland

Abstract. The most massive halos of matter in the Universe grow via accretion and merger events throughout cosmic times. These violent processes generate shocks at many scales and induce large-scale bulk and turbulent motions. These processes inject kinetic energy at large scales, which is transported to the viscous dissipation scales, contributing to the overall heating and virialisation of the halo, and acting as a source of non-thermal pressure in the intra-cluster medium. Characterizing the physical properties of these gas motions will help us to better understand the assembly of massive halos, hence the formation and the evolution of these large-scale structures. We base this characterization on the study of the X-ray and Sunyaev-Zel'dovich effect brightness fluctuations. Our work relies on three complementary samples covering a wide range of redshifts, masses and dynamical states of clusters. We present the results of our X-ray analysis for the low redshift sample, X-COP, and a subsample of higher redshift clusters. We investigate the derived properties according to the dynamical state of our clusters, and the possibility of a self-similar behaviour based on the reconstructed gas motions power-spectra and the correlation with various morphological indicators.

1 Introduction

The baryonic content of galaxy clusters is largely dominated by the hot gas that constitutes the intracluster medium (ICM). The processes governing the growth, as well as the physics of the ICM, introduces perturbations at various scales within the gas. The resulting turbulent motions transport kinetic energy, which cascades to the dissipation scale and contributes to the virialisation of the halo through non-thermal heating of the ICM [e.g. 1].

Direct measurements of these turbulent processes can be achieved using spatially resolved X-ray spectroscopy [2, 3], and will be fully enabled in the future with the upcoming XRISM and Athena missions [4, 5]. These turbulent processes are expected to induce fluctuations in the thermodynamic properties of the ICM, that should be detectable in the related observables [6]. For instance, the X-ray surface brightness and Sunyaev-Zel'dovich distortion scale respectively with the squared density and pressure, integrated along the line of sight. Studies on the nearby Coma and Perseus clusters [7–9] have demonstrated the feasibility of this approach in both X-ray and SZ, and was already extended to a small sample ($N = 10$) to derive statistical trends [10].

*e-mail: sdupourque@irap.omp.eu

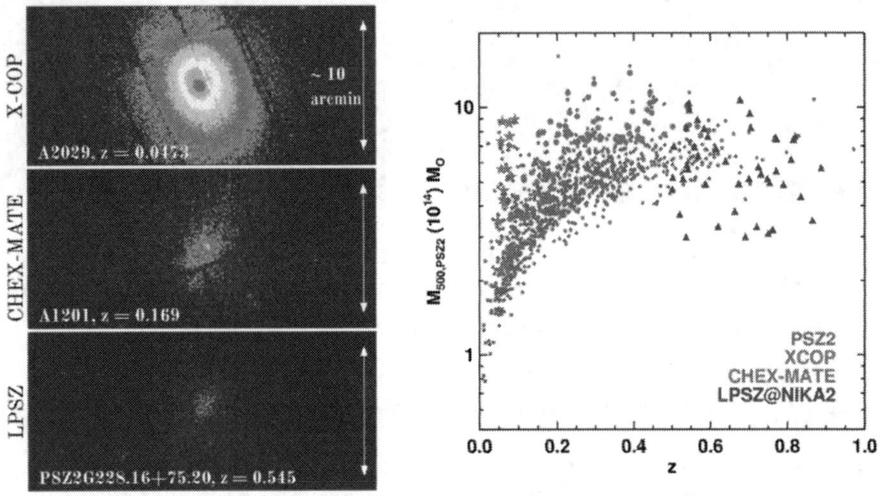

Figure 1. Left: XMM-Newton image (0.7-1.2 keV band) of A2029, A1201 and PSZ2G228.16+75.20 which belong to X-COP, CHEX-MATE and LPSZ@NIKA2, respectively. **Right:** Distribution of the X-COP, CHEX-MATE and LPSZ@NIKA2 samples alongside the PSZ2 catalogue, in the mass-redshift plan.

With this work, we aim at characterizing the X-ray surface brightness and SZ distortion fluctuations for a large cluster sample ($N > 150$) spanning a large range of redshifts, masses and dynamical states. This should allow us to better constrain the properties and impact of turbulence in the ICM, considering the assembly of massive halos.

2 Data

We rely on three samples of clusters built over the Planck [11, 12] and the Atacama Cosmology Telescope [ACT, 13] cluster catalogues:

- The XMM Cluster Outskirts Project [X-COP, 14], a SZ selected sample based on the Planck catalogue, designed to study the outer regions of 12 nearby ($z < 0.1$) galaxy clusters.
- The Cluster HEritage project with XMM-Newton [CHEX-MATE, 15] : a SZ selected sample, based on the Planck catalogue, of 118 clusters designed to obtain an accurate vision of the statistical properties of the cluster population at low to intermediate redshifts ($0.1 < z < 0.5$).
- The NIKA2 SZ Large Program [LPSZ@NIKA2, 16], a SZ-selected sample from both Planck and ACT catalogues, including 50 intermediate to high redshift clusters ($0.5 < z < 0.9$) designed to perform evolution studies and cosmological analysis.

In this paper, we present our preliminary results on the full X-COP sample, and pilot sub-samples of 8 and 4 clusters for CHEX-MATE and LPSZ@NIKA2, respectively.

3 Method

3.1 Mean profile determination

To determine the mean X-ray surface brightness emission from the cluster, an analytical elliptical model is fitted on the photon count image. We first estimate the centroid of emission and the ellipticity using a principal component analysis in a region of $0.5R_{500}$, centred on the emission peak. Then we extract the total photon count in concentric elliptical annuli to obtain a radial surface brightness profile with a 10" binning. The parameter distributions of a β-Model [17] fitting this profile is determined using Bayesian inference, assuming that the number of counts in each annulus follows a Poisson distribution.

3.2 Fluctuation map and 2D power spectrum

The surface brightness image I is directly dependent on the 3D density profile n_e, which can be decomposed into a mean component n_0 and relative fluctuations δ:

$$I(x,y) = \int_{-\infty}^{+\infty} \Lambda(T) n_e^2(\vec{r})\, dz = \int_{-\infty}^{+\infty} \Lambda(T) n_0^2(\vec{r})(1+\delta(\vec{r}))^2\, dz \tag{1}$$

By linearizing for δ and dividing by the previously determined mean count image I_0, we obtain the map of surface brightness fluctuations, J [see 7]:

$$J(x,y) = \frac{I(x,y)}{I_0(x,y)} = 1 + 2\int_{-\infty}^{+\infty} \eta(\vec{r})\delta(\vec{r})\, dz; \quad \eta = \frac{n_0^2(\vec{r})}{\int_{-\infty}^{+\infty} n_0^2(\vec{r}) dz} \tag{2}$$

The associated power spectrum, latter referred to as P_{2D}, is defined as the squared complex modulus of $\mathcal{F}_{2D}\{J\}$, the 2D Fourier transform of the fluctuation map. Assuming that the fluctuation field is isotropic and that the emissivity profile is spherically symmetric, the 2D power spectrum is only a function of $k_\rho = ||\vec{k}_\rho||$:

$$P_{2D}(k_\rho) = |\mathcal{F}_{2D}\{J\}(k_\rho)|^2 = \left| \int J(\vec{\rho}) e^{-2i\pi \vec{k}_\rho \cdot \vec{\rho}} d^2\vec{\rho} \right|^2 \tag{3}$$

For each cluster, the azimuthally averaged power spectrum, P_{2D}, is computed following the method introduced by [Arévalo & al, 18]. It uses Mexican hat filtering at various scales and handles properly masked data (e.g., source exclusions, gaps, and borders effects). The uncertainties due to the shot noise are estimated by computing P_{2D} for 100 Poisson realizations of the image I_0. The power spectrum is also affected by the uncertainty related to the sample variance, which we consider in the following way: assuming that the velocity field is Gaussian, and as in this context, the density fluctuations field is proportional to velocity fluctuations [19, 20], the standard deviation of the 3D power spectrum, P_{3D}, is proportional to the power spectrum itself. As the projection operations from P_{3D} to P_{2D} are linear, we assume that the variance of P_{2D} follows the same law. We therefore added to the P_{2D} variance a term of $(P_{2D}/3)^2$.

3.3 Projection and fitting of the 3D power spectrum

To recover the properties of the inherent 3D density fluctuations power spectrum, we first express the previously obtained 2D power spectrum as a function of the P_{3D}. After introducing the power spectrum of the profile P_η, of the PSF P_{PSF}, the power spectrum of the Poisson noise P_{Poisson} and by designating the 2D convolution with the $**$ operator, we can show that the observed P_{2D} is linked to the P_{3D} with the following dependence :

$$P_{2D}(k_\rho) = P_{\mathrm{Poisson}} + 4P_{\mathrm{PSF}} \int \left\{ P_\eta * *P_{3D} \right\} (k_\rho, k_z)\, dk_z \tag{4}$$

We use a simple power law representation, $P_{3D} = 10^A \left((k_\rho^2 + k_z^2)^{1/2} / k_{\mathrm{ref}} \right)^{-\alpha}$, where the normalization, 10^A [kpc^6], and slope, α, are free parameters. The pivot is set to $k_{\mathrm{ref}} = 10^{-3}\mathrm{kpc}^{-1}$. This model is projected using Eq. (4) and fitted on the measured P_{2D} for each cluster.

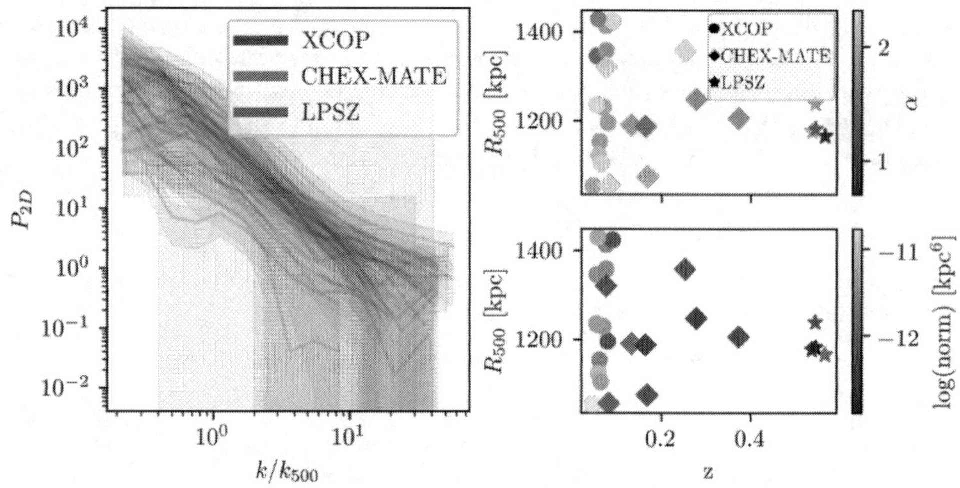

Figure 2. Left: 2D power spectra, P_{2D} [kpc^4], for the set of clusters we are using (See. Sec. 2) plotted as a function of the normalized scale, k/k_{500}. The shaded envelopes correspond to the 1σ dispersion about each individual P_{2D}. **Right:** 3D power spectrum best-fit slope α (11/3 corresponds to a Kolmogorov cascade) and normalization in the R_{500}-z plane.

4 Preliminary results

4.1 Power spectra for a pilot sub-sample

The 2D power spectra, computed for our pilot sample, are shown in the left of Fig. 2, rescaled to $k_{500} = 1/R_{500}$, and exhibit a strong power-law behaviour in this test sample. The spectra do not seem to vary significantly from one sample to another. Thus, we do not see any trend that could correlate with the redshift (though the statistics of our intermediate and higher redshift sub-samples are very limited). The best values of our fit of P_{3D} are displayed in the $R_{500} - z$ plane in Fig. 2, right panel. From this preliminary analysis, we do not see any particular trend of the slope of P_{3D} with the characteristic size or redshift of our clusters, while the normalization decreases with the redshift. This effect will be investigated in a future article. Nevertheless, as the error propagation has not yet been done on these quantities, we cannot draw any solid conclusion about these results.

4.2 Correlation with morphological indicators

To search for links between the dynamical state of the clusters and the role of turbulence in the ICM, we investigate the correlation between our 2D-spectra best parametrization and a set of morphological indicators. We use the following subset of indicators: the centroid shift w, the eccentricity η (from moments) and η_2 (from elliptical fitting), the power ratios $P_{i=1-4}/P_0$, the concentration c ad the Gini coefficient G; see [Lovisari & al, 21]. As the maximum of the characteristic amplitude of the P_{2D} is strongly linked to the 1D Mach number [20], we compute the Spearman rank correlation coefficient between the maximum value of $A_{2D} = \sqrt{2\pi k^2 P_{2D}(k)}$ and the various morphological indicator. The Spearman coefficient, r, is shown in Fig. 3 for the X-COP sample. We observe weak positive correlations for the power-ratios and centroid shifts, which is expected as they reflect the fluctuations at various scales and the disturbances of the surface brightness profile.

 https://doi.org/10.1051/epjconf/202225700015

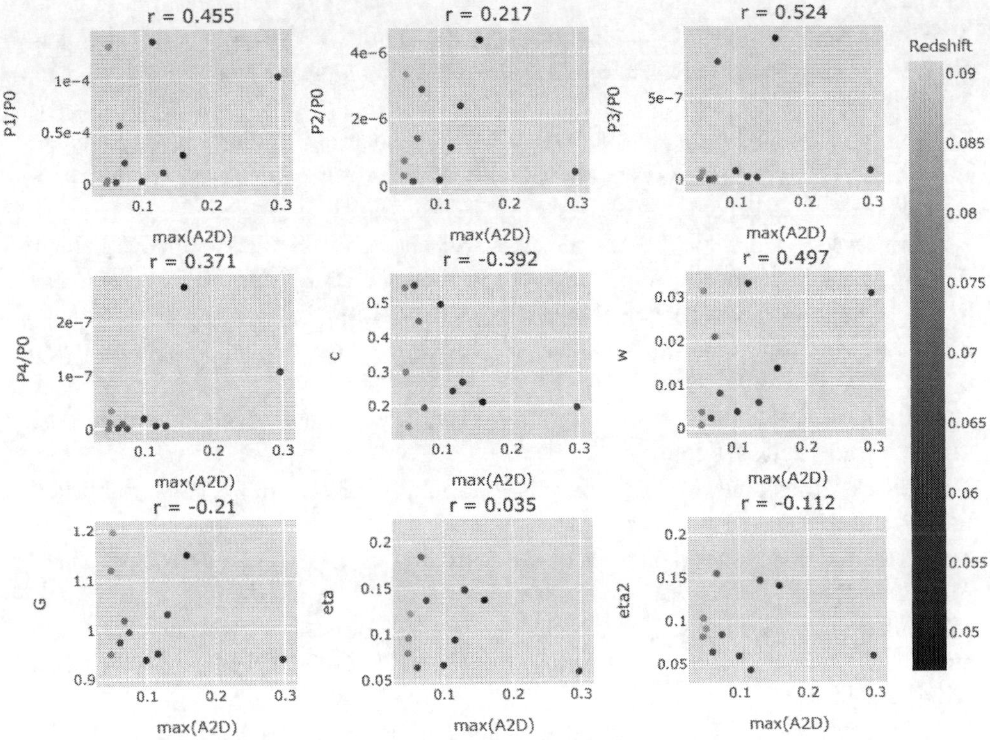

Figure 3. Morphological indicators w.r.t. the maximum of A_{2D} for the X-COP sample, with the corresponding Spearman coefficient

5 Perspectives

We presented here a preliminary analysis of the power spectrum of brightness fluctuations for a sample of galaxy clusters. We will test models with higher levels of complexity for P_{3D} by including, for instance, the injection scale starting the underlying turbulent cascade. We plan to investigate the correlation with morphological indicators, reflecting the dynamical state of each of our clusters, directly in comparison to the parametrization of P_{3D}, to establish the dependencies between the 3D power spectrum and the dynamical state of the cluster. We will extend our study to the whole CHEX-MATE and LPSZ@NIKA2 samples. We also plan to extend this work to the fluctuations of the SZ signal (directly related to the fluctuations of the ICM pressure), making use of the Planck, ACT and NIKA-2 data.

References

[1] F. Vazza, M. Angelinelli, T.W. Jones, D. Eckert, M. Brüggen, G. Brunetti, C. Gheller, Monthly Notices of the Royal Astronomical Society: Letters **481**, L120 (2018)
[2] H. Böhringer, N. Werner, The Astronomy and Astrophysics Review **18**, 127 (2010)
[3] The Hitomi Collaboration, Nature **535**, 117 (2016)
[4] XRISM Science Team, arXiv:2003.04962 [astro-ph] (2020), arXiv: 2003.04962

[5] K. Nandra, D. Barret, X. Barcons, A. Fabian, J.W.d. Herder, L. Piro, M. Watson, C. Adami, J. Aird, J.M. Afonso et al., arXiv:1306.2307 [astro-ph] (2013), arXiv: 1306.2307
[6] A. Simionescu, J. ZuHone, I. Zhuravleva, E. Churazov, M. Gaspari, D. Nagai, N. Werner, E. Roediger, R. Canning, D. Eckert et al., Space Science Reviews **215**, 24 (2019)
[7] E. Churazov, A. Vikhlinin, I. Zhuravleva, A. Schekochihin, I. Parrish, R. Sunyaev, W. Forman, H. Böhringer, S. Randall, Monthly Notices of the Royal Astronomical Society **421**, 1123 (2012)
[8] I. Zhuravleva, E. Churazov, P. Arevalo, A.A. Schekochihin, S.W. Allen, A.C. Fabian, W.R. Forman, J.S. Sanders, A. Simionescu, R. Sunyaev et al., Monthly Notices of the Royal Astronomical Society **450**, 4184 (2015), arXiv: 1501.07271
[9] R. Khatri, M. Gaspari, Monthly Notices of the Royal Astronomical Society **463**, 655 (2016), arXiv: 1604.03106
[10] I. Zhuravleva, S.W. Allen, A. Mantz, N. Werner, The Astrophysical Journal **865**, 53 (2018), publisher: IOP Publishing
[11] The Planck Collaboration, Astronomy & Astrophysics **571**, A29 (2014), publisher: EDP Sciences
[12] The Planck Collaboration, Astronomy & Astrophysics **594**, A13 (2016), publisher: EDP Sciences
[13] M. Mallaby-Kay, Z. Atkins, S. Aiola, S. Amodeo, J.R. Bond, E. Calabrese, G.E. Chesmore, S.K. Choi, K.T. Crowley, O. Darwish et al., arXiv:2103.03154 [astro-ph] (2021), arXiv: 2103.03154
[14] D. Eckert, S. Ettori, E. Pointecouteau, S. Molendi, S. Paltani, C. Tchernin, Astronomische Nachrichten **338**, 293 (2017), arXiv: 1611.05051
[15] The CHEX-MATE Collaboration, arXiv:2010.11972 [astro-ph] (2021), arXiv: 2010.11972
[16] F. Mayet, R. Adam, P. Ade, P. André, A. Andrianasolo, M. Arnaud, H. Aussel, I. Bartalucci, A. Beelen, A. Benoît et al., EPJ Web of Conferences **228**, 00017 (2020), arXiv: 1911.03145
[17] A. Cavaliere, R. Fusco-Femiano, Astronomy and Astrophysics **500**, 95 (1976)
[18] P. Arévalo, E. Churazov, I. Zhuravleva, C. Hernández-Monteagudo, M. Revnivtsev, Monthly Notices of the Royal Astronomical Society **426**, 1793 (2012), publisher: Oxford Academic
[19] I. Zhuravleva, E.M. Churazov, A.A. Schekochihin, E.T. Lau, D. Nagai, M. Gaspari, S.W. Allen, K. Nelson, I.J. Parrish, The Astrophysical Journal **788**, L13 (2014), publisher: IOP Publishing
[20] M. Gaspari, E. Churazov, D. Nagai, E.T. Lau, I. Zhuravleva, Astronomy and Astrophysics **569**, A67 (2014)
[21] L. Lovisari, W.R. Forman, C. Jones, S. Ettori, F. Andrade-Santos, M. Arnaud, J. Démoclès, G.W. Pratt, S. Randall, R. Kraft, The Astrophysical Journal **846**, 51 (2017), arXiv: 1708.02590

Dust Emission in Galaxies at Millimeter Wavelengths

Cooling of star forming regions in NGC6946

G. Ejlali[1,*], *R.* Adam[2], *P.* Ade[3], *H.* Ajeddig[4], *P.* André[4], *E.* Artis[5], *H.* Aussel[4], *A.* Beelen[6], *A.* Benoît[7], *S.* Berta[8], *L.* Bing[9], *O.* Bourrion[5], *M.* Calvo[7], *A.* Catalano[5], *I.* de Looze[17,18], *M.* De Petris[10], *F.-X.* Désert[11], *S.* Doyle[3], *E. F. C.* Driessen[8], *M.* Galametz[4], *F.* Galliano[4], *A.* Gomez[12], *J.* Goupy[7], *A. P.* Jones[6], *A.* Hughes[13], *S.* Katsioli[15,16], *F.* Kéruzoré[5], *C.* Kramer[14], *B.* Ladjelate[14], *G.* Lagache[9], *S.* Leclercq[8], *J.-F.* Lestrade[19], *J.-F.* Macías-Pérez[5], *S. C.* Madden[4], *A.* Maury[4], *P.* Mauskopf[3,20], *F.* Mayet[5], *A.* Monfardini[7], *M.* Muñoz-Echeverría[5], *A.* Nersesian[15,17], *L.* Perotto[5], *G.* Pisano[3], *N.* Ponthieu[11], *V.* Revéret[4], *A. J.* Rigby[3], *A.* Ritacco[6,21], *C.* Romero[22], *H.* Roussel[23], *F.* Ruppin[25], *K.* Schuster[8], *S.* Shu[26], *A.* Sievers[14], *M. W. S. L.* Smith[3], *F. S.* Tabatabaei[1,24], *C.* Tucker[3], *E. M.* Xilouris[15], and *R.* Zylka[8]

[1]School of Astronomy, Institute for Research in Fundamental Sciences (IPM), Larak Garden, 19395-5531 Tehran, Iran
[2]LLR (Laboratoire Leprince-Ringuet), CNRS, École Polytechnique, Institut Polytechnique de Paris, Palaiseau, France
[3]School of Physics and Astronomy, Cardiff University, Queen's Buildings, The Parade, Cardiff, CF24 3AA, UK
[4]AIM, CEA, CNRS, Université Paris-Saclay, Université Paris Diderot, Sorbonne Paris Cité, 91191 Gif-sur-Yvette, France
[5]Univ. Grenoble Alpes, CNRS, Grenoble INP, LPSC-IN2P3, 38000 Grenoble, France
[6]Institut d'Astrophysique Spatiale (IAS), CNRS, Université Paris Sud, Orsay, France
[7]Institut Néel, CNRS, Université Grenoble Alpes, France
[8]Institut de RadioAstronomie Millimétrique (IRAM), Grenoble, France
[9]Aix Marseille Univ, CNRS, CNES, LAM, Marseille, France
[10]Dipartimento di Fisica, Sapienza Università di Roma, Piazzale Aldo Moro 5, I-00185 Roma, Italy
[11]Univ. Grenoble Alpes, CNRS, IPAG, 38000 Grenoble, France
[12]Centro de Astrobiología (CSIC-INTA), Torrejón de Ardoz, 28850 Madrid, Spain
[13]Université de Toulouse, UPS-OMP,F-31028 Toulouse, France; CNRS, IRAP, Av. du Colonel Roche BP 44346, F-31028 Toulouse cedex 4, France
[14]Instituto de Radioastronomía Milimétrica (IRAM), Granada, Spain
[15]National Observatory of Athens, Institute for Astronomy, Astrophysics, Space Applications and Remote Sensing, Ioannou Metaxa and Vasileos Pavlou GR-15236, Athens, Greece
[16]Department of Astrophysics, Astronomy & Mechanics, Faculty of Physics, University of Athens, Panepistimiopolis, GR-15784 Zografos, Athens, Greece
[17]Sterrenkundig Observatorium Universiteit Gent, Krijgslaan 281 S9, B-9000 Gent, Belgium
[18]Department of Physics and Astronomy, University College London, Gower Street, London, WC1E 6BT, UK
[19]LERMA, Observatoire de Paris, PSL Research University, CNRS, Sorbonne Université, UPMC, 75014 Paris, France
[20]School of Earth and Space Exploration and Department of Physics, Arizona State University, Tempe, AZ 85287, USA

*e-mail: gejlali@ipm.ir

[21]Laboratoire de Physique de l'École Normale Supérieure, ENS, PSL Research University, CNRS, Sorbonne Université, Université de Paris, 75005 Paris, France
[22]Department of Physics and Astronomy, University of Pennsylvania, 209 South 33rd Street, Philadelphia, PA, 19104, USA
[23]Institut d'Astrophysique de Paris, Sorbonne Université, CNRS (UMR7095), 75014 Paris, France
[24]Instituto de Astrofísica de Canarias, Vía L'actea S/N, 38205 La Laguna, Spain
[25]Kavli Institute for Astrophysics and Space Research, Massachusetts Institute of Technology, Cambridge, MA 02139, USA
[26]Caltech, Pasadena, CA 91125, USA

Abstract. Interstellar dust plays an important role in the formation of molecular gas and the heating and cooling of the interstellar medium. The spatial distribution of the mm-wavelength dust emission from galaxies is largely unexplored. The NIKA2 Guaranteed Time Project IMEGIN (Interpreting the Millimeter Emission of Galaxies with IRAM and NIKA2) has recently mapped the mm emission in the grand design spiral galaxy NGC6946. By subtracting the contributions from the free-free, synchrotron, and CO line emission, we map the distribution of the pure dust emission at $1.15mm$ and $2mm$. Separating the arm/interarm regions, we find a dominant $2mm$ emission from interarms indicating the significant role of the general interstellar radiation field in heating the cold dust. Finally, we present maps of the dust mass, temperature, and emissivity index using the Bayesian MCMC modeling of the spectral energy distribution in NGC6946.

1 Introduction

The IMEGIN project (PI: Suzanne Madden) is a NIKA2 guaranteed time large project (200 hours) on the 30m IRAM telescope, studying the interstellar medium (ISM) of 22 nearby galaxies. The NIKA2 camera, with a $6.5'$ field of view and resolution of $11.1''$ at $1.15mm$ and $17.6''$ at $2mm$ ([1]) brings us well-resolved, ($\sim kpc$ scale) studies to map full galaxies at these wavelengths for the first time. Here we present the results of a pilot study of the nearby ($D = 6.8Mpc$) grand-design spiral NGC6946, observed almost face-on ($i = 38°$), with a wealth of complementary millimeter, radio, atomic observations as well as molecular lines.

2 Observations

NGC6946 was observed in 2020 and 2021 for a total of 23 hours, reaching rms values of of $0.8mJy/beam$ at $1.15mm$ ($3''$ pixel) and $0.27mJy/beam$ at $2mm$ ($4''$ pixel), with 5% calibration error [1]. The scans were reduced and maps reconstructed using PIIC/GILDAS software[1] (Figure 1).

3 Emission at millimeter wavelengths

Continuum emission at mm wavelengths consists of three main components, 1) thermal emission from dust, 2) thermal free-free emission from ionized gas, and 3) nonthermal synchrotron emission from cosmic rays propagating through the magnetic field [2]. Apart from these, continuum emission might be affected by line emission of molecular gas, such as CO.

We use the thermal and nonthermal radio maps of NGC6946 that have been constructed from the Hα recombination line observations [3]. For pixels with flux values more than 3σ

[1]http://www.iram.es/IRAMES/mainWiki/PIIC

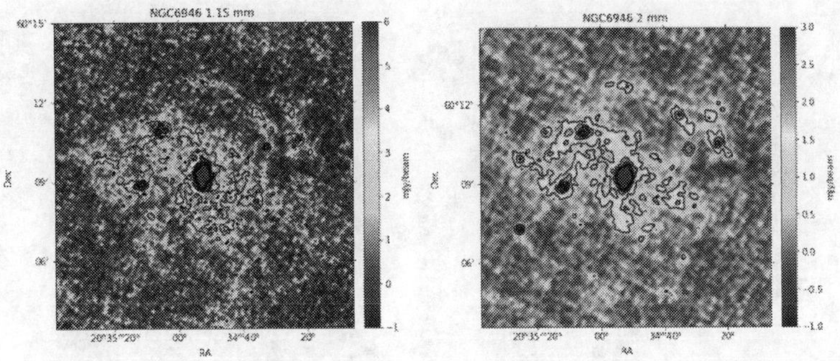

Figure 1. Observed maps of NGC6946 at 1.15*mm* (*left*) and 2*mm* (*right*) at angular resolution of 12″ and 18″ respectively.

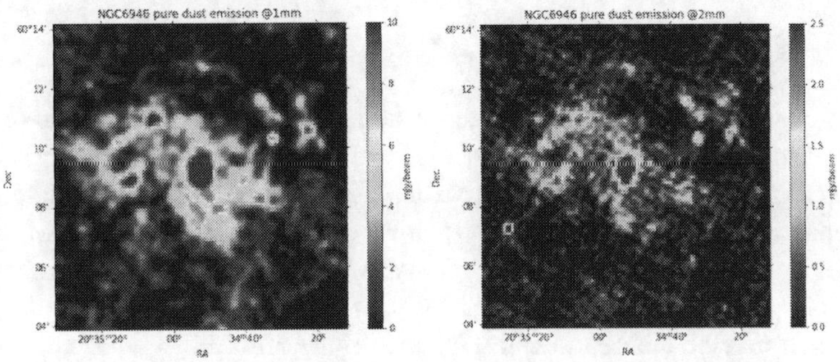

Figure 2. Pure dust emission at 1*mm* (*left*) and 2*mm* (*right*) at angular resolution of 18″.

rms, we find that 5.5% and 28.8% of integrated flux at 1.15*mm* and 2*mm* is due to the radio continuum components.

Next, we take into account the contamination by molecular lines, as the NIKA2 1.15*mm* band includes emission from the CO(2-1) line. We use the CO(2-1) map observed at IRAM as a part of HERACLES project [4]. We subtract the CO(2-1) emission from the observed emission at 1.15*mm* following formulas explained in [5], which uses the transmission function of NIKA2 [1] to convert spectral intensity to pseudo-continuum units. The CO(2-1) emission accounts for 14.1% of the integrated flux of NIKA2 1.15*mm* emission in NGC6946.

The maps of pure dust emission are shown in Figure 2. 80.4% and 63.6% of the integrated flux observed over NGC6946 in the NIKA2 1.15*mm* and 2*mm* bands, respectively, corresponds to pure dust emission (for pixels with larger than 3σ rms observed flux).

4 Regional Analysis

To study the spatial properties of the ISM and dust in NGC6946, we investigate emission in the central, arm and interarm regions separately, making use of a modified version of the mask introduced in [6]. Our modification includes adding a small part of the western arm and a large portion of the interarm region, which is missed due to the limited field of view of SOFIA in [6]. The relative contribution of the different types of emission from each of these three regions in NGC6946 is shown in Figure 3.

It can be seen in Figure 3 that the arm-to-interarm ratio is larger for thermal (free-free) radio emission than the nonthermal (synchrotron) emission. The synchrotron emission is

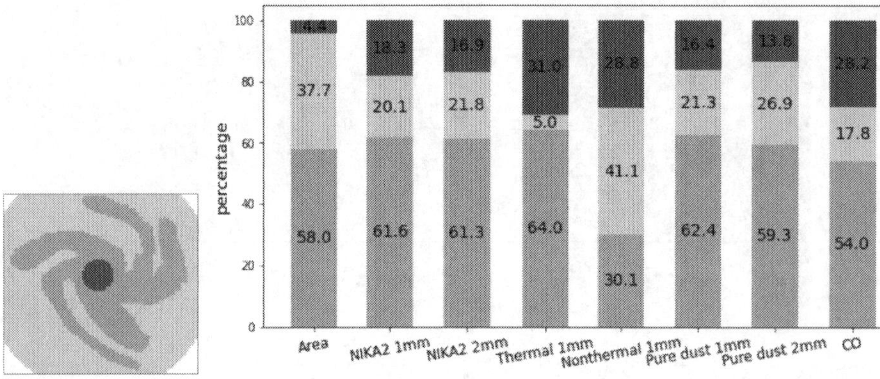

Figure 3. *Left*: The mask we used to distinguish between central (red), arm (green) and interarm (yellow) regions in NGC6946. *Right*: Percentage of contribution of each region to the indicated emissions.

caused by cosmic ray electrons that can propagate large distances in the interarm regions and into the diffuse disc. In addition, a slightly larger contribution of dust emitting at 1.15*mm* compared to 2*mm* in the arm regions is noticeable. The 2*mm* emission is a better tracer of cold dust, compared to the 1.15*mm* emission which can trace warmer dust. Main heating sources of warm dust are the star forming regions, which follow the spiral structure of the galaxy in the arms. On the other hand, colder dust can be heated by the interstellar radiation field (ISRF), which may be the main heating source in the diffuse disc and interarm region. These speculations are investigated further by modeling the dust properties spatially in the galaxy.

5 Spectral Energy Distribution (SED) modeling

To study the physical properties of dust and their spatial variations across NGC6946, we model the SED of this galaxy from centimeter radio to FIR wavelengths. We use the far infrared *Herschel* data and the VLA and Effelsberg data to fully constrain the SED (Table 1). All of the maps were first brought to the same resolution of 18″ using a Gaussian PSF and grid size of 6″. We also subtracted the CO(2-1) emission from NIKA2 1.15*mm* map before including in the SED modeling.

To model the radio continuum SED, we use a power law model (Equation 1), which depends on three parameters, namely A_1 indicating the thermal free-free fraction, A_2 indicating the nonthermal synchrotron fraction, and α which is the synchrotron spectral index [7].

$$S_\nu^{\mathrm{RC}} = A_1 \nu^{-0.1} + A_2 \nu^{-\alpha} \tag{1}$$

To model the thermal emission from dust, we assume a modified black body (MBB) model which constraints dust mass M, dust temperature T and dust emissivity index β as

$$S_\nu^{\mathrm{dust}} = \kappa_0 \left(\frac{\nu}{\nu_0}\right)^\beta \times \frac{M}{D^2} \times B_\nu(T), \tag{2}$$

where $B_\nu(T)$ is the Planck function and we adopt $\kappa_0 = 0.04 m^2/kgr$ [2]. Addition of these two models, $S_\nu^{RC} + S_\nu^{dust}$, with a set of six free parameters $\{A_1, A_2, \alpha, T, M, \beta\}$ construct our model.

To fit this model to our data (Table 1), we use a Bayesian Markov Chain Monte Carlo (MCMC) approach making use of the emcee Python package [11]. This method generated a

Table 1. Summary of data used for resolved SED modeling.

Wavelength	telescope	original resolution	reference
20*cm*	VLA & Effelsberg	15"	[8]
6*cm*	VLA & Effelsberg	15"	[8]
3.6*cm*	VLA & Effelsberg	18"	[8]
1.15*mm*	IRAM 30m NIKA2	12"	This work
2*mm*	IRAM 30m NIKA2	18"	This work
70μm	Herschel PACS	6"	[9],[10]
100μm	Herschel PACS	8"	[9],[10]
160μm	Herschel PACS	12"	[9],[10]
250μm	Herschel SPIRE	18"	[9],[10]

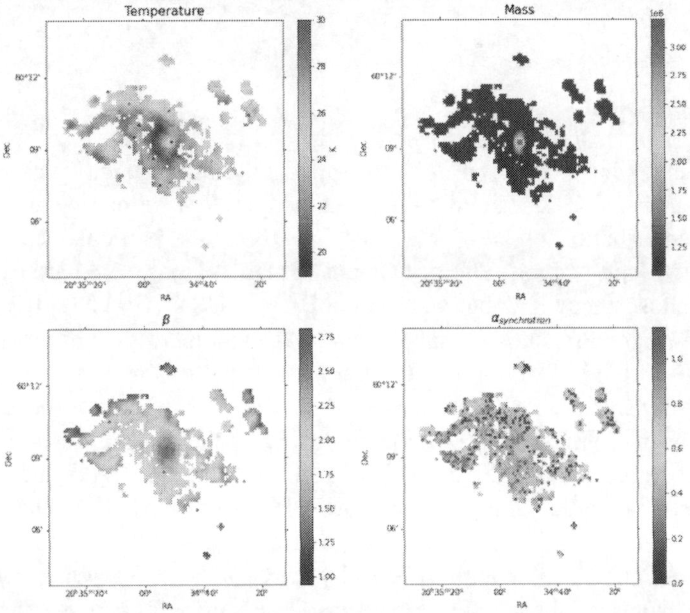

Figure 4. Maps of dust mass M, dust temperature T, dust emissivity index β and synchrotron spectral index α as a result of SED modeling in each $6''$ pixel. The maps are at $18''$ resolution.

model library that encompasses all different combinations of model parameters. The likelihood function of each model in the library is an indicator of how close that model is to the observed data. The Bayesian MCMC approach then derives the posterior distribution of each parameter using a specified number of walkers in the parameter space. The median of the posterior probability distribution function is then used as the reported result.

We fit our model to the observed flux densities of each pixel of the map and report the six free parameters for each pixel. The maps of the 4 main dust properties which we find through the SED modeling are demonstrated in Figure 4.

6 Discussion

We studied the correlation of dust emission with the CO(2-1) emission, as a tracer of the molecular gas. For this purpose, we only use pixels with flux values larger than 3σ RMS of both maps and separated by one resolution element. We find a strong correlation (Pearson coefficient = 0.978) between dust emission at 1.15*mm* and CO(2-1) for pixels located in the central region of NGC6946. A similar trend has been reported in the central region of the Milky Way [12]. On the other hand, points belonging to arm and interarm region show a much weaker correlation (Pearson coefficient=0.600) between 1.15*mm* emission of dust and molecular gas. This observation brings us to conclude that dust and molecular gas have a common heating source in the central region, whereas they are being heated by different heating sources in the disc. Both the diffuse interstellar radiation field and old stars play important roles in heating the dust in the disk, while they are not energetic enough to bring molecular gas to thermal equilibrium with dust.

We report an anti-correlation between T and β for galactocentric radii larger than $\sim 1.5kpc$ in NGC6946. The fact that the anti-correlation breaks down for the inner $\sim 1.5kpc$ brings us to conclude that this is not caused by degeneracies in our model and both β and T are well constrained.

Acknowledgements

We would like to thank the IRAM staff for their support during the campaigns. The NIKA2 dilution cryostat has been designed and built at the Institut Néel. In particular, we acknowledge the crucial contribution of the Cryogenics Group, and in particular Gregory Garde, Henri Rodenas, Jean Paul Leggeri, Philippe Camus. This work has been partially funded by the Foundation Nanoscience Grenoble and the LabEx FOCUS ANR-11-LABX-0013. This work is supported by the French National Research Agency under the contracts "MKIDS", "NIKA" and ANR-15-CE31-0017 and in the framework of the "Investissements d'avenir" program (ANR-15-IDEX-02). This work has benefited from the support of the European Research Council Advanced Grant ORISTARS under the European Union's Seventh Framework Programme (Grant Agreement no. 291294). This work was supported by the Programme National "Physique et Chimie du Milieu Interstellaire" (PCMI) of CNRS/INSU with INC/INP co-funded by CEA and CNES. F.R. acknowledges financial supports provided by NASA through SAO Award Number SV2-82023 issued by the Chandra X-Ray Observatory Center, which is operated by the Smithsonian Astrophysical Observatory for and on behalf of NASA under contract NAS8-03060.

References

[1] L. Perotto, N. Ponthieu, J.-F. Macías-Pérez, *et al.*, Astron. Astrophys. **637**, A71 (2020)
[2] F.S. Tabatabaei, *et al.* Astron. Astrophys. **555**, A128 (2013b)
[3] F.S. Tabatabaei, *et al.* Astron. Astrophys. **552**, A19 (2013a)
[4] A.K Leroy *et al.* The Astronomical Journal, **137**, 4670-4696 (2009)
[5] E. Drabek *et al.* Mon. Not. R. Astron. Soc. **426**, 23-39 (2012)
[6] F. Bigiel, *et al.*, The Astrophysical Journal, **903**, 30 (2020)
[7] F.S. Tabatabaei *et al.* The Astrophysical Journal, **836**, 185 (2017)
[8] Beck, R. Astron. Astrophys. 470, 539 (2007)
[9] Kennicutt *et al.* PASP, 123, 1347 (2011)
[10] C. J. R. Clark *et al.*, Astron. Astrophys. **609**, A37 (2018)
[11] D. Foreman-Mackey *et al.*, Pub. of the Astronomical Society of the Pacific, **925**, 306-312 (2013)
[12] D. J. Eden *et al.*, Monthly Notices of the Royal Astronomical Society, **498**, 5936–5951(2020)
[13] R. Adam *et al.*, Astron. Astrophys. **609**, A115 (2018)
[14] M. Calvo *et al.*, Journal of Low Temperature Physics, **184**, 816 (2016)
[15] O. Bourrion *et al.*, Journal of Instrumentation, **11**, 11001 (2016)

Observations with KIDs Interferometer Spectrum Survey (KISS)

A. Fasano[1*], A. Catalano[2], J. F. Macías-Pérez[2], M. Aguiar[4,5], A. Beelen[1], A. Benoit[3], A. Bideaud[3], J. Bounmy[2], O. Bourrion[2], G. Bres[3], M. Calvo[3], J. A. Castro-Almazán[4,5], P. de Bernardis[6], M. De Petris[6], A. P. de Taoro[4,5], M. Fernández-Torreiro[4,5], G. Garde[3], R. Génova-Santos[4,5], A. Gomez[7], M. F. Gómez-Renasco[4,5], J. Goupy[3], C. Hoarau[2], R. Hoyland[4,5], G. Lagache[1], J. Marpaud[2], M. Marton[2], A. Monfardini[3], M. W. Peel[4,5], G. Pisano[8], N. Ponthieu[9], R. Rebolo[4,5], S. Roudier[2], J. A. Rubiño-Martín[4,5], D. Tourres[2], C. Tucker[8], and C. Vescovi[2]

[1]Aix Marseille Univ, CNRS, CNES, LAM, Marseille, France
[2]Univ. Grenoble Alpes, CNRS, LPSC/IN2P3, 38000 Grenoble, France
[3]Univ. Grenoble Alpes, CNRS, Grenoble INP, Institut Néel, 38000 Grenoble, France
[4]Instituto de Astrofísica de Canarias, C/Vía Láctea, E-38205 La Laguna, Tenerife, Spain
[5]Universidad de La Laguna, Dept. Astrofísica, E-38206 La Laguna, Tenerife, Spain
[6]Dipartimento di Fisica, Sapienza Università di Roma, Piazzale Aldo Moro 5, I-00185 Roma, Italy
[7]Centro de Astrobiología (CSIC-INTA), Torrejón de Ardoz, E-28850, Madrid, Spain
[8]Astronomy Instrumentation Group, University of Cardiff, The Parade CF24 3AA, UK
[9]Univ. Grenoble Alpes, CNRS, IPAG, 38400 Saint Martin d'Hères, France

Abstract. We describe the preliminary on-sky results of the KIDs Interferometer Spectrum Survey (KISS), a spectral imager with a 1 deg field of view (FoV). The instrument operates in the range 120–180 GHz from the 2.25 m Q-U-I JOint TEnerife telescope in Teide Observatory (Tenerife, Canary Islands), at 2 395 m altitude above sea level. Spectra at low resolution, up to 1.45 GHz, are obtained using a fast (3.72 Hz mechanical frequency) Fourier transform spectrometer, coupled to a continuous dilution cryostat with a stabilized temperature of 170 mK that hosts two 316-pixel arrays of lumped-element kinetic inductance detectors. KISS generates more than 3 000 spectra per second during observations and represents a pathfinder to demonstrate the potential for spectral mapping with large FoV. We give an overall description of the spectral mapping paradigm and we present recent results from observations, in this paper.

1 Introduction

In the millimeter (mm) astronomy domain, there is a strong demand from the scientific community to develop multi-band instruments for component separation, foreground characterization, cosmic microwave background spectral distortions and line intensity mapping [1]. An interesting instrumental candidate to meet these demands is the exploitation of Fourier transform spectrometers (FTSs) and, in particular, the Martin-Puplett interferometer (MPI)

*Corresponding author: Alessandro Fasano, alessandro.fasano@lam.fr configuration [2, 3]. The case of ground-based experiments has an additional requirement: the atmospheric fluctuations have to be addressed. For this purpose, FTSs have to be coupled with fast detectors, and kinetic inductance detectors (KIDs) are the fastest available in large format arrays at mm-wavelength [4, 5].

In this scientific framework, we have developed the KIDs Interferometer Spectrum Survey (KISS), which uses two arrays of KIDs coupled to an MPI. The reference source of such a differential FTS configuration can be set between the image of the de-focused sky and a cryogenics stage, see [3] for more details. KISS allows us to exploit a wide instantaneous field of view (1 deg) and a spectral resolution as fine as 1.45 GHz in the 120–180 GHz frequency band. The instrument is installed on the 2.25-meter Q-U-I JOint TEnerife (QUIJOTE) telescope at the Teide Observatory, in Tenerife and has been operational since February 2019. KISS is also the pathfinder of the new CarbON CII line in the post-rEionization and ReionizaTiOn epoch project (CONCERTO) [6].

In this paper, we give an overall description of the spectral mapping paradigm and we present recent results from the last year of observations. In Section 2, we describe the necessity for fast scanning strategy and we explain the spectral mapping technique. Section 3 presents the preliminary on-sky results with point sources in spectral mapping reduction and we address the data reduction of the KISS project.

2 Spectral mapping with FTS fast scanning

Ground-based experiments have to deal with a particular additional issue with respect to space-borne telescopes. Despite the ease to debug and modify the hardware of telescopes on the ground, even after the deployment, an additional source of contamination is present: the atmosphere, both in terms of absorption and emission. This limits the frequency range over which they can be used. The emission has two main effects: 1) it increases the background (photon) noise with respect to what is expected for satellite experiments, and 2) leads to low-frequency noise induced by low-frequency drifts in the atmospheric emission. For the first, it requires an increase of the necessary integration time for a given signal-to-noise ratio. However, this requirement is compensated by the lower cost of a ground-based facility. For reducing the latter, there are specific observation strategies for different instruments that prevent the introduction of systematic errors in the signal. This problem has to be carefully addressed to avoid affecting the accuracy of the calibration. In the case of FTS, the solution is to acquire the full interference pattern, the so-called "interferogram", on times scales shorter than those of the atmospheric fluctuations.

State-of-the-art detectors in the mm domain reach a white noise level dominated component is the photon noise, which is the intrinsic noise derived by the random nature of the incoming photons. The best noise performance is matched if this condition is verified and the detectors are named photon-noise limited. The second higher noise component is produced predominantly by the electronic noise. In addition, there is the noise coming from the state fluctuation of the superconducting particles, called generation-recombination noise, in the KID case [7]. The slow fluctuations of the atmosphere are translated to a 1/f noise in the frequency domain, whose intersection with the white noise forms a knee shape. The frequency of the intersection, the knee frequency, is particularly important for ground-based experiments. The recording of every single interferogram at stable atmospheric absorption (in other words, faster than the 1/f knee) is a prior requirement, in the specific KISS case. We use the so-called "fast scanning" technique for interferometric pattern sampling: we record the interferogram on the fly while continuously oscillating the roof mirror, which introduces the optical path difference (OPD) while pointing the telescope on-sky (see [8] for a description of the phase and amplitude modulation alternative techniques, not suitable for our utilization). The typical 1/f atmospheric knee is expected to be at ~1 Hz [9], in the heritage of the New IRAM KID Array 2 (NIKA2) [10] at the Institut de Radioastronomie Millimétrique (IRAM) telescope in Pico Veleta (Spain). This, coupled with the requirement on the spectroscopic resolution, leads to the acquisition of hundreds to thousands of samples on time scales of 0.25 seconds. To achieve this performance and ensure a large field of view (FoV), we need large arrays of fast detectors and readout as is the case for the KIDs. To meet these requirements KISS integrates two full interferograms at 3.72 Hz (0.27 s). The presence of a double interferogram is due to the adopted sampling technique of the interferogram (see [11] for a detailed discussion): we double-sample the interferogram by measuring slightly above the zero optical path difference (ZPD) point that corresponds to the null interference. The first and second interferograms are the result of the forward and backward mirror motion, respectively, as shown in Fig. 1.

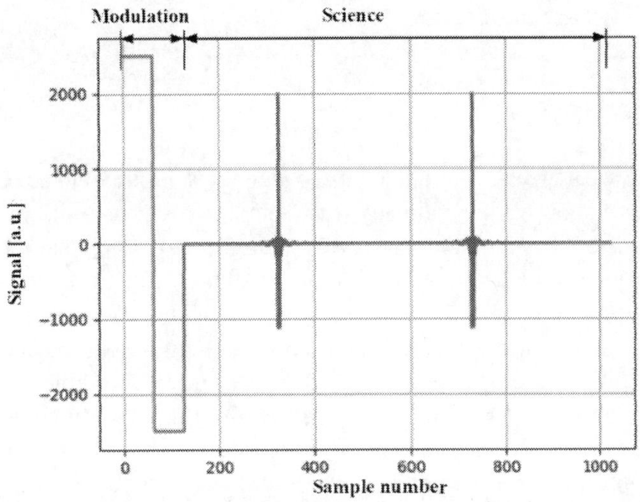

Figure 1. KISS data block recorded at 3.72 Hz: frequency signal as a function of the sample number. There are a total of 1024 data samples: the first 128 are dedicated to the modulation and the remaining ones are to the forward and backward interferogram acquisition. The signal is the result of the median value subtraction (i.e., by scaling to 0 Hz the median value), which represents the photometry, i.e., the signal integrated over the electromagnetic bandwidth.

In addition, we have introduced an electrical modulation before each interferogram recording, as shown in Fig. 1. This readout modulation is adopted to improve the sky signal reconstruction accuracy for types of instruments for which a fast sampling frequency is required, both to remove atmospheric fluctuations and to perform full spectroscopic measurements on each sampled sky position. Kinetic inductance detectors are superconducting resonators that are read out by injecting a reference signal in the circuit and recording the output one. The physical quantity that can be directly translated into an astrophysical signal is the resonance frequency, which is different for each single KID. The optical load changes this electrical parameter and by recording it we retrieve the astrophysical signal. The raw data from the KID read-out system is the complex signal (I, Q) from which we can compute the phase signal ($\phi = \arctan 2(I/Q)$). Before each on-sky scan, the single KID is optimized (tuned) for the background at the observation time. The electrical modulation introduced before each interferogram sampling allow us to follow the background evolution, reducing the downsides from the evolved optical load that introduces imperfection on the initial tuning approximation. In addition, it allows us to convert the raw ϕ to the resonance frequency: for a detailed discussion see [12].

This whole method is translated into cubes of data composed of the x-y positions on the sky plus the electromagnetic information: we retrieve a whole interferogram per sky position. Starting from the raw interferogram data presented above we go to maps of the sky as follows: 1) using the modulation we calibrate the signal in frequency shift, 2) the data corresponding to the forward and backward interferograms are separated; 3) the interferograms are cleaned from low-frequency drifts and the ZPD is obtained, by regridding the interferograms on the OPD and identifying the maximum or minimum signal; 4) the interferograms in the time domain are projected into maps using all detectors, and 5) the final spectral maps are obtained from the Fourier transform of the interferograms accounting for the ZPD. The result is a set of N maps at the N bins of the frequency bandwidth.

In this paper, we present uncalibrated signals, which are to be intended as resonance frequency ones, as previously mentioned.

3 On-sky results

In this section, we show early results in uncalibrated signals of Jupiter and the calibrated flux of Venus. In previous works, we have already demonstrated the possibility of exploiting an FTS as a pure photometer. In particular, observing a bright and wide source like the Moon [13] which acted as a quick-look demonstrator for our challenging instrument, and calibrating Venus with Jupiter observations [12].

We exploited Jupiter to calibrate our detector response in an astrophysical signal and obtain the first scientific result, as shown in Fig. 2. We measure the brightness temperature of Venus at a frequency that has not been measured before. We obtain a brightness temperature of $T^{\mathrm{KISS}}_{b,\mathrm{Venus}} = 338 \pm 27\,\mathrm{K}$ at 150 GHz (the central frequency of the 120–180 GHz KISS band and the incertitude represents the 1-sigma level of confidence). We also show the "Bellotti" model [14] as a solid blue line for the Venus brightness temperature and its extrapolation of a simple power-law as a dashed line at mm wavelengths. In solid cyan line, we represent the best-fit power law. We observe that the Bellotti model underestimates the power law at high frequency, for more details see [12].

The next challenge is to obtain the spectral mapping of the two planets. The coupling between KISS and the Q-U-I JOint TEnerife (QUIJOTE) telescope results in 7 arcmin (at the central electromagnetic frequency of 150 GHz) full width at half maximum (FWHM). Venus and Jupiter have respectively $338 \pm 27\,\mathrm{K}$ and $174.1 \pm 0.2\,\mathrm{K}$ [15] in thermodynamic temperatures, and 66 arcsec and 50 arcsec maximum intrinsic apparent diameters when they are at the closest approach in the orbit. We are, therefore, affected by a strong dilution factor, which is defined geometrically as $D = \Omega/2\pi\sigma^2$, where Ω is the source solid angle and σ is the sigma of the single-pixel $\sigma = FWHM/2\sqrt{2\log(2)}$. In conclusion, the planet signal is diluted within the beam and it results in few kelvins at the different apparent diameters, in the KISS case. This was a major issue for the early observations because we experienced the absence of bright and large-enough primary calibrators to accurately constrain the pointing model at the beginning of the commissioning phase. Despite these difficulties, we managed to detect and systematically observe these two cited planets.

After the first spectral mapping with the Moon presented in [3], we introduce here the results with spectral mapping capability with point sources with KISS which can be considered as the proper first light of the experiment. Figure 3 shows the spectral mapping of Jupiter, projecting the spectra at the given sky position per electromagnetic bin. We observe how,

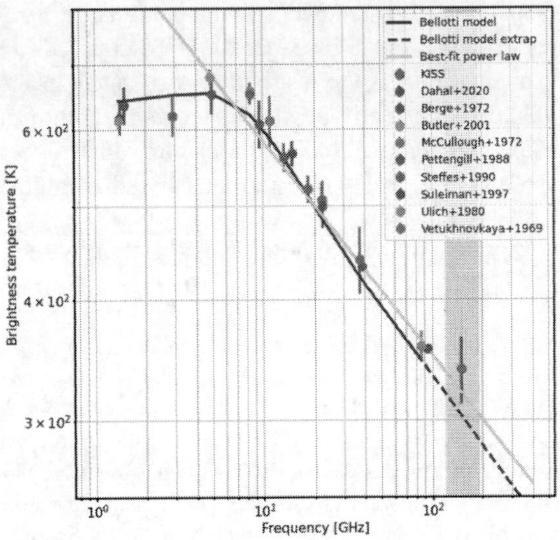

Figure 2. Venus brightness temperature as a function of frequency. We present a collection of radio and mm wavelength measurements including those of KISS. The blue and cyan lines represent two models for the Venus brightness temperature. [12]

despite the large size of the map needed due to the large FoV, the source is well identified in the center with a signal-to-noise ratio (SNR) of ~10.

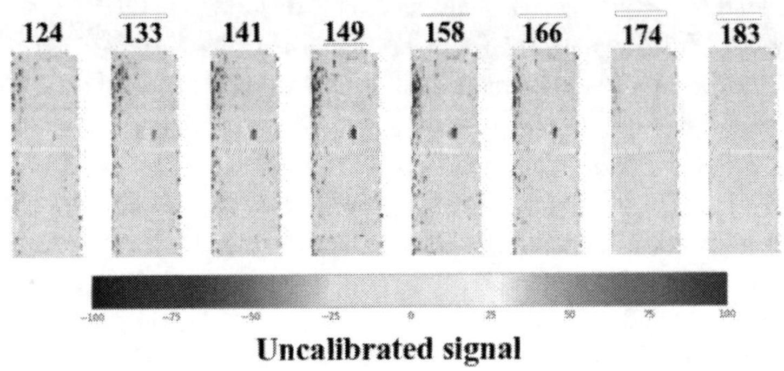

Figure 3. Uncalibrated spectral mapping of Jupiter with 8 GHz resolution in frequency. The large bin has been adopted to increase the sensitivity. We show the mean value from all detectors. The central frequency in GHz is given at the top of each map.

The presented spectral maps are the result of a portion of the original (typical) 90x90 arcmin^2 sky coverage, obtained in 10 minutes of total integration time per observation. We can calibrate the spectral response by exploiting the integrated spectrum of such observations: in other words, we obtain a calibration factor between the detector response and the astrophysical signal per electromagnetic bin.

4 Conclusions

We have demonstrated the capability of instantaneously mapping large areas of the sky (one-degree diameter FoV) in spectroscopic mode by exploiting FTS coupled to KIDs.

KISS has played a fundamental role as the test bed for dedicated spectral mapper large-telescope, CONCERTO, which has currently passed its technical commissioning phase, as reported in [16]. Despite its pioneering and challenging nature, it has paved the way for the technological advancement of the mm observations ground-based at multi-wavelength large FoV.

In this paper, we have reported the integrated Venus brightness temperature and the uncalibrated spectral maps of Jupiter.

Acknowledgements. The KID arrays described in this paper have been produced at the PTA Grenoble microfabrication facility. This work has been partially supported by the LabEx FOCUS ANR-11-LABX-0013, the European Research Council (ERC) under the European Union's Horizon 2020 research and innovation program (project CONCERTO, grant agreement No 788212), and the Excellence Initiative of Aix-Marseille University-A*Midex, a French "Investissements d'Avenir" program. This work has been partially funded by the Spanish Ministry of Science under the project AYA2017-84185-P. Based on observations made with the first QUIJOTE telescope (QT-1), operated on the island of Tenerife in the Spanish Observatorio del Teide of the Instituto de Astrofisica de Canarias.

References

[1] K. Basu, M. Remazeilles, J. B. Melin et al., Experimental Astronomy, (2021)
[2] D. H. Martin and E. Puplett, Infrared Physics **10**, 105-109 (1970)
[3] A. Fasano, M. Aguiar, A. Benoit et al., JLTP **199**, 529-536 (2020)
[4] A. Catalano, A. Bideaud, O. Bourrion et al., A&A **641**, A179 (2020)
[5] S. Masi, P. de Bernardis, A. Paiella et al., JCAP **7**, 3 (2003)
[6] Concerto Collaboration, P. Ade, M. Aravena et al., A&A **642**, 13 (2020)
[7] J. Zmuidzinas, Annual Review of Condensed Matter Physics **3**, 169 (2012)
[8] M. De Petris, S. De Gregori, M. Decina et al., MNRAS **429**, 849–858 (2013)
[9] A. Ritacco, N. Ponthieu, A. Catalano et al., A&A **599**, A34 (2017)
[10] R. Adam, A. Adane, P. Ade et al., A&A **609**, A115 (2018)
[11] L. D. Spencer, Dissertation, *Spectral characterization of the HERSCHEL SPIRE photometer*, (University of Lethbridge, Canada, 2005), ADS bibcode 2005MsT..........1S
[12] A. Fasano, J. F. Macías-Pérez, A. Benoit, A. et al., A&A accepted, eprint arXiv:2109.03145
[13] A. Fasano, M. Aguiar, A. Benoit et al., EPJ Web of Conferences **228**, (2020)
[14] A. Bellotti, Dissertation, *Jupiter: a study of atmospheric composition, structure, and dynamics using microwave techniques*, (Georgia Institute of Technology, 2018), http://hdl.handle.net/1853/61122
[15] Planck Collaboration, Y. Akrami, M. Ashdown et al., A&A **607**, A122 (2017)
[16] A. Monfardini, A. Beelen, A. Benoit et al., JLTP submitted, eprint arXiv:2106.14028

Velocity dispersion vs cluster mass: A new scaling law with THE THREE HUNDRED clusters

Antonio Ferragamo[1,*], *Marco* De Petris[1], *Gustavo* Yepes[2,3], *Juan* Macías-Pérez[4], *Weiguang* Cui[5,3], and *Alejandro* Jiménez-Muñoz[4]

[1]Dipartimento di Fisica, Sapienza Universitá di Roma, Piazzale Aldo Moro 5, I-00185 Roma, Italy
[2]Departamento de Física Téorica, Módulo 15, Facultad de Ciencias, Universidad Autónoma de Madrid, E-28049 Madrid, Spain
[3]Centro de Investigación Avanzada en Física Fundamental (CIAFF), Facultad de Ciencias, Universidad Autónoma de Madrid, E-28049 Madrid, Spain
[4]Univ. Grenoble Alpes, CNRS, LPSC-IN2P3, 53, avenue des Martyrs, 38000 Grenoble, France
[5]Institute for Astronomy, University of Edinburgh, Edinburgh EH9 3HJ, United Kingdom

Abstract. The Planck Collaboration has shown that the number of clusters as a function of their mass and redshift is an extremely powerful tool for cosmological analyses. However, the true cluster mass is not directly measurable. Among the possible approaches, clusters mass could be related to different observables via self similar scaling law. These observables are related to the baryonic components of which a cluster is composed. However, the theoretical relations that allow the use of these proxies often are affected by observational and physical biases, which impacts on the determination of the cluster mass. Fortunately, cosmological simulations are an extremely powerful tool to assess these problems. We present our calibration of the scaling relation between mass and velocity dispersion of galaxy members from the study of the simulated clusters of THE THREE HUNDRED project with mass above $10^{13} M_{\odot}$. In order to investigate the presence of a redshift dependence, we analyzed 16 different redshifts between $z = 0$ and $z = 2$. Finally, we investigated the impact of different AGN feedback models.

1 Introduction

According to the hierarchical bottom-up scenario, galaxy clusters (GC) are the last structure to form, and are, therefore, the perfect tracers of the evolution of structures along the history of the Universe. They reside in the deepest potential wells of the Large Scale Structure (LSS), generate by accretion of Dark matter (DM) [1], therefore their number as a function of their mass is very sensitive to the underlying Cosmology [2]. Although GC are dominated by DM, they include several forms and phases of the baryonic matter. Indeed, the galactic component is surrounded by the so called Intra-Cluster Medium (ICM) composed by warm and hot gas and non-thermal plasma. Each component is characterised by its own physical process. Therefore we can measure different quantities, such as the X-ray temperature, The integrated Compton parameter or the velocity dispersion of the galaxies, by observing different regions of the electromagnetic spectrum. Under the assumption of relaxed systems,

*e-mail: ferragamoantonio@gmail.com

these observables can be directly related to the cluster mass trough scaling relation. Since GC formation is gravitationaly driven starting from initial density fluctuations, they evolve self-similarly [3] and this allow us to approximate these scaling relation to power laws. This property has been shown to be fundamental for the use of clusters in cosmology, as their mass is not directly measurable. In fact, it is crucial to constrain the scaling relations as accurately and precisely as possible. For this purpose, cosmological simulations play a fundamental role being the perfect laboratories in which it is possible to calibrate and to test new scaling relations understanding possibly biases and physical limitations.

This paper is organised as follows. In section 2 we describe the simulation and the data-set we used for our analysis. In sec. 3 we present the results of the scaling relation $\sigma - M$. Finally in section 4 we presents our conclusions.

2 The Three Hundred

The dataset of simulated GCs that we used in this study is selected within The Three Hundred (The300) [4]. This is a zoomed re-simulation of the 324 most massive Lagrangian region of a MultiDark Planck 2 simulation (MDPL2) [5]. The MDPL2 consists in a $1h^{-1}$Gpc volume populated with 3840^3 DM particles of mass $1.5 \times 10^9 h^{-1}\mathrm{M}_\odot$, consistent with the cosmology of Planck 2015 data release [6] ($h = 0.678$, $n = 0.96$, $\sigma_8 = 0.823$, $\Omega_\Lambda = 0.693$, $\Omega_m = 0.307$ and $\Omega_b = 0.048$).

The region of a radius of $15h^{-1}$Mpc around the center each of the 324 selected Lagrangian regions are then populated with gas particles with initial mass of $2.36 \times 10^8 h^{-1}\mathrm{M}_\odot$ and they are re-simulated with several codes. In this work we focus our analysis in the results of the Gadget-X [7, 8] and GIZMO-SIMBA [9] hydrodynamical simulations codes that implement different feedback mechanisms.

2.1 Dataset

Within each region the bounded structures (halos and sub-halos) are identified by using the Amiga Halo Finder (AHF) [10]. For our purpose, we consider only halos with total mass $M_{200} \geq 10^{13} h^{-1}\mathrm{M}_\odot$ and with at least 5 DM sub-halos or galaxies that are not contaminated by low resolution particles. The sub-halos are structure of at least 20 DM particles with mass above $10^{11} h^{-1}\mathrm{M}_\odot$. Whereas, in analogy with [11] (hereafter M13), we call the galaxy with more than ~ 20 stars ($M_* \gtrsim 10^9 h^{-1}\mathrm{M}_\odot$) within the sub-halo. This definition of galaxy assures the selection of all the sub-halos and many more with mass lower than $10^{11}\mathrm{M}_\odot$.

Finally, in order to study the possible redshift evolution of the $\sigma - M$ scaling relation we selected 16 redshifts between $z = 0$ and $z = 2$ within the 128 snapshots stored by The300.

The final selection of GC contains more than 1000 clusters at each redshift both in the Gadget-X and GIZMO catalogs.

3 Scaling relation results

In analogy with the previous works in literature on the dynamical scaling relation [e.g. 11–13], we characterised the parameters of a power law relation

$$\frac{\sigma_{200}}{\mathrm{km\,s^{-1}}} = A \left(\frac{h(z)\, M_{200}}{10^{15}\,\mathrm{M}_\odot} \right)^{\alpha}, \tag{1}$$

where $h(z)$ is the Hubble factor as a function of redshift, while σ_{200} and M_{200} are the velocity dispersion and the mass of the cluster within the radius R_{200}, respectively. For all the clusters

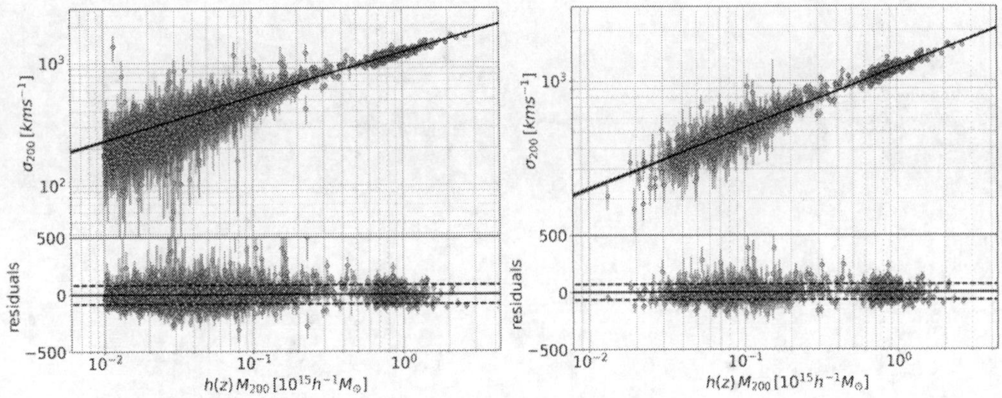

Figure 1. Sub-halos velocity dispersion σ_{200} as a function of total mass $h(z)M_{200}$, at z = 0, for the Gadget-X simulation run. The right and left top panels show the cases $n_{gal} \geq 5$ and $n_{gal} \geq 10$, respectively. The solid black line represents the best-fitting relation. In the bottom panels there are depicted the corresponding residuals.

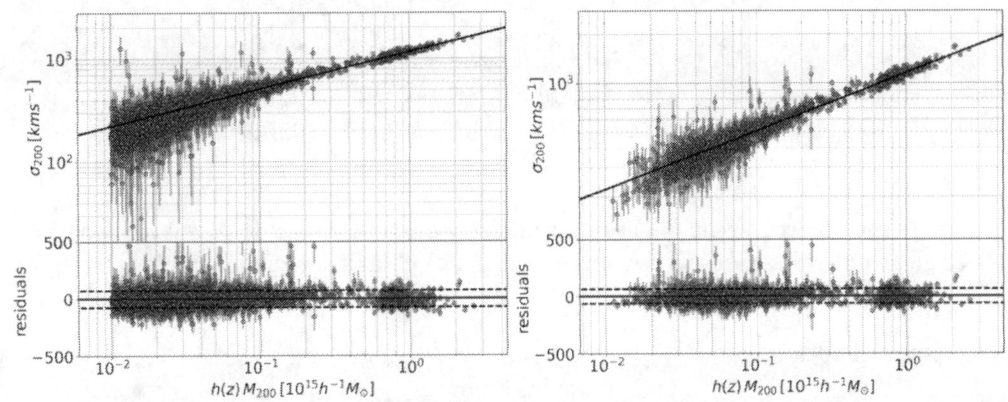

Figure 2. Galaxies velocity dispersion σ_{200} as a function of total mass $h(z)M_{200}$, at z = 0, for the Gadget-X simulation run. The right and left top panels show the cases $n_{gal} \geq 5$ and $n_{gal} \geq 10$, respectively. The solid black line represents the best fitting relation. In the bottom panels there are depicted the corresponding residuals.

described in sec. 2.1 we used the Biweight estimator [14] to estimate the one-dimensional velocity dispersion $\sigma_{200} = \sigma_{200}^{3D}/\sqrt{3}$. textbfThe uncertainty on the velocity dispersion in each main projection are estimated via bootstrap and than propagated to obtain the error an the one-dimensional dispersion. Subsequently, we performed the fit with a linear function in the relation in the logarithmic space by using the python routine `scipy.optimize.curve_fit`.

In order to investigate the impact of the number of galaxy members on the scaling relation, we performed the entire analysis taking into account two different thresholds for the minimum number of galaxy members: $n_{gal} \geq 5$ and $n_{gal} \geq 10$. Figures 1 and 2 show the fitted relations for the Gadget-X run in the two cases at redshift z = 0 for sub-halos and galaxies, respectively. We see that the slope of the best fit is less steep when the minimum number of galaxy members is 10. Furthermore, the scatter is higher in the case with $n_{gal} \geq 5$. We can explain these evidence with the combinations of two effects. From one side the scatter increases because lower the number of galaxies higher the uncertainties on the velocity

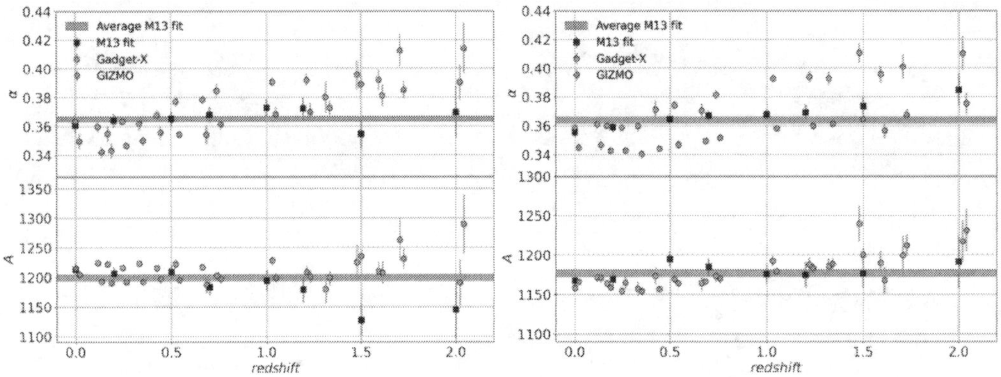

Figure 3. Best fit parameters α (top panels) and A (bottom panels) as a function of redshift, for sub-halos (left panels) and galaxies (right panels). Light blue and coral dots represent the results from the Gadget-X and Gizmo analyses, respectively. The M13 result are represented by the black squares, while the grey band represents their weighted mean.

Table 1. Best fit parameter of equation 2 for sub-halos

	A	α	β
Gadget-X	1216.2 ± 0.6	0.358 ± 0.002	0.027 ± 0.003
GIZMO-SIMBA	1198.8 ± 0.7	0.341 ± 0.002	0.026 ± 0.002

dispersion. From the other side the cluster sample in the case with $n_{gal} \geq 5$ is affected by selection effects. For these reasons we present the results obtained with the sample $n_{gal} \geq 10$.

3.1 Redshift dependence

In figure 3 we show the parameter of the $\sigma - M$ relation along the redshift between $z = 0$ and $z = 2$ that we took into account for sub-halos and galaxies in the left and right panel, respectively. The black squares are the M13 results (slope in the top panel and the normalization in the central panel) for the AGN sample only at the 10 investigated redshifts, while the shaded regions represent the weighted mean of the 10 redshifts. In light blue and orange we plotted the best fit values obtained with the Gadget-X and the GIZMO runs of the The300. Since the Dianoga simulation and the Gadget-X implement the same baryonic physics recipe, in this section we only focus on the comparison between the M13 and Gadget-X results. The comparison with GIZMO is reported in sec. 3.2. In the cases of sub-halos and galaxies, the two results are in an excellent agreement at $z < 0.8$. Whereas, going towards higher redshift we observe an increasing value of the slope that is not present in M13 results. differently from the normalization where any redshift dependence is noticeable.

The GIZMO best fits show the same behaviour for both slope and normalization. For this reason we decided to take into account the redshift dependence of the slope in our parametrization of the scaling relation by fitting a linear relation between the redshift and the slope. In Tables 1 and 2 we list the parameters of the $\sigma - M$ relation

$$\frac{\sigma_{200}}{\mathrm{km\,s^{-1}}} = A \left(\frac{h(z)\, M_{200}}{10^{15}\,\mathrm{M_\odot}} \right)^{(\alpha+\beta z)} . \tag{2}$$

Table 2. Best fit parameter of equation 2 for galaxies

	A	α	β
Gadget-X	1172.7 ± 0.8	0.357 ± 0.002	0.029 ± 0.002
GIZMO-SIMBA	1171.5 ± 0.6	0.340 ± 0.001	0.016 ± 0.001

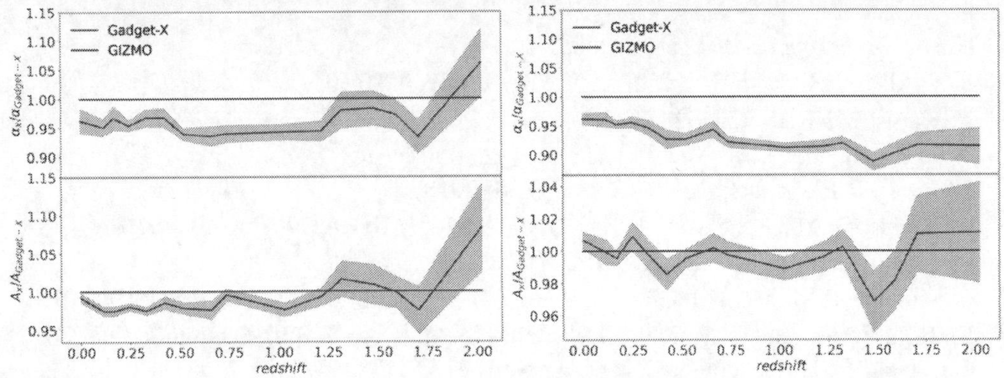

Figure 4. Ratio of the best fit parameters from GIZMO and Gadget-X runs (red line), for sub-halos (left panel) and galaxies right panel). The slope, α, and the normalization, A, are represented in top and bottom panels, respectively.

3.2 Gadget-X vs GIZMO-SIMBA

It is interesting to report that in Fig. 3 the GIZMO parameters are mostly lower than the Gadget-X. The ratio between parameters obtained from the Gadget-X and GIZMO runs are shown in Fig. 4. On one hand, the slope α from GIZMO is $\sim$ 5% lower than the slope obtained in the Gadget-X fits in the whole redshift range for the sub-halos, whereas for the galaxies the offset increases with the redshift passing from 5% at $z = 0$ to a $\sim$ 10% at $z = 2$. On the other hand, the bias in the normalization, A, is much smaller for the sub-halos and completely absent for the galaxies. This evidence can be explained with the differences in the baryonic physics recipes between the Gadget-X and the GIZMO-SIMBA runs of The300, in particular in the differences in the AGN feedback, dramatically affecting the less massive halos, which can explain the shallower slope of GIZMO relations [15].

4 Conclusions

In this work we presented the calibration of a scaling relation between the velocity dispersion and the mass of GCs. For this purpose we used the synthetic clusters from the The300 simulation with mass $M_{200} \geq 10^{13}\mathrm{M}_{\odot}$, at 16 different redshifts between $z = 0$ and $z = 2$, from two different runs, Gadget-X and GIZMO-SIMBA. Due of the similarities between the Gadget-X run of The300 and the simulation used by M13 (in particular with the AGN feedback model), we performed a direct comparison between the results of this work and M13. In analogy with M13, we constrained the scaling relation by using different tracers such as sub-halos and galaxies with thresholds in total ($M_{tot} \geq 10^{11}\mathrm{M}_{\odot}$) and stellar mass ($M_{*} \geq 10^{9}\mathrm{M}_{\odot}$). While we are consistent with M13 results up to z=0.8, our scaling law shows a different slope at high redshifts. In fact, we observed a dependence of the slope of the relation with the redshift. Furthermore, we investigated the effect of different model of AGN feedback on the parameters of $\sigma - M$. By comparing the best fits from Gadget-X and GIZMO-SIMBA, we observed that the slope obtained with GIZMO is a ~ 5% smaller than the Gadget-X one for both tracers and in the whole range of redshifts. We asses this result to the different implementations of the AGN feedback.

References

[1] V. Springel, MNRAS **364**, 1105 (2005)
[2] S.W. Allen, A.E. Evrard, A.B. Mantz, ARA&A **49**, 409 (2011)
[3] N. Kaiser, MNRAS **222**, 323 (1986)
[4] W. Cui, A. Knebe, G. Yepes, F. Pearce, C. Power, R. Dave, A. Arth, S. Borgani, K. Dolag, P. Elahi et al., MNRAS **480**, 2898 (2018)
[5] A. Klypin, G. Yepes, S. Gottlöber, F. Prada, S. Heß, MNRAS **457**, 4340 (2016)
[6] Planck Collaboration XIII, A&A **594**, A13 (2016)
[7] G. Murante, P. Monaco, M. Giovalli, S. Borgani, A. Diaferio, MNRAS **405**, 1491 (2010)
[8] E. Rasia, S. Borgani, G. Murante, S. Planelles, A.M. Beck, V. Biffi, C. Ragone-Figueroa, G.L. Granato, L.K. Steinborn, K. Dolag, ApJl **813**, L17 (2015)
[9] R. Davé, D. Anglés-Alcázar, D. Narayanan, Q. Li, M.H. Rafieferantsoa, S. Appleby, MNRAS **486**, 2827 (2019)
[10] S.R. Knollmann, A. Knebe, ApJs **182**, 608 (2009)
[11] E. Munari, A. Biviano, S. Borgani, G. Murante, D. Fabjan, MNRAS **430**, 2638 (2013)
[12] A.E. Evrard, J. Bialek, M. Busha, M. White, S. Habib, K. Heitmann, M. Warren, E. Rasia, G. Tormen, L. Moscardini et al., ApJ **672**, 122-137 (2008)
[13] A. Saro, J.J. Mohr, G. Bazin, K. Dolag, ApJ **772**, 47 (2013)
[14] T.C. Beers, K. Flynn, K. Gebhardt, AJ **100**, 32 (1990)
[15] W. Cui, arXiv e-prints arXiv:2111.01889 (2021), 2111.01889

Cosmic dust evolution: The challenges for NIKA2

Frédéric Galliano[1,*] and the DustPedia collaboration

[1]AIM, CEA, CNRS, Université Paris-Saclay, Université Paris Diderot, Sorbonne Paris Cité, F-91191 Gif-sur-Yvette, France

Abstract. I summarize a recent study aimed at constraining dust evolution timescales in nearby galaxies. We were able to quantify the efficiency of dust condensation in SN II ejecta, grain growth in the ISM and dust destruction by SN II blast waves. Around Solar metallicity, our results are consistent with what we know in the Milky Way, that is dust production is dominated by grain growth. We however show that this changes with metallicity. The dust content in very-low-metallicity systems is dominated by grains that have formed in SN II ejecta. I end this article with a discussion about the potential of NIKA2 to study dust evolution in galaxies.

1 Introduction

Interstellar dust is a key physical ingredient of galaxies, obscuring star formation, regulating the heating and cooling of the gas, and building-up chemical complexity. Its properties are however difficult to estimate, as there are important uncertainties about the constitution of the grain mixture in the diffuse Interstellar Medium (ISM) of the Milky Way, where dust models are calibrated. In addition, dust properties are known to drastically evolve as a function of their local physical conditions (gas density, radiation field, shocks) and as a function of the history of the galaxy (enrichment of the ISM by successive waves of star formation). It is thus crucial to quantify this evolution.

Cosmic dust evolution models attempt to account for the balance between the main dust evolution processes, in a global way (*cf.* [1] for a review). They however rely on simple descriptions of the evolution mechanisms and depend on the value of uncertain tuning parameters, such as sticking coefficients, binding energies, sputtering yields, *etc.* We have attempted to provide an empirical estimate of these tuning parameters, using a sample of $\simeq 800$ nearby galaxies.

2 Galaxy sample and SED Modeling

In order to constrain a dust evolution model, we need to estimate the following global quantities, for a large sample of galaxies: the dust mass, $M_{\rm dust}$, the gas mass, $M_{\rm gas}$, the stellar mass, $M_\star$, the metallicity, Z, and the Star Formation-Rate (SFR). We have used the DustPedia sample [2]. It provides homogeneous multi-wavelength photometry [3], with ancillary H I and

*e-mail: frederic.galliano@cea.fr

CO observations. The stellar mass and SFR have been derived by [4], using ultraviolet-to-mm Spectral Energy Distribution (SED) fitting. The metallicity have been estimated with a consistent calibration by [5].

One of the limitations of DustPedia is its selection of objects larger than 10′, excluding most local dwarf galaxies. Yet, the latter are crucial to constrain dust evolution, as they sample the low-metallicity regime corresponding to the early stages of the dust build-up. To overcome this deficiency, we have added to our sample the sources from the Dwarf Galaxy Survey (DGS, [6–9]). In total we have ≃ 800 objects covering a large metallicity range.

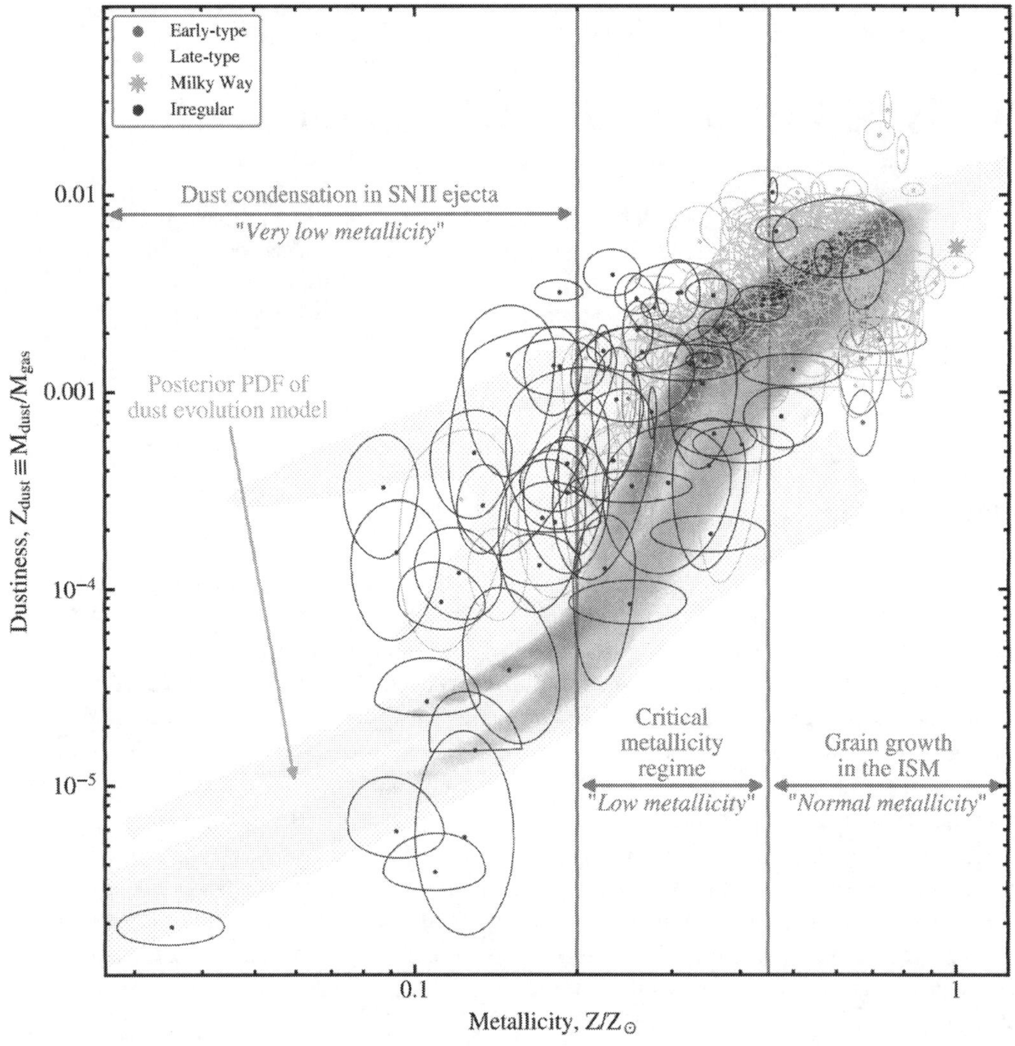

Figure 1. The Skewed Uncertainty Ellipses (SUEs) represent the observed dustiness, as a function of metallicity, of each galaxy in our sample, color-coded according to its type [10]. The yellow-to-red density contours in the background represent the posterior probability density of our dust evolution model, inferred from the observations of galaxies (*i.e.* the SUEs). We have highlighted the three main dust production regimes in magenta.

We have performed infrared SED modeling of the observed photometry, using the hierachical Bayesian code HerBIE [11]. This model provides a rigorous account of the different sources of uncertainties and is efficient at recovering the intrinsic scatter and correlations of

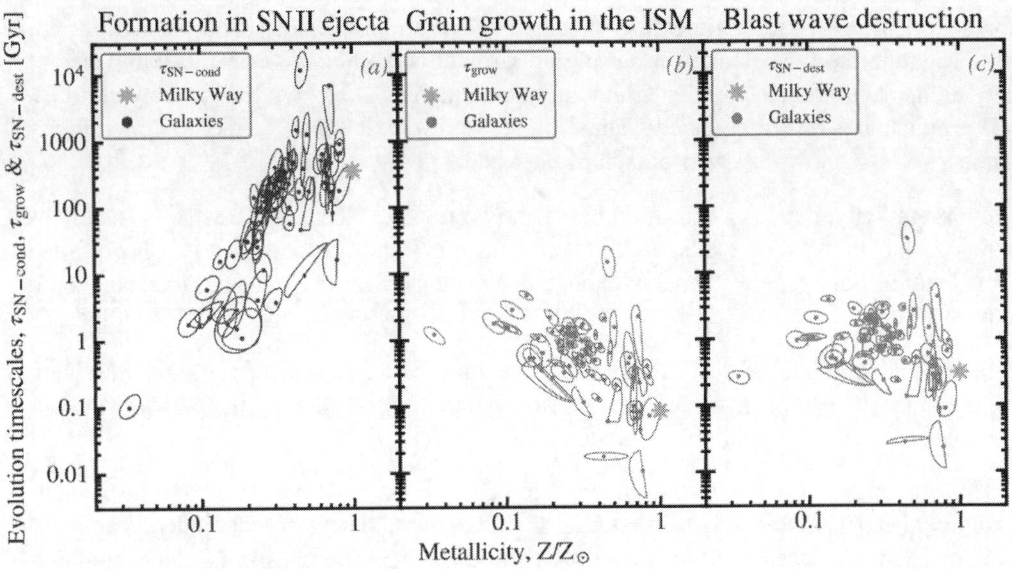

Figure 2. In each panel, the SUEs represent the inferred dust evolution timescales of the galaxies for which all constraints are available, as a function of metallicity [10]. Panel *(a)* shows the timescale of dust condensation in SN II ejecta, panel *(b)*, the timescale of grain growth in the ISM, and panel *(c)*, the timescale of dust destruction by SN II blast waves.

the studied parameters. We have used the properties of the THEMIS dust model [12]. The results are the Skewed Uncertainty Ellipses (SUEs) in figure 1 [10]. We have displayed the relation between the metallicity and the dustiness (*i.e.* the dust-to-gas mass ratio).

3 Constraining dust evolution timescales

The estimated M_{dust}, M_{gas}, $M_{\star}$, Z and SFR of each galaxy are then used to constrain the parameters of our cosmic dust evolution model. This model, for a given Star Formation History (SFH), computes the evolution, as a function of time, of the masses of stars, gas, heavy elements and dust. At each time step, new stars are formed, depleting the gas. At the end of their lifetime (which depends on their initial mass), these stars return to the ISM a fraction of their gas, newly-formed elements, and dust seeds. The three dust evolution processes we account for (on top of astration, which is straightforward) are: the condensation of grain seeds in SN II ejecta, grain growth in the ISM and dust destruction by SN II blast waves. The efficiency of these different processes are vigorously debated in the literature and they can vary by more than one order of magnitude, depending on the study.

We fit our dust evolution models to our observed estimates in a hierarchical Bayesian way, assuming that each galaxy has a particular SFH (parameterized by its timescale, amplitude, age, infall and outlfow rates), but that the dust evolution efficiencies are universal (*i.e.* they have the same value for each galaxy). The posterior probability distribution of our dust evolution model is represented as contours in figure 1.

From our inferred dust evolution efficiencies, we can derive dust evolution timescales. These are shown in figure 2. Although we only have one estimate of the three efficiencies for our whole sample, each galaxy has its own set of evolution timescales, because these depend also on the SFR, M_{gas}, Z, *etc.* The most important conclusions from these trends are the following.

At very low metallicity, the timescale for dust condensation in SN II is shorter than that of grain growth in the ISM. This is because grain growth is inefficient when the abundance of heavy elements in the gas is low. Grains in these young systems are thus mainly stardust. SN II production is rather inefficient. This is the reason why, in figure 1, very low-metallicity galaxies, lie on the low-dustiness branch of the trend.

At low metallicity, there is a transition between two regimes. The dustiness rises steeply as grain growth in the ISM becomes rapidly more efficient. We are around what [13] have called the *critical metallicity*. Now, there are enough heavy element atoms in the gas to accrete onto grain surfaces.

At normal metallicity, the dust production is dominated by grain growth in the ISM, with a high efficiency. Although we did not use the constraints provided by the Milky Way, our timescales are consistent with the Galactic ones.

This study also provides methodological insights. First, it demonstrates the instrumental role of dwarf galaxies in understanding dust evolution. These objects indeed sample a metallicity range characterized by a dust production regime different from Solar-metallicity systems. The constraint on the low efficiency of dust production by SN II ($\gtrsim 0.03\ M_\odot$/SN in our article) relies entirely on dwarfs. Second, our work is the first rigorous fit of individual dust evolution models to actual observations of galaxies. Most studies before us were simply overlaying dust evolution tracks on top of observed scaling relations. Yet, such a visual comparison is inconsistent in many cases, as it permits interpreting two different quantities of a galaxy with two different ages. Lastly, our results emphasize that grain growth is fast, provided the metallicity is high enough ($Z \gtrsim 0.4\ Z_\odot$). It thus provides a natural mechanism to explain the existence of dusty objects at very high redshifts ($z \gtrsim 6$), only a few 100 Myrs after reionization.

4 Prospective: The potential of NIKA2

The dust evolution study we have just presented was limited because we analyzed galaxies as a whole. The dustiness we estimated for each galaxy was representative of their average conditions, but we know that local grain evolution, at the scale of molecular clouds, is important. Spatially-resolved studies are thus the necessary next step to constrain local dust evolution, that will provide empirical constraints on the effects of gas density and radiation field intensity on the grain properties.

NIKA2 is an instrument providing unprecedentedly deep images of nearby galaxies, at 1.2 and 2 mm, with angular resolutions $12''$ and $18''$, respectively. It provides unique constraints on the grain properties at spatial scales of a few hundreds of parsecs, close to the size of giant molecular complexes. Figure 3 represents the typical SED of a nearby star-forming galaxy. We see that NIKA2 samples an important spectral domain, where the following physical processes can be probed.

Thermal dust emission, at long-wavelengths, is poorly constrained. Yet, it is crucial to properly estimate dust masses, as it is dominated by cold, massive regions. We have represented the potential presence of a "submm excess", which is likely the result of the inaccuracy of contemporary dust models (*cf.* [1] for a review). The detailed characterization of dust emission will likely account for this excess. NIKA2 might be instrumental in addressing the origin of this enigmatic effect.

The Anomalous Microwave Emission (AME) is the cm emission shown in red. It originates in fastly spinning small grains [14]. The exact nature of the carriers of this process is still

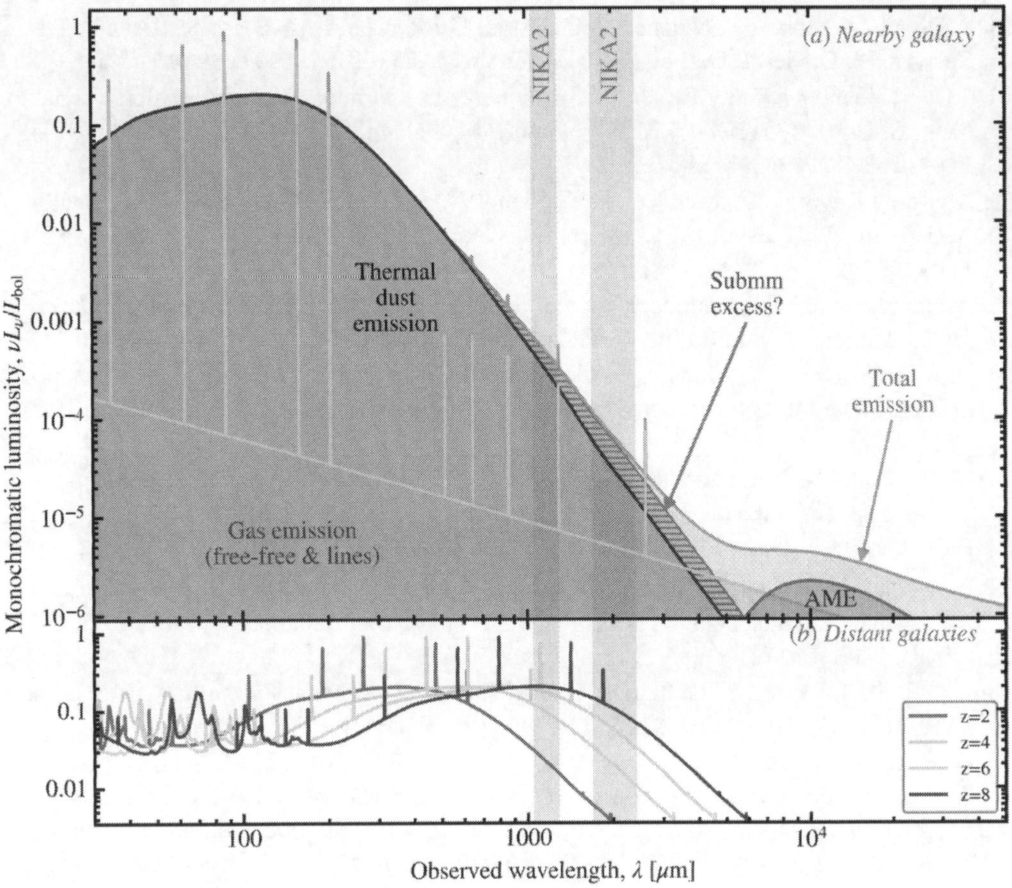

Figure 3. We display the physical processes probed by NIKA2, in the nearby Universe (panel *a*), and at high redshift (panel *b*). We have represented the tpyical SED of a star-forming galaxy [1].

debated. NIKA2 observations provide the data to properly constrain the baseline required to quantify the AME. NIKA2 data will thus open the possibility to make maps of AME intensity in external galaxies and understand how it correlates with other dust properties, which should give us clues about the nature of its carriers.

The consortium of the NIKA2 guaranteed program, IMEGIN, is currently tackling these challenges (*cf.* [15] and [16] in this volume).

References

[1] F. Galliano, M. Galametz, A.P. Jones, ARA&A **56**, 673 (2018)

[2] J.I. Davies, M. Baes, S. Bianchi, A. Jones, S. Madden, M. Xilouris, M. Bocchio, V. Casasola, L. Cassara, C. Clark et al., PASP **129**, 044102 (2017), `1609.06138`

[3] C.J.R. Clark, S. Verstocken, S. Bianchi, J. Fritz, S. Viaene, M.W.L. Smith, M. Baes, V. Casasola, L.P. Cassara, J.I. Davies et al., A&A **609**, A37 (2018), `1708.05335`

[4] A. Nersesian, E.M. Xilouris, S. Bianchi, F. Galliano, A.P. Jones, M. Baes, V. Casasola, L.P. Cassarà, C.J.R. Clark, J.I. Davies et al., A&A **624**, A80 (2019), `1903.05933`

[5] P. De Vis, A. Jones, S. Viaene, V. Casasola, C.J.R. Clark, M. Baes, S. Bianchi, L.P. Cassara, J.I. Davies, I. De Looze et al., A&A **623**, A5 (2019), 1901.09040

[6] S.C. Madden, A. Rémy-Ruyer, M. Galametz, D. Cormier, V. Lebouteiller, F. Galliano, S. Hony, G.J. Bendo, M.W.L. Smith, M. Pohlen et al., PASP **125**, 600 (2013), 1305.2628

[7] A. Rémy-Ruyer, S.C. Madden, F. Galliano, V. Lebouteiller, M. Baes, G.J. Bendo, A. Boselli, L. Ciesla, D. Cormier, A. Cooray et al., A&A **582**, A121 (2015), 1507.05432

[8] A. Rémy-Ruyer, S.C. Madden, F. Galliano, S. Hony, M. Sauvage, G.J. Bendo, H. Roussel, M. Pohlen, M.W.L. Smith, M. Galametz et al., A&A **557**, A95 (2013), 1309.1371

[9] A. Rémy-Ruyer, S.C. Madden, F. Galliano, M. Galametz, T.T. Takeuchi, R.S. Asano, S. Zhukovska, V. Lebouteiller, D. Cormier, A. Jones et al., A&A **563**, A31 (2014), 1312.3442

[10] F. Galliano, A. Nersesian, S. Bianchi, I. De Looze, S. Roychowdhury, M. Baes, V. Casasola, L.P. Cassará, W. Dobbels, J. Fritz et al., A&A **649**, A18 (2021)

[11] F. Galliano, MNRAS **476**, 1445 (2018), 1801.06660

[12] A.P. Jones, M. Köhler, N. Ysard, M. Bocchio, L. Verstraete, A&A **602**, A46 (2017)

[13] R.S. Asano, T.T. Takeuchi, H. Hirashita, A.K. Inoue, Earth, Planets, and Space **65**, 213 (2013), 1206.0817

[14] A.C. Bell, T. Onaka, F. Galliano, R. Wu, Y. Doi, H. Kaneda, D. Ishihara, M. Giard, PASJ **71**, 123 (2019), 1910.01265

[15] G. Ejlali, R. Adam, P. Ade, H. Ajeddig, P. André, E. , Artis, H. Aussel, A. Beelen, A. Benoît, S. Berta et al., *Dust Emission in Galaxies at Millimeter Wavelengths. Cooling of star forming regions in NGC6946*, in *Proceedings of the International Conference entitled "mm Universe @ NIKA2", Rome (Italy), June 2021, EPJ Web of conferences*, edited by M. De Petris, A. Ferragamo, F. Mayet (2021)

[16] S. Katsioli, R. Adam, P. Ade, H. Ajeddig, P. André, E. , Artis, H. Aussel, A. Beelen, A. Benoît, S. Berta et al., *Exploring the millimetre emission in nearby galaxies: analysis of the edge-on galaxy NGC 891*, in *Proceedings of the International Conference entitled "mm Universe @ NIKA2", Rome (Italy), June 2021, EPJ Web of conferences*, edited by M. De Petris, A. Ferragamo, F. Mayet (2021)

The hydrostatic mass bias in The Three Hundred clusters

Giulia Gianfagna[1,2,*], *Elena* Rasia[3,4], *Weiguang* Cui[5], *Marco* De Petris[1], and *Gustavo* Yepes[6]

[1]Dipartimento di Fisica, Sapienza Università di Roma, Piazzale Aldo Moro 5, I-00185 Rome, Italy
[2]INAF, Istituto di Astrofisica e Planetologia Spaziali, via Fosso del Cavaliere 100, 00133 Rome, Italy
[3]IFPU - Institute for Fundamental Physics of the Universe, Via Beirut 2, 34014 Trieste, Italy
[4]INAF Osservatorio Astronomico di Trieste, via Tiepolo 11, I-34131, Trieste, Italy
[5]Institute for Astronomy, University of Edinburgh, Royal Observatory, Edinburgh EH9 3HJ, UK
[6]Departamento de Física Teórica and CIAFF, Módulo 8, Facultad de Ciencias, Universidad Autónoma de Madrid, 28049 Madrid, Spain

Abstract. The assumption of Hydrostatic equilibrium (HE) is often used in observations to estimate galaxy clusters masses. We use a set of almost 300 simulated clusters from The Three Hundred Project, to estimate the cluster HE mass and the bias deriving from it. We study the dependence of the bias on several dynamical state indicators across a redshift range from 0.07 to 1.3, finding no dependence between them. Moreover, we focus our attention on the evolution of the HE bias during the merger phase, where the bias even reaches negative values due to an overestimation of the mass with HE.

1 Introduction

Galaxy clusters are the most massive, gravitationally bound structures in the Universe and they are a powerful tool in cosmology. In particular, their mass is fundamental for the estimation of cosmological parameters. In observations, one way to estimate this mass is through X-rays and SZ (Sunyaev-Zeldovich) effect, from which the temperature, density and pressure profiles of the hot gas between the galaxies (Intra Cluster Medium, ICM) are extracted, then the assumption of hydrostatic equilibrium (HE) has to be made in order to estimate their mass [6]. This method usually leads to an underestimation of the mass, as shown in numerical simulations [8]. The lack of spherical symmetry and non-thermal motions in the gas lead to a violation of the HE assumption. These are usually present in the object outskirts, in major-merger events and in general in disturbed clusters [7, 14, 17, 19]. The presence of a correlation between hydrostatic mass bias and cluster dynamic state is still a matter of debate [7, 17], in this work we seek to understand this in detail by looking directly to the bias change along the evolution of major-merger events.

2 The Three Hundred Project

The clusters in this work are the most massive clusters in The Three Hundred Project [1], they come from a set of 324 Lagrangian regions from the dark-matter (DM) only simulation MultiDark [2]. We analyse 9 redshifts in the [0.07, 1.32] range.

*e-mail: giulia.gianfagna@uniroma1.it

THE THREE HUNDRED set is a volume-limited mass-complete sample, the objects masses M_{500}[1] are larger than $6.5 \times 10^{14} M_{\odot}$. Initial conditions are generated at the redshift $z = 120$ by refining the mass resolution in the central region and degrading it in the outer part with multiple levels of mass refinement. The high-resolution dark-matter particle mass is equal to $m_{DM} = 1.27 \times 10^{9} h^{-1} M_{\odot}$, while the initial gas mass is equal to $2.36 \times 10^{8} h^{-1} M_{\odot}$.

We consider the resimulated cluster sample with the hydrodynamical code GADGET-X [3], which includes radiative processes, gas-cooling, star formation, supernovae thermal feedback, chemical evolution and enrichment, super massive black holes and their feedback.

This simulation assumes a standard cosmological model according to the 2016 Planck results [5]: $h = 0.6777$ for the reduced Hubble parameter, $n = 0.96$ for the primordial spectral index, $\sigma_8 = 0.8228$ for the amplitude of the mass density fluctuations in a sphere of $8h^{-1}$ Mpc co-moving radius, $\Omega_\Lambda = 0.692885$, $\Omega_m = 0.307115$, and $\Omega_b = 0.048206$ respectively for the density parameters of dark energy, dark matter, and baryonic matter.

3 Methods

In this section we show all the basics to estimate the hydrostatic masses and their bias with respect to the real mass.

3.1 The Hydrostatic Equilibrium

The hydrostatic equilibrium assumption assumes that the cluster is in equilibrium due to the balance between the gas thermal pressure and the gravitational force. The HE mass, namely the total mass inside a sphere of radius r, can be written as

$$M_{\mathrm{HE,SZ}}(<r) = -\frac{r^2}{G\rho_{\mathrm{g}}(r)} \frac{\mathrm{d}P_{\mathrm{th}}(r)}{\mathrm{d}r}, \tag{1}$$

it is based on the assumptions of spherical symmetry of the system and purely thermal gas pressure. In Eq.(1) G is the gravitational constant, ρ_{g} and P_{th} are the gas density and the thermal pressure of the gas. Assuming the equation of state of an ideal gas, it follows

$$M_{\mathrm{HE,X}}(<r) = -\frac{r k_{\mathrm{B}} T(r)}{G\mu m_p} \left[\frac{\mathrm{d}\ln \rho_{\mathrm{g}}(r)}{\mathrm{d}\ln r} + \frac{\mathrm{d}\ln T(r)}{\mathrm{d}\ln r} \right]. \tag{2}$$

The real mass M_{true} of a simulated cluster can be easily computed by summing all the dark matter, stars and gas particle masses inside an aperture radius. The mass bias, at a specific radius, is defined as $b_{\mathrm{SZ/X}} = (M_{\mathrm{true}} - M_{\mathrm{HE,SZ/X}})/M_{\mathrm{true}}$. It can be either positive, when the HE mass is underestimating the cluster true mass, or negative, when there is an overestimation. A null bias suggests that the true mass is perfectly described by HE.

3.2 The ICM profiles

The 3D radial profiles of pressure, temperature and gas density are estimated dividing the cluster in logarithmically equispaced radial bins, starting from the center (the place of the minimum potential well [7]). The profiles are fitted in the radial range [0.2-3]R_{500}.

[1] The mass with the subscript 500 is indicating the mass of a sphere with radius R_{500} whose density is 500 times the critical density of the Universe $\rho_c = 3H(a)^2/8\pi G$, where $H(a)$ is the Hubble function.

The 3D gas density profiles are estimated as the total gas mass in the spherical shell, divided by the shell volume. We use the Vikhlinin model [9] to fit these profiles

$$\rho_g(r) = \rho_0^2 \frac{(r/r_c)^{-a}}{(1+(r/r_c)^2)^{3b-a/2}} \frac{1}{(1+(r/r_s)^c)^{e/c}} + \frac{\rho_{02}^2}{(1+(r/r_{c2})^2)^{3b_2}}, \quad (3)$$

where c is fixed at 3 and $e < 5$. The other 8 parameters are left free.

For the mass-weighted temperature profile we consider only particles with temperature $kT > 0.3$keV. The analytical model was introduced again by Vikhlinin in [9]

$$T(r) = T_0 \times \frac{x+\tau}{x+1} \times \frac{(r/r_t)^{-a}}{(1+(r/r_t)^b)^{c/b}} \quad (4)$$

with $x = (r/r_{\text{cool}})^{a_{\text{cool}}}$. All the 8 parameters are free to vary.

To model the radial pressure profile we make use of the generalized Navarro-Frenk-White (gNFW) model [10]

$$P(r) = \frac{P_0}{x^c(1+x^a)^{\frac{b-c}{a}}} \quad (5)$$

with $x = r/r_s$ a dimensionless radial distance normalised to the scale radius $r_s = R_{500}/c_{500}$, where c_{500} is the concentration parameter. This model has 5 free parameters.

The hydrostatic masses are estimated using the fitted profiles, instead of the numerical ones, to avoid the impact of fluctuations on the profiles (see [8] for a deeper discussion).

3.3 Dynamical state

We divide the clusters depending on their dynamical state thanks to two 3D estimators [11]: $f_s = M_{\text{sub}}/M$, the fraction of cluster mass included in substructures; $\Delta_r = |\mathbf{r}_\delta - \mathbf{r}_{\text{cm}}|/R$, the offset between the central density peak, and the centre of mass of the cluster, normalized to the aperture radius. Each of the estimators is computed at the three nominal overdensities of 200, 500 and 2500. Both indicators are lower than 0.1 for relaxed cluster, larger than 0.1 for disturbed cluster and all the other cases are defined as hybrid clusters [12, 13].

The information given separately by these indicators can be joint through the so called relaxation parameter $\chi_{\text{DS}} = \sqrt{2/[(\Delta_r/0.1)^2 + (f_s/0.1)^2]}$ [13]. For a relaxed cluster $\chi_{DS} \geq 1$.

4 Results

We study the HE bias dependencies on redshift, mass and dynamical state at the overdensities R_{2500}, R_{500} and R_{200}.

4.1 Bias correlations

The bias dependence on the redshift is presented in Fig. 1. The disturbed clusters have the largest dispersion and have negative biases, independently from the overdensity radius. Near the cluster center, the deviations from HE are stronger than in the other overdensities [8, 14], however the bias at R_{2500} shows the largest scatter. No bias dependence on the redshift is evident inside the errors, in agreement with [15, 16].

The bias as a function of the relaxation parameter is represented in Fig.2, the logarithmic scale is used to inflate the data distribution in the small χ_{DS} range. The fit helps to spot any correlation, reducible to linear due to the narrow χ_{DS} scale. The Pearson correlation coefficients are 0.14 (p-value 0.02) for b_{SZ} and 0.23 (p-value 0.0001) for b_X at z=1.32; 0.01

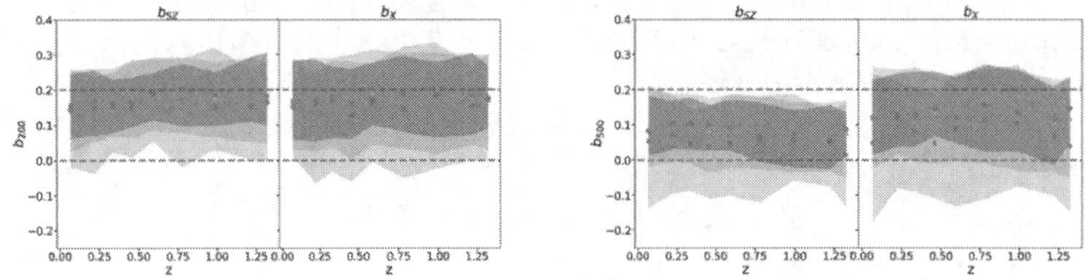

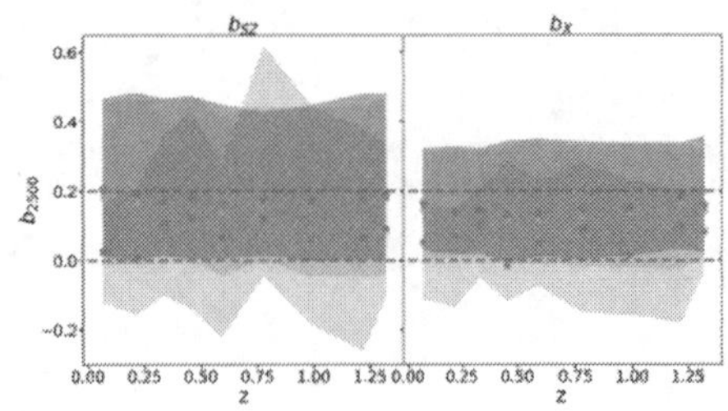

Figure 1. The redshift evolution of b_{SZ} (left panels) and b_X (right panels). The median values of the bias for all clusters, relaxed and disturbed are represented with yellow triangles, magenta diamonds and blue squares respectively. The shaded regions represent the 16th and 84th percentiles. The bias estimated at R_{200}, R_{500} and R_{2500} are represented in the top left, top right and bottom panels respectively. The dashed lines represent 0 and 0.2 biases.

(p-value 0.91) for b_{SZ} and 0.01 (p-value 0.81) for b_X at z=0.07. Fixing the confidence level at 0.05, we cannot say that the biases are correlated at z=0.07 (p-value > 0.05), instead at $z = 1.32$ there is indeed a mild correlation. At the other redshifts we have p-value > 0.05, except for $z = 0.46$, where the correlation coefficient points again to a mild correlation. However, the connection between the biases and the dynamical state can be affected by the use of the best fit model of the profiles, as this usually leads to a smoothing of any possible fluctuation. Also, the $b - \chi_{DS}$ dependency can be influenced by the thresholds chosen to determine the dynamical state of each cluster [13]. Both these issues will most likely tend to reduce the potential correlation. Moreover, also in this case, the disturbed clusters show a wider dispersion with respect to the relaxed ones [7, 17].

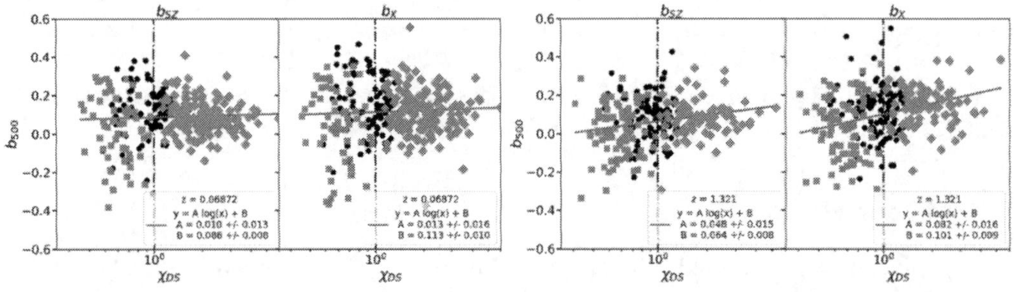

Figure 2. The biases are shown as a function of the relaxation parameter χ_{DS}. The intermediate, relaxed and disturbed are represented with black dots, magenta diamonds and blue squares respectively. In the left panel we show the quantities at $z = 0.07$ and in the right one at $z = 1.3$.

Finally, the bias dependence on the cluster mass at R_{500} is represented in Fig.3. We show the biases at 4 redshifts (1.32, 0.78, 0.33, 0.07), and we find no dependence of the bias on

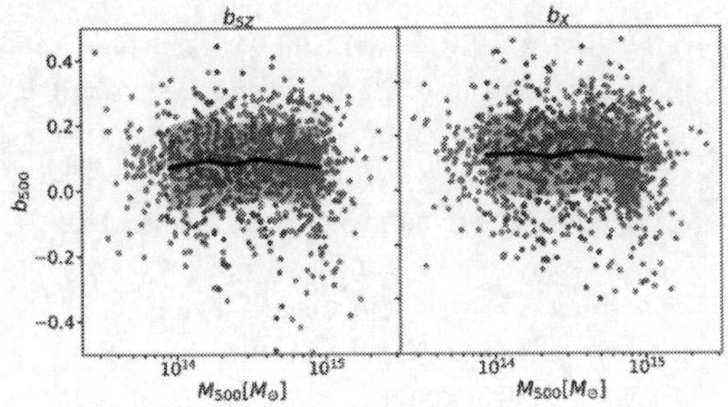

Figure 3. The biases are represented as a function of the cluster total mass at R_{500}. The redshifts 1.32, 0.78, 0.33, 0.07 are represented in red, blue, magenta and green respectively. The black line and the shaded region represent the binned median and 16th-84th percentiles.

the total mass of the clusters, as in [7, 14, 15, 18]. The same happens also for R_{2500} and R_{200}. However, the mass range considered is not wide enough to make definitive statements.

4.2 Hydrostatic bias and merging events

Hydrodynamical simulations allow for tracking the entire history of a cluster, identifying the merging event and the moment it takes place. We define a merger as a very rapid increase in the halo mass. Using the fractional mass change $\Delta M/M = (M_f - M_i)/M_i$, we consider major-merger, with $\Delta M/M \geq 1$, a merger that takes place in half of the dynamical time, with a 100% mass increase [19]. The masses in this case are referring to R_{200}. The dynamical time is defined as $t_d = \sqrt{(3/4\pi)(1/200G\rho_{\rm crit})}$, the critical density $\rho_{\rm crit}$ depends only on the cosmology, so the dynamical time does not depend on the cluster features, but evolves only with redshift. As in [19], the merger event can be described introducing 4 characteristic redshifts: $z_{\rm before}$, the moment right before the merger, when the main object is still in equilibrium; $z_{\rm start}$, the cluster starts growing in mass and the merger begins; $z_{\rm end}$, the merger disturbing effects end; $z_{\rm after}$, the end of the whole merger phase, when the cluster returns to equilibrium. We call merger phase the time between $z_{before} - z_{after}$ and merger event between $z_{start} - z_{end}$.

In our sample 12 clusters experience a major-merger. In the top panel of Fig. 4 the stacking of the two biases is represented as a function of $\Delta t/t_{\rm dyn}$, which is defined as $\Delta t/t_{\rm dyn} = (t_{(\rm before-1)+i} - t_{(\rm before-1)})/t_{\rm dyn}$, where $t_{\rm dyn}$ is the dynamical time and $t_{\rm before-1}$ corresponds to the analysed redshift right before z_{before}, namely the time before the whole merger phase.

Before the merger phase (first point in Fig. 4) the bias is in agreement with the typical relaxed clusters values. Right after z_{before}, the biases increase until ~0.3 and stay almost constant till $z_{\rm end}$. After the end of the merger event, where the secondary object is completely within the R_{200} of the main object, the relaxation process begins and the bias starts to quickly decrease, until $z_{\rm after}$ where a minimum is reached, with negative values. Here the profiles of the thermodynamic quantities (see Section 3.2) become steeper, due to the internal moving substructure, leading to an increase in the HE mass (see Eqs 1,2) and a decrease in the bias. After the end of the merger phase, the bias value is again compatible with the one of the relaxed clusters. This behaviour can be explained also looking at the stacked profile of the relaxation parameter, Fig.4 - bottom panel. The value of $\chi_{\rm DS}$ at $z_{\rm before}$ is close to 1, pointing to a general relaxed state. At $z_{\rm start}$, it starts to decrease, so the clusters become disturbed and after $z_{\rm end}$ it increase again, even if not reaching again the original relaxation state, as [19].

5 Conclusions

In this work the 3D profiles of temperature, pressure and gas density of a set of almost 300 cluster are analysed. These clusters are extracted from The Three Hundred simulations. From the profiles, we recover the HE masses and their biases. The bias dependencies on

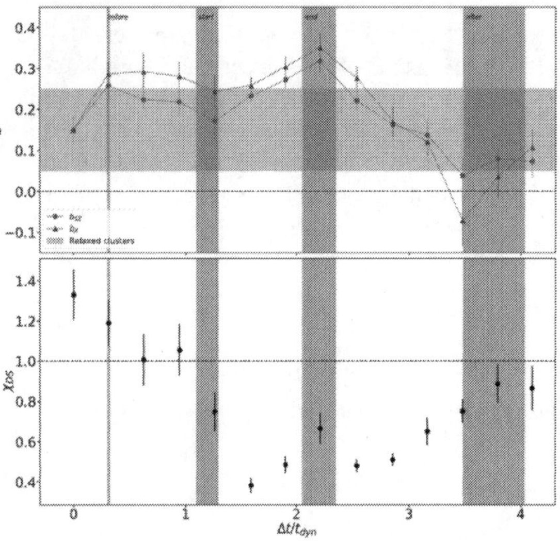

Figure 4. Top panel: the stacked biases of 12 events at R_{200} (medians and standard deviations) are represented as a function of the difference in times from $t_{\text{before}-1}$, divided by the dynamical time. The SZ bias is represented in purple, while the X one in blue. The magenta shaded region shows the bias ranges for the relaxed clusters at R_{200}. Central panel: the stacked relaxation parameter estimated at R_{200}, the means and the standard deviations are represented as a function of $\Delta t/t_{dyn}$. The black shaded areas represent the 1σ regions of the averaged time before, after, start and end of the merger.

redshift, dynamical state and mass are studied. We do not find correlation between the bias and these quantities, in agreement with other simulations. We also find that the very disturbed systems tend to have a negative bias, namely an overestimation of the mass with HE.

Moreover, we analysed the bias in 12 major-mergers. Before the merger, the objects are on average relaxed. During the merger, the main object gets more and more disturbed, due to the secondary object which is being absorbed, translating into a steepening in the thermodynamic profiles. This causes an overestimation of the mass with HE, i.e. the bias reaches negative values. As soon as the merger phase ends, the cluster gets back to a more relaxed state, and the bias assumes the typical values of relaxed clusters.

References

[1] Cui W., Knebe A., Yepes G. et al., MNRAS **480**, 2898-2915, 3 (2018)
[2] Klypin A., Yepes G., Gottlöber S. et al., MNRAS **457**, 4340-4359, 4 (2016)
[3] Rasia E., Borgani S., Murante G. et al., ApJ **813**, L17, 1 (2015)
[4] Sembolini F., Yepes G., De Petris M. et al., MNRAS **429**, 323-343, 1 (2013)
[5] Planck Collaboration, A&A **594**, A13 (2016)
[6] Kravtsov A. V. and Borgani S., ARA&A **50**, 353-409, 1 (2012)
[7] Ansarifard S., Rasia E., Biffi V. et al., A&A **634**, A113 (2020)
[8] Gianfagna G., De Petris M., Yepes G. et al., MNRAS **502**, 5115-5133, 4 (2021)
[9] Vikhlinin A., Kravtsov A., Forman W. et al., ApJ **640**, 691-709, 2 (2006)
[10] Nagai D., Kravtsov A. V. and Vikhlinin A., ApJ **668**, 1-14, 1 (2007)
[11] Neto A. F., Gao L., Bett P. et al., MNRAS **381**, 1450-1462 (2007)
[12] Cialone G., De Petris M., Sembolini F. et al., MNRAS **477**, 139–152, 1 (2018)
[13] De Luca F., De Petris M., Yepes G. et al., MNRAS **504**, 5383-5400, 4 (2021)
[14] Pearce F. A., Kay S. T., Barnes D. J. et al., MNRAS **491**, 1622-1642, 2 (2019)
[15] Le Brun A. M. C., McCarthy I. G., Schaye J. et al., MNRAS **466**, 4442–4469 (2017)
[16] Salvati L., Douspis M. and Aghanim N., A&A **614**, A13 (2018)
[17] Rasia E., Meneghetti M., Martino R. et al., New Journal of Physics **14**, 055018, 5 (2012)
[18] Barnes D. J., Vogelsberger M., Pearce F. A. et al., MNRAS **506**, 2533–2550, 2021
[19] Contreras A. et al., in preparation

Generalised scalar-tensor theories of gravity and pressure profiles of galaxy clusters

Balakrishna S. Haridasu[1,2,3,*], *Purnendu* Karmakar[4], *Marco* De Petris[4,5,6], *Vincenzo F.* Cardone[5,6], and *Roberto* Maoli[4,5]

[1]SISSA-International School for Advanced Studies, Via Bonomea 265, 34136 Trieste, Italy
[2]INFN, Sezione di Trieste, Via Valerio 2, I-34127 Trieste, Italy
[3]IFPU, Institute for Fundamental Physics of the Universe, via Beirut 2, 34151 Trieste, Italy
[4]Dipartimento di Fisica, Sapienza Universitá di Roma, P.le Aldo Moro 2, 00185, Roma, Italy
[5]INAF - Osservatorio Astronomico di Roma, Via Frascati 33, 00040, Monteporzio Catone, Roma, Italy
[6]INFN- Sezione di Roma 1, P.le Aldo Moro 2, 00185, Roma, Italy

Abstract. In the current proceedings, we summarise the results presented during the mm Universe@NIKA2 conference, taken from our main results in [1]. We test the Degenerate higher-order scalar-tensor(DHOST) theory as a generalised platform for scalar-tensor theory at galaxy cluster scales to predict in such static systems small scale modification to the gravitational potential. DHOST theory is not only a good alternative to ΛCDM for the background evolution but also predicts small-scale modification to the gravitational potential in static systems such as galaxy clusters. With a sample of 12 clusters with accurate X-ray Intra Cluster Medium (ICM) data (X-COP project) and Sunyaev-Zel'dovich (SZ) ICM pressure (Planck satellite), we place preliminary constraints on the DHOST parameter (Ξ_1) defining the deviation from GR. Moreover, we also collect a few supplementary analyses we have performed during the course: i) Gaussian process reconstruction without parametric assumptions, ii) P_{SZ}-only data analysis not aided by the X-ray data. Finally, we present possible extensions to the current work which may benefit from future high sensitivity and spatial resolution observations.

1 Introduction

Einstein developed the standard general relativity (GR) by considering gravity as warps and curves in the fabric of geometric space-time, and that geometry of space-time is described only by a metric. GR can explain the majority of the cosmological observations, and requires a cosmological constant to explain the acceleration in the cosmological distance, which is however phenomenological. Therefore, if we plan to keep the basic metric structure because of the large success of GR, and add one scalar degrees of freedom to explain the acceleration, then the degenerate higher-order scalar-tensor (DHOST) theory is the best possible platform to test the largest classes of scalar-tensor theories [2, 3].

The additional degree of freedom must be suppressed at small scales by a screening mechanism to fit with the observations [4]. The radial derivative of modified Newtonian potential for the DHOST theory in the galaxy cluster, which produces the gravitational waves propagating at the velocity of light ($c_g = c$), is

*e-mail: sandeep.haridasu@sissa.it

$$\frac{d\Phi(r)}{dr} = \frac{G_{\rm N}^{\rm eff} M_{\rm HSE}(r)}{r^2} + \Xi_1 G_{\rm N}^{\rm eff} \frac{d^2 M_{\rm HSE}(r)}{dr^2}\,, \tag{1}$$

while assuming the hydrostatic equilibrium (HSE), where $M_{\rm HSE}(r)$ is the total mass within the radial distance r. Both, Ξ_1 and $G_{\rm N}^{\rm eff}$ parameters track the departure of DHOST theory from GR, which modify the ICM thermal pressure profile as

$$P^{\rm th}(r) = P^{\rm th}(0) - \int_0^r \rho_{\rm gas}(\tilde{r}) \frac{d\Phi(\tilde{r})}{dr} {\rm d}\tilde{r}\,, \tag{2}$$

where $\rho_{\rm gas}(r)$ is the gas density. Our formalism is a straight-up implementation of the so called *forward*-method, where the pressure profile is computed while assuming empirical profiles for the mass, $M_{\rm HSE}(r)$, and the electron density, $n_e(r)$, radial profiles which in our case are the standard NFW [5] and simplified Vikhlinin parametric model [6], respectively. Vikhlinin profile provides a more general parametric model in comparison to the β or the double β profiles and we also validate that the former produces variation from the multiscale fitting in [7], of no more than $\sim 5\%$ for the $n_e(r)$. In this context, several previous works have implemented similar approach either using stacked clusters and/or having complementary weak lensing data [8–10].

Throughout the current discourse we assume $H_0 = 70$ km/s Mpc^{-1}and $\Omega_{\rm m} = 0.3$. R_{500} and M_{500} carry usual definitions, i.e, total mass M_Δ within the radius R_Δ, with a mean mass density Δ times the critical density ($\rho_c(z) = 3H^2(z)/8\pi G_{\rm N}^{\rm eff}$). Here $G_{\rm N}^{\rm eff} = \tilde{\gamma}_{\rm N} G_{\rm N}$, which represents the running of the Planck mass on large scales.

2 Main Results

Firstly, we figure that only 8 clusters out of the 12 available clusters in the dataset prefer the NFW mass profile in a comparison through information criteria (see Fig 2, of [7]). We refer to the other 4 clusters (A644, A1644, A2319 and A2255) as non-NFW clusters for brevity. As a consequence of which, we find that the non-NFW clusters fit the data within the DHOST scenario with statistical preferences reaching $\Delta \log(\mathcal{B}) >> 25$ (further details in [1] and [7]), which imply a very high preference yet providing very low values of the parameter Ξ_1, in complete disagreement with the theoretical limits [11–13]. In Figure 1, we show the fits of the GR and DHOST scenarios for A644 (non-NFW) and A1795 cluster and the 1σ dispersion to provide a comparison. We show the posteriors for Ξ_1 obtained using the individual cluster and the joint constraint in left panel of Figure 2 and a comparison with earlier constraints in the right panel.

As the main result of our analysis we updated the limits on the modified gravity parameter $\Xi_1 = -0.030 \pm 0.043$, which is a much more stringent constrain in comparison to the earlier result of $\Xi_1 \sim -0.028^{+0.23}_{-0.17}$, reported in [8] using a similar formalism with stacked X-ray cluster profile. In a more conservative case where we utilise only 4 clusters (A1795, A3158, RXC1825 and ZW1215) having mass estimates consistent with [7], we find a less stringent $\Xi_1 = -0.061 \pm 0.074$, yet twice as tighter constraint than the earlier analysis. These 4 conservative clusters are restricted within a redshift range of $0.0597 < z < 0.0766$. In the right panel of Figure 2, we show a comparison of the earlier constraints, our results and the theoretical limits. As can be seen, our results are in excellent agreement with the latter and an improvement for the constraints on Ξ_1. The lower theoretical limit is marked by the requirement of stable static solutions of non-relativistic stars [11, 14] and the upper limits are obtained on the based on the minimum mass required for hydrogen burning in low mass red dwarfs [12, 13].

An added advantage of our current analysis is that having individual clusters spread over a redshift range of $0.04 < z < 0.1$, we are able to present for the first time the necessary

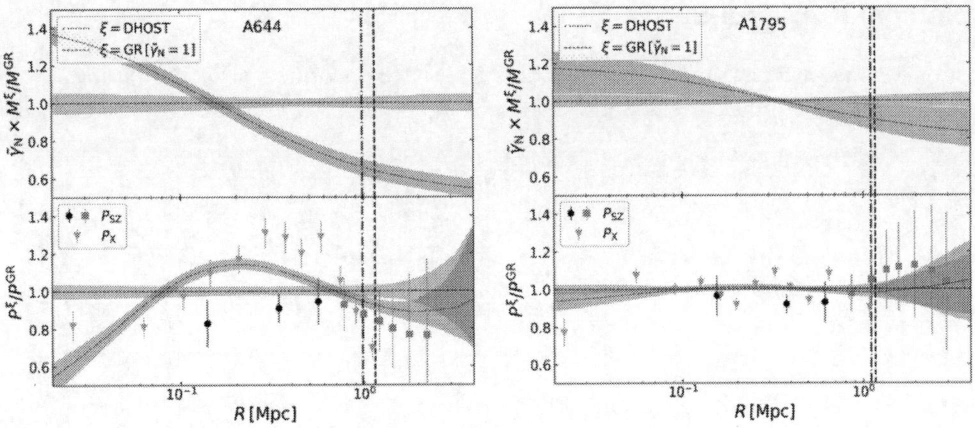

Figure 1. Mass (*upper*) and pressure (*lower*) radial profiles for the cluster A644 (*Left*) and A1795 (*Right*) in the cases of GR (blue) and DHOST (red), normalised to GR. The former cluster is an example case of the 4 non-NFW clusters, while the latter is one of the 4 clusters we retain in estimating our final conservative result. The vertical dashed (GR) and dot-dashed (DHOST) lines show R_{500}. The 3 inner P_{SZ} data points (black) are excluded in the main analysis.

assessment of time-evolution of the parameter $\Xi_1(a \equiv 1/(1+z))$. We utilise a simple first order Taylor expansion of $\Xi_1(a)$ as the scale factor, $a \to 1$ and perform a post-MCMC analysis on the inferred posteriors of the Ξ_1 obtained for individual clusters. Using 8 clusters we find a mild deviation ($\lesssim 2\sigma$) from a constant behaviour for Ξ_1, at present is however driven only by two clusters (A85 and A2142). Note however that the current redshift range used to assess possible time dependence is limited (see also [1]), and the advantage of extending the analysis over a larger dataset that covers a wider redshift range is clear.

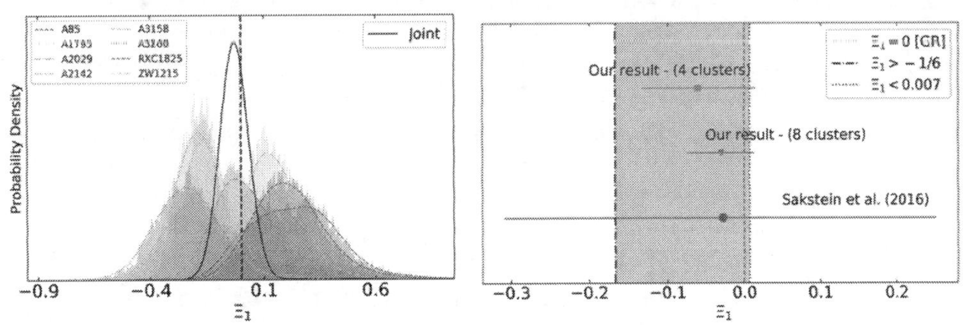

Figure 2. *Left*: Posterior distribution of the parameter Ξ_1, over-plotted by smooth Gaussian kernel density profiles. The vertical dashed line marks the GR ($\Xi_1 = 0$) case. The black curve represents our joint constraint using 8 clusters. *Right*: Constraints on the parameter Ξ_1, from an earlier analysis of 58 stacked clusters (blue) [8] and our results with 8 clusters (green) and conservative 4 clusters (orange). The grey shaded region shows the allowed region for the same parameter obtained from theoretical limits [11–13].

3 Supplementary Results

In this section we present a few supplementary analyses we have performed, also constituting possible future extension of the current work.

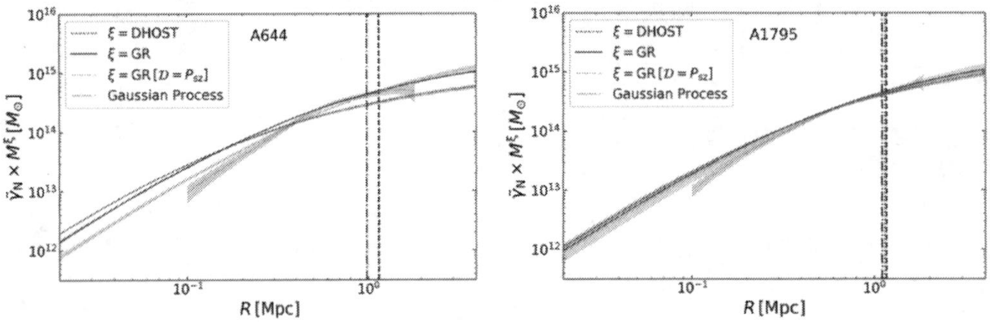

Figure 3. Mass radial profiles with 1σ dispersion for clusters A644 (*left*) and A1795 (*right*) for the GR (blue), DHOST (red) and GP (green). A644 and A1795 represent one of the non-NFW cluster and one of the 4 conservative clusters, respectively. The vertical dashed, dot-dashed and dotted lines show the R_{500} in GR, DHOST and GP, respectively. The R_{500} from the GP case coincides very well with that in GR case. The mass profiles obtained utilising only the P_{SZ} data for the GR case are shown in orange. Here the GR case corresponds to $\tilde{\gamma}_N = 1$.

3.1 Gaussian Process

In addition to the parametric analysis, we also perform a non-parametric model-independent analysis based on the Gaussian Process (GP) formalism, please see [15] and references therein for details on the method. In essence, once could reconstruct the underlying functional form $[f(x_i)]$ for a collection of Gaussian data points at $\mathbf{x} \equiv \{x_1, x_2, ..x_N\}$, assumed belonging to the same process, which now represents a GP. This is implemented by modelling a covariance $K(x_i, x_j)$, among the data points, instead of functional form for the process itself. For this purpose we independently reconstruct the electron density and the P_{SZ} pressure profiles assuming the squared-exponential (SE) [1] kernel. Note that here we cannot extrapolate the profile beyond the range of the available data, as GP tends to retrieve broad prior regions, with no constrain on the posterior. We intend the analysis with GP as an alternate verification for the fits based on the profile themselves as the model-independent method is agnostic with no assumptions made for the electron density profile and the mass profile. Therefore, a parametric method based on assumed empirical profiles is expected to serve better if it is in a good agreement with the GP method.

In Figure 3, we show the mass reconstructions obtained using the NFW profile in both the GR (blue) and DHOST (red) scenarios and the GP (green) reconstructed result. For the GP method we firstly find a lowering of the mass profile in the inner regions, which is an immediate consequence of not excluding the inner most 3 data-points of P_{SZ} data (see [7, 16]). However, being a non-parametric estimation and that it only follows the data points, we find the GP to be in better agreement towards the outskirts of the mass profile. In the case of the A644, a non-NFW cluster, we find that the model-independent reconstruction is in better agreement with the GR case than with the DHOST scenario, which is a clear indication of our results for

[1]The functional form for the SE kernel is $K(x_i, x_j) = \sigma_f^2 \exp\left[(x_i - x_j)^2/2l^2\right]$, where $\{\sigma_f, l\}$ are now free parameters inferred by optimising a log-marginal likelihood.

the latter being biased when utilising the NFW mass profile. The $R_{500} = 1.15 \pm 0.06$ [Mpc] and $M_{500} = 4.57^{+0.73}_{-0.65}$ [$10^{14}\, M_{\odot}$] estimated using GP here are in excellent agreement with our constraints in the GR case. In the case of A1795 however, the deviation between the GR and the DHOST cases is minimal and is also reflected in the GP reconstruction. Once again the GP based $R_{500} = 1.12 \pm 0.04$ [Mpc] and $M_{500} = 4.24^{+0.045}_{-0.040}$ [$10^{14}\, M_{\odot}$] are in excellent agreement with the parametric method assuming the NFW profile.

3.2 P_{SZ} only analysis

Indeed, one of the major advantages in our analysis is that we perform the joint-fit to the T_X(or P_X) and the P_{SZ} data, which provides a better estimation of the mass profile to the outskirts ($R \sim 1.2$ Mpc) of the clusters. In the screening mechanism, the GR is recovered in the innermost region of the galaxy cluster, while the gravity is gradually deviates from GR in the outskirt of the cluster. Given that, the inclusion of the P_{SZ} data actually provides a very suitable opportunity to asses the mass profile in a larger radial range and hence the modifications to gravity.

To contrast our constraints against the combined analysis of $P_{SZ} + P_X$, we also perform an MCMC analysis using only the P_{SZ} data (without excluding the 3 inner most data points). We also show the inferred mass profiles (orange) of the same in Figure 3. One can immediately notice that for the cluster A644, all the 4 contrasted scenarios are almost discordant at the face-value. This in turn indicates that a more careful assessment has to be made when combining the data sets and/or trying to assess the modifications to physics using this cluster. On the other hand, the A1795 cluster shows that all the approaches are in very good agreement, especially in the outskirts of the cluster. However, comparing the P_{SZ} parametric fit and the GP based reconstruction we find that the former tends to show higher masses, also in agreement with the $P_X + P_{SZ}$ analysis. The results for the posterior distributions on Ξ_1 are summarised in Appendix C. of [1].

Within the joint analysis however, one needs to check internally for a consistency between the X-ray and SZ data, see Appendix A of [16] for more details. Therefore in a similar approach we introduce an intrinsic dispersion ($\sigma_{P,\rm int}$) as a free parameter which is scaled by the modelled pressure ($\sigma_{P,\rm int} \sim P^{\rm model}\sigma_{\rm int}$) (Appendix D of [17]), added in quadrature to the error and inferred through the MCMC analysis. Alongside the dispersion, we also include a rescaling parameter η, such that $P_X^{\rm data} \to \eta \times P_X^{\rm data}$. We show the posterior distributions of these two consistency parameters in Figure 4 for the A644 and A1795 clusters. Clearly, both clusters show a good agreement between P_{SZ} and P_X data, while A644 shows a larger dispersion within the P_X data points. This in turn provides additional correlations with M_{500} and increases the error-bars, however note also that we fix $\eta = 1$ in the main analysis. Please refer to original analysis in [7, 16, 17], where these arguments are elaborated.

4 summary

As we have shown here, having the galaxy cluster physics well-mapped in a larger radial range and having several of these objects also distributed along the redshift will be crucial to assess the constraints on the Ξ_1 in individual clusters and its time evolution. We intend to explore the same with soon to be available future data. For this goal, NIKA2 SZ observations could be helpful to extend this analysis.

References

[1] B.S. Haridasu, P. Karmakar, M. De Petris, V.F. Cardone, R. Maoli (2021), 2111.01101
[2] M. Crisostomi, K. Koyama, G. Tasinato, JCAP **1604**, 044 (2016), 1602.03119

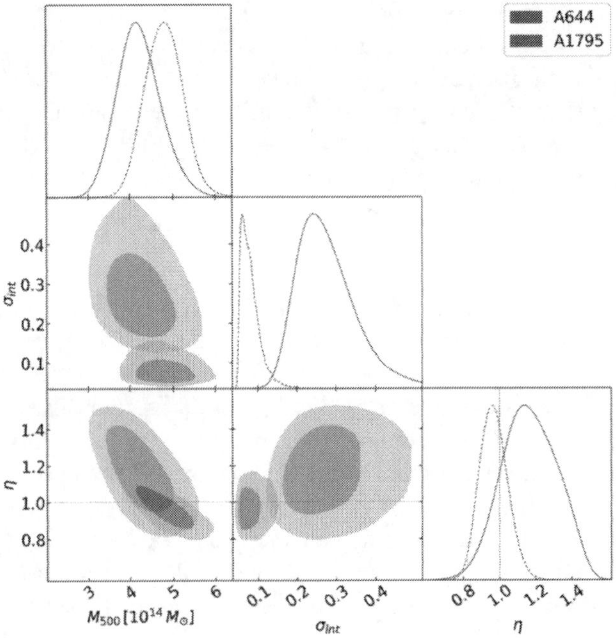

Figure 4. We compare here the posteriors for the intrinsic dispersion in the X-ray data and the scaling parameter. The dashed grey line marks the case of $\eta = 1$, which implies good agreement between the P_X and P_{SZ} data.

[3] J. Ben Achour, D. Langlois, K. Noui, Phys. Rev. **D93**, 124005 (2016), `1602.08398`

[4] E. Babichev, C. Deffayet, Class. Quant. Grav. **30**, 184001 (2013), `1304.7240`

[5] J.F. Navarro, C.S. Frenk, S.D.M. White, Astrophys. J. **462**, 563 (1996), `astro-ph/9508025`

[6] A. Vikhlinin, A. Kravtsov, W. Forman, C. Jones, M. Markevitch, S.S. Murray, L. Van Speybroeck, Astrophys. J. **640**, 691 (2006), `astro-ph/0507092`

[7] S. Ettori, V. Ghirardini, D. Eckert, E. Pointecouteau, F. Gastaldello, M. Sereno, M. Gaspari, S. Ghizzardi, M. Roncarelli, M. Rossetti, Astron. Astrophys. **621**, A39 (2019), `1805.00035`

[8] J. Sakstein, H. Wilcox, D. Bacon, K. Koyama, R.C. Nichol, JCAP **1607**, 019 (2016), `1603.06368`

[9] H. Wilcox et al., Mon. Not. Roy. Astron. Soc. **452**, 1171 (2015), `1504.03937`

[10] A. Terukina, L. Lombriser, K. Yamamoto, D. Bacon, K. Koyama, R.C. Nichol, JCAP **04**, 013 (2014), `1312.5083`

[11] R. Saito, D. Yamauchi, S. Mizuno, J. Gleyzes, D. Langlois, JCAP **1506**, 008 (2015), `1503.01448`

[12] J. Sakstein, Phys. Rev. **D92**, 124045 (2015), `1511.01685`

[13] J. Sakstein, Phys. Rev. Lett. **115**, 201101 (2015), `1510.05964`

[14] I.D. Saltas, I. Sawicki, I. Lopes, JCAP **05**, 028 (2018), `1803.00541`

[15] B.S. Haridasu, V.V. Luković, M. Moresco, N. Vittorio, JCAP **10**, 015 (2018), `1805.03595`

[16] V. Ghirardini et al., Astron. Astrophys. **621**, A41 (2019), `1805.00042`

[17] V. Ghirardini, S. Ettori, D. Eckert, S. Molendi, F. Gastaldello, E. Pointecouteau, G. Hurier, H. Bourdin, Astron. Astrophys. **614**, A7 (2018), `1708.02954`

The Three Hundred project: Contrasting clusters galaxy density in hydrodynamical and dark matter only simulations

A. Jiménez Muñoz[1], *J.F.* Macías-Pérez*[1], *W.* Cui[4], *M.* De Petris[3], *A.* Ferragamo[3], and *G.* Yepes[2]

[1]Univ. Grenoble Alpes, CNRS, Grenoble INP, LPSC-IN2P3, 53, avenue des Martyrs, 38000 Grenoble, France
[2]Departamento de Física Teórica and CIAFF, Módulo 8, Facultad de Ciencias, Universidad Autónoma de Madrid, 28049 Madrid, Spain
[3]Dipartimento di Fisica, Sapienza Universitá di Roma, Piazzale Aldo Moro, 500185 Roma, Italy
[4]Institute for Astronomy, University of Edinburgh, Edinburgh, UK

Abstract.
Cluster number count is a major cosmological probe for the next generation of cosmological large scale-structure surveys like the one expected from the Euclid satellite mission. Cosmological constraints will be mainly limited by the understanding of the selection function (SF), which characterize the probability of detecting a cluster of a given mass and redshift. The SF can be estimated by injecting realistic simulated clusters into the survey and re-applying the detection procedure. For this purpose we intend to use The Three Hundreds project, a 324 cluster sample simulated with full-physics hydrodynamical re-simulations. In this paper we concentrate on the study of the distribution of member galaxies in the cluster sample. First, we study possible resolution effects by comparing low and high resolution simulations. Finally, accounting for the latter we derive the density profiles of the member galaxies and discuss their evolution with cluster mass and redshift.

1 Introduction

The next generation of large-scale structure survey instruments, like the Euclid satellite [1] and the Rubin observatory [2], will use cluster abundance as a major cosmological probe to constraint the parameters of the ΛCDM cosmological model. In practice, one needs to compare the total number of observed clusters in a given solid angle region, Ω, to the predicted cluster counts as a function of mass and redshift [3]. For this, it is necessary to account for the probability of finding a cluster at a certain mass and redshift in this solid angle, which is known as the Selection Function (SF). The SF is an intrinsic characteristic of the cluster catalogue and depends on the cluster finder algorithm, the observational and quality cuts, and the characteristics of the survey. In order to estimate the SF, it is needed to use simulated data sets. These can be either a full mock galaxy catalogue including clusters and field galaxies, or individual simulated clusters which are injected in the observed galaxy

*e-mail: macias@lpsc.in2p3.fr

catalogue. Various ways of computing the SF related to different levels of refinement and related to the cluster finder used have been developed [e.g. 3–6].

Large sky volume is needed to find massive objects like galaxy clusters. However, it is difficult to produce simulations including both dark matter and baryons with enough resolution for these large volumes. A solution to this is the 'zoom' simulations, as adopted by The Three Hundreds [7]. For this project, a large cosmological volume is simulated by N-body dark-matter-only simulation, and only in the regions where a galaxy cluster is found, full-physics simulations are performed. For having enough statistics it is necessary to run at least hundreds of independent simulations. We describe in this paper the main optical and IR properties of the Three Hundreds clusters in the prospect of using them for computing the SF of future cluster catalogues like the one expected for the Euclid satellite.

2 The Three Hundreds project

The Three Hundreds project consists of a 324 cluster sample simulated with full-physics hydrodynamical re-simulations. They have been produced starting from the dark-matter-only MultiDark Simulations [MDPL2 8], which consists in a 1 h^{-1} Gpc cube containing 3840^3 dark matter (DM) particles with a mass of 1.5 x 10^9 $h^{-1}M_\odot$ each. The ROCKSTAR halo finder [9] is ran on this simulation to find dark matter haloes. A total of 324 spherical regions were extracted from the halo finder results, selecting as center for these regions the position of the most massive halo at redshift $z = 0$. The radius of each spherical region is 15 h^{-1} Mpc that is much larger than the virial radius of the central cluster. The phase space initial conditions for the 324 selected regions are used to perform the 'zoom' re-simulations. For the study presented here, The Three Hundreds collaboration has ran these simulations using the GADGET-X [10] code, which can generate both dark matter only or hydrodynamical simulations. The GADGET-X code[**?**] is based on a modern version of the Smoothed Particle Hydrodynamics (SPH) first proposed by [11, 12].

Depending on the physics and the resolutions we consider the following The Three Hundreds simulations in this paper:

1. ***LR DMONLY:*** Dark-matter-only simulations with a dark matter particle resolution of 1.5 x 10^9 $h^{-1}M_\odot$. A total of 324 regions have produced.
2. ***HR DMONLY:*** Dark-matter-only simulations at high resolution. With respect to *LR DMONLY*, it has twice particles per dimension, thus, 7680^3, with eight times less mass per particle i.e., 1.8 x 10^8 h^{-1} $M_\odot$ each. For these simulations we only have 68 regions. Hence, when comparing with the low resolution simulations with the *HR DMONLY* ones, we will use the 68 regions for which we have high resolution data.
3. ***LR HYDRO:*** Full-physics hydrodynamics zoom simulations at low resolution. These simulations are at the same resolution as the *LR DMONLY* simulations and 324 regions are available.
4. ***HR HYDRO:*** Physics as LR HYDRO but at high resolution as HR DMONLY. Only one region has been simulated because to time constraints. The latter has been mainly used to identify resolution problems, which manifest in the luminosity function of cluster galaxy members.

We concentrate only on the most massive and central cluster of the 15 h^{-1} Mpc region, also called host halo. This is for ensuring large enough clusters that can host a significant

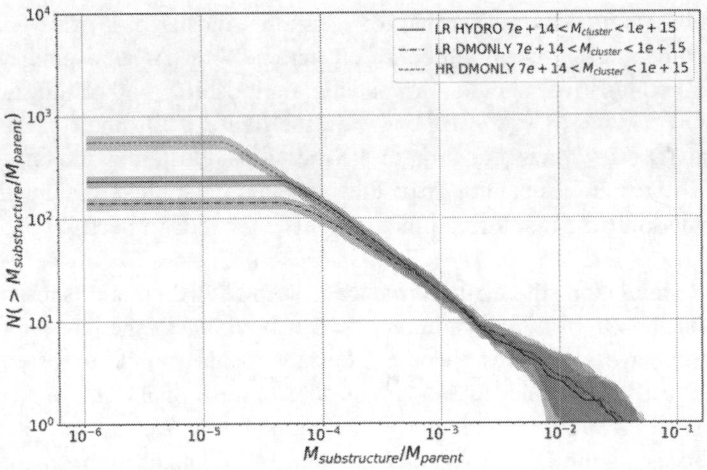

Figure 1. 3D cumulative satellite galaxy mass function at $z = 0$ and 7×10^{14} $M_\odot$ < M < 1×10^{15} $M_\odot$. In blue, red and black, for *LR DMONLY*, *HR DMONLY* and and *LR HYDRO* simulations, respectively. Each line corresponds to the mean value of 3D galaxy mass functions of the individual clusters in the bin. The shaded regions are the 1σ uncertainties coming from the standard deviation across clusters in the mass bins.

number subhalos or substructures, implying a higher galaxy statistics. **A subhalo is defined as a gravitationally bound structure either to the host halo or to another subhalo. The subhalo were identified using a the Amiga's Halo Finder (AHF) algorithm (see [17]) as discussed in [7]. To define a galaxy, first we consider a mass threshold, for the subhalo mass, for each of three simulations. This translates into a mass resolution limit. Notice that we consider the particle mass instead of the number of particles as our threshold because for the different simulations the particle mass is different. Thus, the same number of particles does not translate into the same subhalo mass. Taking into account the particle mass and the simulation resolution we consider a minimum mass of 4×10^{10} $M_\odot$. In the case of the hydrodynamical simulations, we need also to consider the mass ratio between the dark matter and the star content of the galaxy. We assume that the mass of a real galaxy is mainly coming from the dark matter halo surrounding the stars and gas, so we also impose that the stellar mass component is not higher than 30% of the total mass. For this analysis the galaxies identified within a subhalo will be associated directly to the host halo. In the following, a galaxy will be called substructure or subhalo depending on the context.**

3 Subhalo Mass Function

The subhalo mass function, which gives the number of substructures above a given mass relative to the halo mass, can be used to study possible resolution issues. Indeed, the lack of mass resolution in the LR simulation will imprint in this cumulative subhalo mass function as a deficit of low-mass subhalos with respect to the HR ones, as discussed in [13]. In practice, we study this cumulative subhalo mass function defined as the total number of subhalos with relative mass (ratio between their mass and their host halo mass) below a given value. We have divided The Three Hundreds cluster sample in various bins in mass and redshift. We selected as narrow bins in mass as possible to ensure that cluster properties within the bin are similar while having a sufficient number of clusters per bin. As an example, we present in Figure 1, the 3D cumulative subhalo mass function at redshift $z = 0$ for cluster with mass in the range 7 x 10^{14} $M_{\odot}$ < M < 1 x 10^{15} $M_{\odot}$. In the figure, each line corresponds to the mean value of the 3D galaxy mass function of the individual clusters in the bin. The shaded regions are the 1σ uncertainties coming from the standard deviation across clusters. For this analysis we consider only the first three simulation sets described in Sect. 2.

For *DMONLY* simulations the main difference is seen for the smaller substructures. We find many more of them from the *HR* simulations. For both mass functions we observe that the cumulative number of structures becomes constant for the smallest masses, indicating that there are no structures smaller than a certain threshold in relative mass (M_{sub}/M_{parent}). This threshold is different for the *LR* and *HR* simulations. We can see that the three simulations are consistent for the more massive substructures, and again the main differences come from the low-mass subhalos. At the same resolution (i.e., the blue and black lines), we find more objects in the hydro simulation, and these objects are less massive. This could be due to the fact that baryonic physics, such as gas cooling processes, avoid the strip out of the particles in these low-mass subhalos. The slope for the *LR HYDRO* simulations is closer to the *HR DMONLY* but still we can see that the black line is slightly above the red one in the low mass regime. This means that improving the resolution in dark-matter-only simulations is not equivalent to adding baryonic physics in the simulations.

To compare the properties of clusters from the three sets of simulations, we define a relative halo mass (M_{sub}/M_{parent}) threshold beyond which the resolution effect does not play a significant role. This threshold is defined from the observed plateau in Figure 1 for each set. Note that when comparing the different simulations we will use the highest mass cut for the simulations involved. The mass cut defined from the DMONLY and HYDRO LR simulations will be called *DMONLY Cut* and *HYDRO Cut* in the following.

4 Subhalo Density Distribution

In this section we compare the subhalo density distribution obtained for the different types of simulations discussed above. First of all, we apply the mass cut discussed above to ensure that resolution effects can be neglected. If we use the *LR DMONLY* we will the use the *DMONLY Cut* and the *HYDRO Cut* otherwise.

4.1 Subhalo density distribution estimation

For each simulation and a given threshold in mass we estimate the subhalo distribution from each cluster in a mass and redshift bin and average it over all the clusters in that bin. Uncertainties are computed from the dispersion across clusters in the same bin. Figure 2 shows the Einasto model fit for the 3D cumulative radial subhalo density profiles for *LR HYDRO* and *HR DMONLY* in black and red respectively, at $z = 0$ for clusters in the mass range 7 x 10^{14} $M_{\odot}$ < M < 1 x 10^{15} $M_{\odot}$. We observe that the subhalo density distribution of the *LR HYDRO* simulations is overall above the *HR DMONLY* except the cluster outskirt. Note that the decrease of the subhalo density distribution above R_{200} is just an artifact as no member subhalos are expected beyond this radius for The Three Hundreds simulations.

To better characterize the difference between the two simulation sets we fit them to an analytical model and check their evolutions with redshift. It has been proven in [14], that for numerical simulations with an accurate resolution, the Navarro-Frenk-White profile differs from the data, and that the Einasto profile provides a better fit. Thus, we use here the Einasto profile defined by [15, 16] as

$$\rho(r) = n_0 \exp\left(\frac{-2}{\alpha}\left[\left(\frac{r}{r_0}\right)^{\alpha} - 1\right]\right), \tag{1}$$

where n_0, r_0 and α are the free parameters describing the model. The fit is performed using a MCMC algorithm. We do not use the data beyond R_{200}, nor radial bins with less than three galaxies (nearby the center of the cluster) for the fit. In practice, we only use the darker regions in the Figure. Note that for the *LR HYDRO* the used region is larger as we have more galaxies in the inner part of the clusters.

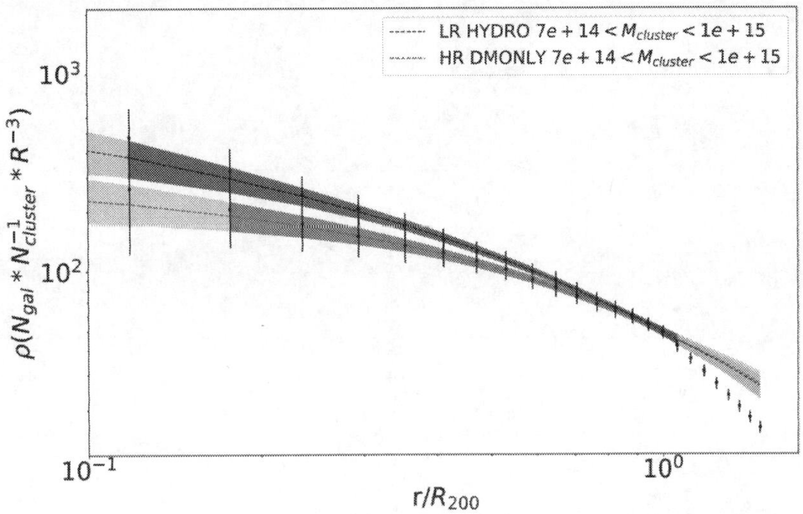

Figure 2. Subhalo density radial profiles for the *HR DMONLY* and the *LR HYDRO* simulations when the *HYDRO Cut* is applied, for $z = 0$ and for cluster mass in the range 7 x 10^{14} $M_\odot$ < M < 1 x 10^{15} $M_\odot$. The dotted lines and the shaded regions correspond to the best-fit Einasto profile with its uncertainties based on the 16th, 50th and 84th percentiles over the posterior distribution of the model parameters.

4.2 Evolution in mass and redshift

We repeated the previous process for all mass and redshift bins. Figure 3 shows the 2D posterior probability distribution for parameters corresponding to cluster in the mass range 7 x 10^{14} $M_\odot$ < M < 1 x 10^{15} $M_\odot$ for several redshifts between 0 and 1. We show the results by pairs of parameters: 1) α and r_0, in the top row, 2) r_0 and n_0, in the middle row, 3) α and n_0, in the bottom row. The shaded colored regions represent the 68 % C.L. uncertainties. In general we observe that α is lower and n_0 greater for the *LR HYDRO* simulations, and r_0 values are consistent for both types of simulations. A lower value of the slope, α, and greater of the normalization, n_0, for a fixed value of r_0 means that the subhalo density profile increases towards the center of the cluster and drops rapidly in the outskirts. This consistent with the difference between *LR HYDRO* and *HR DMONLY* simulations observed in Figure 2. At high redshift this behaviour is accentuated as α decreases with redshift, while n_0 increases. The fact that n_0 increases with redshift can be due to that at high redshift there is more fragmentation, and then a larger number of galaxies is giving a higher density value.

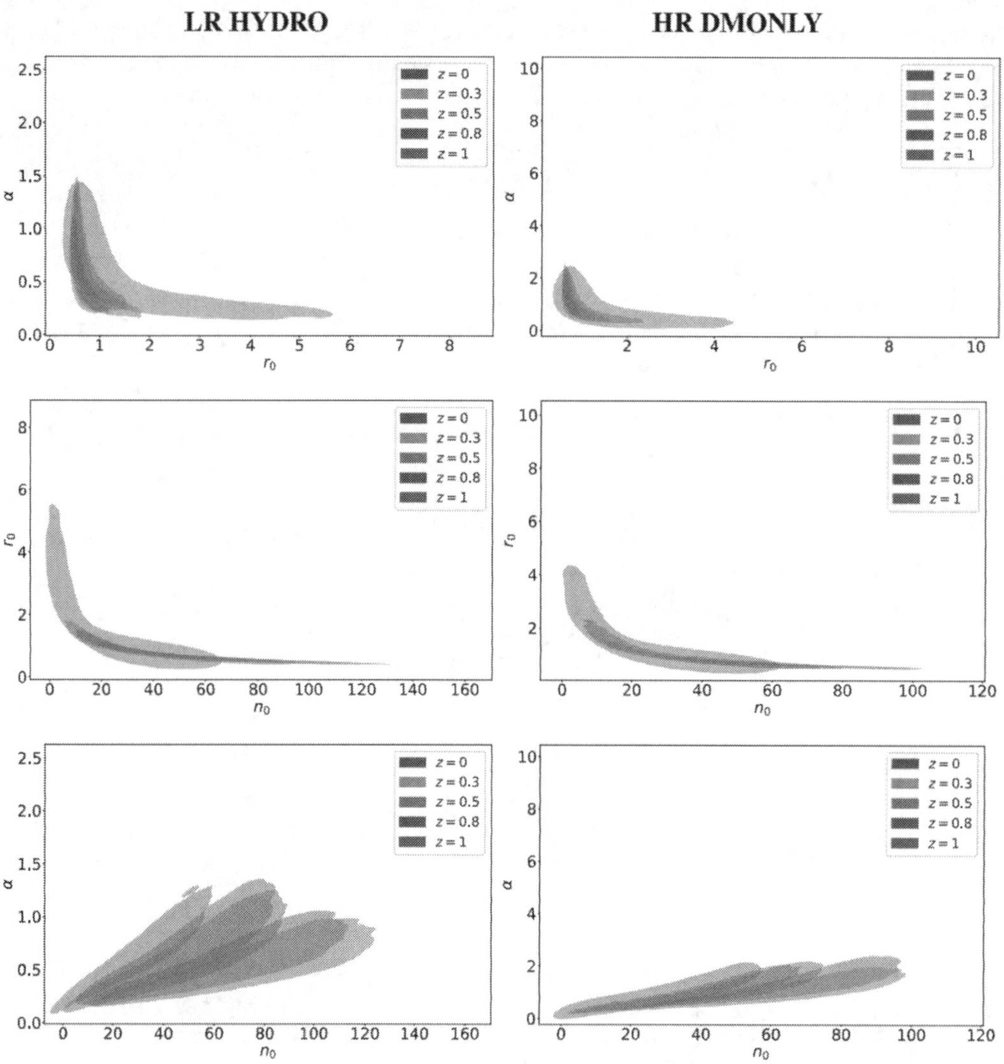

Figure 3. Evolution with redshift of 2D probability distributions for the n_0, α and r_0 parameters of the Einasto model. In the left and right columns we show the *LR HYDRO* and *HR DMONLY* simulations, respectively. The shaded regions correspond to the 68 % C.L.

5 Conclusions

The determination of the cluster selection function will be crucial to perform cluster cosmology with the next generation of large-scale structure experiments as the Euclid satellite and the Rubin observatory. To compute the SF, large simulations with realistic cluster properties are needed. At this respect, we have investigated here the properties of the clusters of The Three Hundred Project. For this we consider low and high resolution dark-matter-only and hydrodynamical simulations.

We have identified a resolution effect in The Three Hundreds low resolution simulations, which does not allow us to study the luminosity function. These effects were identified previously by [13] but with a much lower statics and much sparse coverage in mass and redshift.

Studying the density mass function for the different sets of simulations we have been able to set subhalo mass thresholds above which resolution effects are negligible for the determination of the subhalo density distribution.

Using a common cut we have determined and compared the subhalo density distribution for the *DMONLY* and *HYDRO* simulations. We find that the latter preserve more structures and in particular towards the center of the cluster. This difference increases with redshift. We have also found that the Einiasto profile is a good fit to the density of subhalos.

References

[1] Laureijs et al., arXiv:1110.3193 (2011).
[2] Abell et al., arXiv:0912.0201 (2009).
[3] Sartoris et al., Monthly Notices of the Royal Astronomical Society, **459**, 1764-1780 (2016)
[4] Maturi et al., Monthly Notices of the Royal Astronomical Society, **485**, 498-512 (2019)
[5] Adam et al., Astronomy & Astrophysics, **122**, 399-407 (2019)
[6] Rykoff et al., The Astrophysical Journal, **785**, 104 (2014)
[7] Cui et al., Monthly Notices of the Royal Astronomical Society **480**, 2898–2915 (2018).
[8] Giocoli et al., Monthly Notices of the Royal Astronomical Society **461**, 209-223 (2016)
[9] Behroozi et al., The Astrophysical Journal, **762**, 109 (2013)
[10] Beck et al., Monthly Notices of the Royal Astronomical Society, **455**, 2110-2130 (2016)
[11] Lucy, L.B., The astronomical journal, **82**, 1013-1024 (1977)
[12] Gingold et al., Monthly notices of the royal astronomical society, **181**, 375-389 (1977)
[13] Dolag et al., Monthly Notices of the Royal Astronomical Society, **399**, 497-514 (2009)
[14] Navarro et al., Monthly Notices of the Royal Astronomical Society, **402**, 21-34 (2010)
[15] Einasto et al., Trudy Astrofizicheskogo Instituta Alma-Ata, **5**, 87-100 (1965)
[16] Navarro et al., Monthly Notices of the Royal Astronomical Society, **349**, 1039-1051 (2004)
[17] Knollmann & Knebe, the Astrophysical Journal Supplement, **182**, 608-624 (2009)

Exploring the millimetre emission in nearby galaxies: Analysis of the edge-on galaxy NGC 891

S. Katsioli[1,2,*], *R.* Adam[3], *P.* Ade[4], *H.* Ajeddig[5], *P.* André[5], *E.* Artis[6], *H.* Aussel[5], *A.* Beelen[7], *A.* Benoît[8], *S.* Berta[9], *L.* Bing[10], *O.* Bourrion[6], *M.* Calvo[8], *A.* Catalano[6], *I.* De Looze[11,12], *M.* De Petris[13], *F.-X.* Désert[14], *S.* Doyle[4], *E. F. C.* Driessen[9], *G.* Ejlali[15], *M.* Galametz[5], *F.* Galliano[5], *A.* Gomez[16], *J.* Goupy[8], *A. P.* Jones[7], *A.* Hughes[7], *F.* Kéruzoré[6], *C.* Kramer[9,17], *B.* Ladjelate[17], *G.* Lagache[10], *S.* Leclercq[9], *J.-F.* Lestrade[18], *J.-F.* Macías-Pérez[6], *S. C.* Madden[5], *A.* Maury[5], *P.* Mauskopf[4,19], *F.* Mayet[6], *A.* Monfardini[8], *M.* Muñoz-Echeverría[6], *A.* Nersesian[11,1], *L.* Perotto[6], *G.* Pisano[4], *N.* Ponthieu[14], *V.* Revéret[5], *A. J.* Rigby[4], *A.* Ritacco[7,19], *C.* Romero[20], *H.* Roussel[21], *F.* Ruppin[22], *K.* Schuster[9], *S.* Shu[23], *A.* Sievers[17], *M. W. L.* Smith[4], *F.* Tabatabaei[15,24], *C.* Tucker[4], *E. M.* Xilouris[1], and *R.* Zylka[9]

[1]National Observatory of Athens, Institute for Astronomy, Astrophysics, Space Applications and Remote Sensing, Ioannou Metaxa and Vasileos Pavlou GR-15236, Athens, Greece

[2]Department of Astrophysics, Astronomy & Mechanics, Faculty of Physics, University of Athens, Panepistimiopolis, GR-15784 Zografos, Athens, Greece

[3]LLR (Laboratoire Leprince-Ringuet), CNRS, École Polytechnique, Institut Polytechnique de Paris, Palaiseau, France

[4]School of Physics and Astronomy, Cardiff University, Queen's Buildings, The Parade, Cardiff, CF24 3AA, UK

[5]AIM, CEA, CNRS, Université Paris-Saclay, Université Paris Diderot, Sorbonne Paris Cité, 91191 Gif-sur-Yvette, France

[6]Univ. Grenoble Alpes, CNRS, Grenoble INP, LPSC-IN2P3, 53, avenue des Martyrs, 38000 Grenoble, France

[7]Université Paris-Saclay, CNRS, Institut d'astrophysique spatiale, 91405, Orsay, France

[8]Institut Néel, CNRS, Université Grenoble Alpes, France

[9]Institut de RadioAstronomie Millimétrique (IRAM), Grenoble, France

[10]Aix Marseille Univ, CNRS, CNES, LAM (Laboratoire d'Astrophysique de Marseille), Marseille, France

[11]Sterrenkundig Observatorium Universiteit Gent, Krijgslaan 281 S9, B-9000 Gent, Belgium

[12]Department of Physics and Astronomy, University College London, Gower Street, London WC1E 6BT, UK

[13]Dipartimento di Fisica, Sapienza Università di Roma, Piazzale Aldo Moro 5, I-00185 Roma, Italy

[14]Univ. Grenoble Alpes, CNRS, IPAG, 38000 Grenoble, France

[15]Institute for Research in Fundamental Sciences-IPM, Larak Garden,19395-5531 Tehran, Iran

[16]Centro de Astrobiología (CSIC-INTA), Torrejón de Ardoz, 28850 Madrid, Spain

[17]Instituto de Radioastronomía Milimétrica (IRAM), Granada, Spain

[18]LERMA, Observatoire de Paris, PSL Research University, CNRS, Sorbonne Université, UPMC, 75014 Paris, France

[19]School of Earth and Space Exploration and Department of Physics, Arizona State University, Tempe, AZ 85287, USA Laboratoire de Physique de l'École Normale Supérieure, ENS, PSL Research University, CNRS, Sorbonne Université, Université de Paris, 75005 Paris, France

*e-mail: s.katsioli@noa.gr

[20]Department of Physics and Astronomy, University of Pennsylvania, 209 South 33rd Street, Philadelphia, PA, 19104, USA
[21]Institut d'Astrophysique de Paris, Sorbonne Université, CNRS (UMR7095), 75014 Paris, France
[22]Kavli Institute for Astrophysics and Space Research, Massachusetts Institute of Technology, Cambridge, MA 02139, USA
[23]Caltech, Pasadena, CA 91125, USA
[24]Instituto de Astrofísica de Canarias, Vía L'actea S/N, 38205 La Laguna, Spain

Abstract. New observations of the edge-on galaxy NGC 891, at 1.15 and 2 mm obtained with the IRAM 30-m telescope and the NIKA2 camera, within the framework of the IMEGIN (Interpreting the Millimetre Emission of Galaxies with IRAM and NIKA2) Large Program, are presented in this work. By using multiwavelength maps (from the mid-IR to the cm wavelengths) we perform SED fitting in order to extract the physical properties of the galaxy on both global and local (~kpc) scales. For the interpretation of the observations we make use of a state-of-the-art SED fitting code, HerBIE (HiERarchical Bayesian Inference for dust Emission). The observations indicate a galaxy morphology, at mm wavelengths, similar to that of the cold dust emission traced by sub-mm observations and to that of the molecular gas. The contribution of the radio emission at the NIKA2 bands is very small (negligible at 1.15 mm and ~ 10% at 2 mm) while it dominates the total energy budget at longer wavelengths (beyond 5 mm). On local scales, the distribution of the free-free emission resembles that of the dust thermal emission while the distribution of the synchrotron emission shows a deficiency along the major axis of the disc of the galaxy.

1 Introduction

Nearby galaxies have provided a great piece of knowledge in the field of galaxy formation and evolution. Their proximity allows us to study galaxies at high resolution and to derive their properties on local (sub-kpc/kpc) scales. The current study is part of the IMEGIN Large Program (P.I.: *S. Madden*), a NIKA2 guaranteed time program of 200 hours allocated telescope time with the aim of mapping 22 nearby galaxies (distances of < 30 Mpc) at 1.15 and 2 mm using the NIKA2 continuum camera at the IRAM 30-m telescope [1–4]. The main objective of the project is to explore the emission at millimetre wavelengths originating from nearby galaxies, a complex wavelength regime with contributions from a mix of different emission mechanisms (free-free, synchrotron, and dust thermal emission). This is the first time that nearby galaxies are being observed in the continuum at mm wavelengths at resolutions of 11.1″ and 17.6″ (~ 0.5 kpc, and ~ 0.8 kpc at a distance of 10 Mpc) at 1.15 and 2 mm respectively. At these scales it is possible to disentangle the contribution of the different emission mechanisms in different environments inside the galaxies (disc, halo, HII regions, etc).

As a pilot study, we analyze the NIKA2 observations of the edge-on galaxy NGC 891, a SA(s)b type galaxy [5]. Due to its proximity (D = 9.6 Mpc; [6]) and its edge-on orientation (i ≅ 89.8°; [7]), NGC 891 is one of the most observed and well studied galaxies of the Local Universe (e.g. [7–10]).

2 Observations and Analysis

The NIKA2 observations of NGC 891 were obtained between October 2019 and January 2020 with a total on-source integration time of 7 hours. The maps of the galaxy at 1.15 and

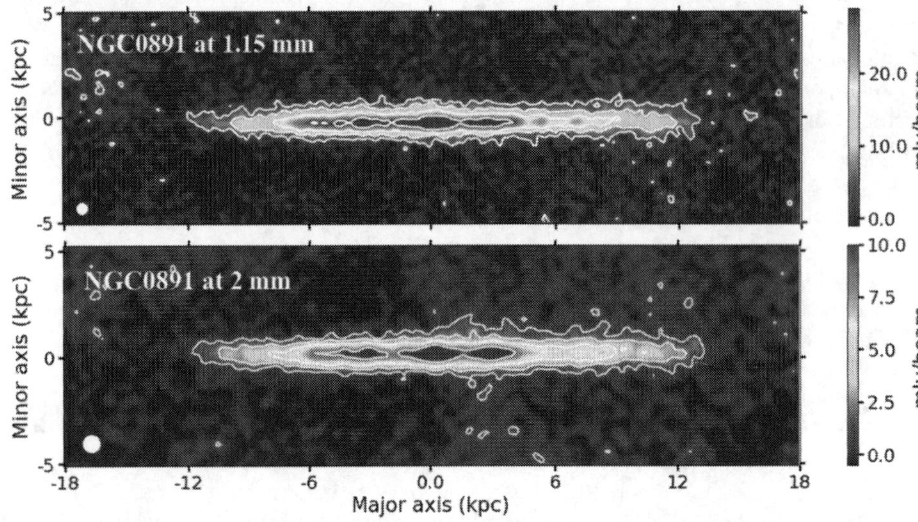

Figure 1. NIKA2 high-resolution millimeter observations of NGC 891 at 1.15 (top panel) and 2 mm (bottom panel). The maps are presented in their native resolution (11.1″ and 17.6″ at 1.15 and 2 mm respectively; indicated by the white circles in each panel) and they are rotated by 22.9° (the position angle of the galaxy). Their sky RMS values are 1.0 mJy beam^{-1} and 0.3 mJy beam^{-1} at 1.15 mm and 2 mm. The surface brightness contours correspond to 3.5, 8, 15 and 30 × RMS.

2 mm were reduced using the GILDAS/PIIC[1] (version of 29.04.2020) software [11] and are shown in Fig. 1 (top and bottom panels respectively). The edge-on orientation of the galaxy and the high resolution of the maps allow us to discern different regions along the major axis with the main characteristics being a peak emission at the nucleus and secondary maxima at ~ 3 kpc either side of the center. These emission features closely resemble the distribution of the cold dust traced by sub-mm observations [12] as well as of the molecular gas [13, 14] which are possibly originating either from a ring structure or limb brightening associated with spiral arms (see [15]).

In our study we performed a Spectral Energy Distribution (SED) analysis of NGC 891 using multiwavelength data ranging from 3.4 μm to 5 cm. We used archival data from the *WISE* telescope (3.4, 4.6, 11.6 and 22.1 μm), the *Spitzer* telescope (3.6, 4.5, 5.8, and 24 μm), as well as the *Herschel* telescope (70, 100, 160, 250, and 350 μm). Along with the NIKA2 observations at 1.15 and 2 mm we also used centimeter data from the *AMI* (2 cm) and *EVLA* (5 cm) radio telescopes. The archival data have been retrieved from the DustPedia[2] and NED[3] databases. The maps were all convolved (using Gaussian Kernels) to the same angular resolution of 25″ corresponding to the resolution of the 350 μm SPIRE map. This resolution corresponds to ~ 1.1 kpc at the distance of NGC 891. All maps have also been regridded into a common reference frame with a pixel size of 8″. Fitting the continuum emission of the galaxy, we had to remove the background sources from the near-IR maps as well as to correct the 1.15 mm map (see [1] for the NIKA2 transmission curves) from the CO(2-1) line contamination following the procedure described in [16]. In order to compute the global fluxes at each wavelength we used an elliptical aperture centered at $\mathrm{RA}_{J2000} = 2^h22^m33^s$, $\mathrm{DEC}_{J2000} = +42°20'53''$ [17] with major and minor axes of 5′ and 48″ respectively and

[1] https://publicwiki.iram.es/PIIC/
[2] http://dustpedia.astro.noa.gr/
[3] https://ned.ipac.caltech.edu/

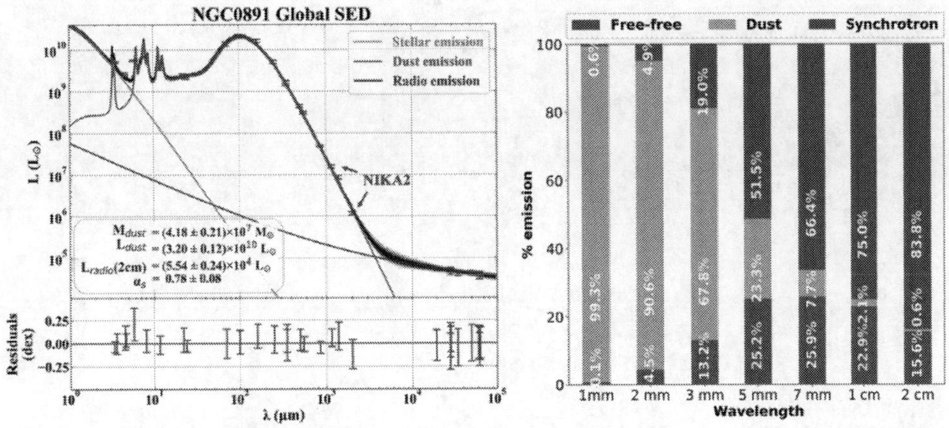

Figure 2. Left panel: the global SED of NGC 891 with multiwavelength data ranging from 3.4 μm to 6.3 cm. The curve passing through the photometric data is the best fitted model with the width of the curve indicating the uncertainty of the model along the spectrum. The emission components of the total SED of the galaxy, stellar, dust and radio, are plotted with orange, red and brown color respectively. The residuals between the observed and the model predicted fluxes are given in the graph below the SED in dex. Right panel: An emission decomposition plot for characteristic wavelengths in the spectral range between 1 mm to 2 cm. The percentage of the dust emission is presented with orange color, while red and blue colors are used for percentages of the free-free and the synchrotron emission respectively.

positional angle of 22.9°, so as to encompass the bulk of the emission originating from the disc of the galaxy. Additional global photometry at radio wavelengths (from *Planck, 100-m Effelsberg, OVRO, WSRT* and *GBT* telescopes) were also culled from the literature.

To interpret the SED we used the SED fitting code HerBIE (Hierarchical Bayesian Inference for dust Emission; [18, 19]). In contrast to the commonly used least-squares fitting method, HerBIE eliminates the noise-induced correlations between the parameters. The code takes into account realistic optical properties and stochastic heating for the dust grains by making use of the THEMIS dust model which is described in detail in [20, 21]. Free-free and synchrotron emission mechanisms are also implemented in the code. The global SED of the galaxy, fitted by the HerBIE code, is presented in the left panel in Fig. 2.

With an angular resolution of 25″ (see Sec. 2) the observations used in this study allowed us to perform spatially resolved SED fitting within the galaxy and create maps of the local physical properties associated with the dust and radio emission. The maps presented in Fig. 3 show the total dust luminosity map (left), the free-free luminosity at 2 cm (middle) and the synchrotron luminosity at 2 cm (right).

3 Results

Modelling the SED of NGC 891 in the wavelength range from 3.4 μm to 6.3 cm allows us to accurately determine the dust content, taking into account emission from the aromatic features as well as from the continuum emission from small and big grains, but also the radio emission composed by the free-free and the synchrotron radiation. Our model predicts a total dust mass of $(4.18 \pm 0.21) \times 10^7$ $M_\odot$, while the mass fraction of the small carbon grains over the total dust mass is 0.10 ± 0.01. The total radio luminosity at 2 cm is $(5.54 \pm 0.24) \times 10^4$ $L_\odot$ with a fraction of 0.16 ± 0.11 originating in free-free radiation. The inferred value of the synchrotron radiation index (α_s) is 0.78 ± 0.08 (in agreement with other studies, e.g. [9]).

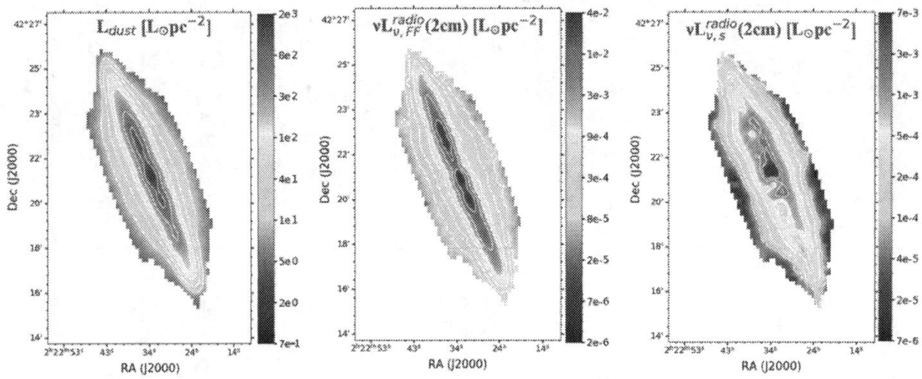

Figure 3. Maps of the model-inferred luminosity for the dust emission (left panel), the free-free luminosity at 2 cm (middle panel) and the synchrotron luminosity at 2 cm (right panel).The colorbar on the right in each panel indicates the luminosity in units of $L_{\odot}pc^{-2}$.

Having modelled the SED we can now study the largely unexplored wavelength range between ~ 1 mm and ~ 2 cm where a mix of emission mechanisms exists. In the right panel of Fig. 2 we present the percentages to the global emission of dust (orange color), free-free (red color) and synchrotron emission (blue color) at 1, 2, 3, 5, 7 mm and 1, and 2 cm. From this plot it is evident that the dust thermal emission dominates to up to ~ 3 mm where the contributions of the free-free and synchrotron emission mechanisms become almost equally significant to the dust emission. At longer wavelengths, these latter emission mechanisms dominate with the synchrotron emission increasing rapidly with values up to ~ 84% at 2 cm, while the free-free emission reaches a maximum contribution of ~ 26% at ~ 7 mm and then drops again.

The spatial decomposition of the emission of the galaxy (see Fig. 3) allows us to determine the places where the different emission mechanisms dominate. From the dust luminosity map (left panel of Fig. 3) it is evident that the bulk of the dust emission comes from the disc of the galaxy, as it is clearly seen also from the NIKA2 observations (Fig. 1), but with significant amount also coming from the dust halo at large distances from the disc. The free-free emission at 2 cm (middle panel of Fig. 3) shows a similar distribution to that of the dust emission, while the synchrotron luminosity (right panel of Fig. 3) shows a different distribution with a deficiency along the galactic plane and with the peak emission originating from the galactic center. One interesting difference between the dust and the radio emission maps is the shape of the halo which, in the dust case has an elliptical shape, while in the radio case (and especially in the synchrotron component) has a more complex "peanut"-like shape.

4 Conclusions

In this study we present new millimetre NIKA2 observations of the edge-on galaxy NGC 891 at 1.15 and 2 mm. Using the NIKA2 observations and also multiwavelength maps (from the mid-IR to the cm wavelengths) we construct the galaxy's SED and perform SED fitting analysis using the HerBIE SED fitting code. The observations indicate a morphology similar to that of the cold dust (detected at sub-mm) and of the molecular gas with the emission peaking at the nucleus of the galaxy and with two secondary maxima at ~ 3 kpc either side of the center. Emission at the NIKA2 bands is dominated by thermal dust, while the radio emission becomes almost equally significant to the thermal emission at $\sim$ 3 mm. At longer wavelengths the radio emission dominates with synchrotron emission reaching up to $\sim$ 84% at 2 cm. On local scales our analysis shows that the bulk of the dust thermal emission and the free-free emission comes from the disc of the galaxy while there is a deficiency, of the synchrotron emission, along the major axis of the disc of the galaxy.

Acknowledgements

We would like to thank the IRAM staff for their support during the campaigns. The NIKA2 dilution cryostat has been designed and built at the Institut Néel. In particular, we acknowledge the crucial contribution of the Cryogenics Group, and in particular Gregory Garde, Henri Rodenas, Jean Paul Leggeri, Philippe Camus. This work has been partially funded by the Foundation Nanoscience Grenoble and the LabEx FOCUS ANR-11-LABX-0013. This work is supported by the French National Research Agency under the contracts "MKIDS", "NIKA" and ANR-15-CE31-0017 and in the framework of the "Investissements d'avenir" program (ANR-15-IDEX-02). This work has benefited from the support of the European Research Council Advanced Grant ORISTARS under the European Union's Seventh Framework Programme (Grant Agreement no. 291294). F.R. acknowledges financial supports provided by NASA through SAO Award Number SV2-82023 issued by the Chandra X-Ray Observatory Center, which is operated by the Smithsonian Astrophysical Observatory for and on behalf of NASA under contract NAS8-03060. This work was supported by the Programme National "Physique et Chimie du Milieu Interstellaire" (PCMI) of CNRS/INSU with INC/INP co-funded by CEA and CNES.

References

[1] L. Perotto, N. Ponthieu, J.-F. Macías-Pérez, *et al.*, Astron. Astrophys. **637**, A71 (2020)
[2] R. Adam *et al.*, Astron. Astrophys. **609**, A115 (2018)
[3] M. Calvo *et al.*, Journal of Low Temperature Physics, **184**, 816 (2016)
[4] O. Bourrion *et al.*, Journal of Instrumentation, **11**, 11001 (2016)
[5] I.S. Konstantopoulos *et al.*, Astrophys. J. **145**, 137 (2013)
[6] D.K. Strickland *et al.*, Astrophys. J. **608**, 829 (2004)
[7] E.M. Xilouris, Y.I. Byun, N.D. Kylafis *et al.*, Astron. Astrophys. **344**, 868 (1999)
[8] T.M. Hughes, M. Baes, J. Fritz, *et al.*, Astron. Astrophys. **565**, A4 (2014)
[9] D.D. Mulcahy, A. Horneffer, R. Beck, *et al.*, Astron. Astrophys. **615**, A98 (2018)
[10] J.H. Yoon, C.L. Martin, S. Veilleux, *et al.*, Mon. Not. Roy. Astron. Soc. **502**, 969 (2021)
[11] R. Zylka, *MOPSIC: Extended Version of MOPSI* (2013)
[12] S. Bianchi & E.M. Xilouris, Astron. Astrophys. **531**, L11 (2011)
[13] N.Z. Scoville, D. Thakkar, J.E. Carlstrom & A.I. Sargent, Astrophys. J. **404**, L59 (1993)
[14] S. Garcia-Burillo, M. Guelin, J. Cernicharo *et al.*, Astron. Astrophys. **266**, 21 (1992)
[15] P.B. Alton, S. Bianchi, R.J. Rand, *et al.*, Astrophys. J. **507**, L125 (1998)
[16] E. Drabek, J. Hatchell, p. Friberg, *et al.*, Mon. Not. Roy. Astron. Soc. **472**, 23 (2012)
[17] T. Oosterloo, F. Fraternali, & R. Sancisi, Astrophys. J. **134**, 1019 (2007)
[18] F. Galliano, Mon. Not. Roy. Astron. Soc. **476**, 1445 (2018)
[19] F. Galliano, A. Nersesian, S. Bianchi, *et al.*, Astron. Astrophys. **649**, A18 (2021)
[20] A.P. Jones, Astron. Astrophys. **555**, A39 (2013)
[21] A.P. Jones, M. Köhler, N. Ysard *et al.*, Astron. Astrophys. **602**, A46 (2017)

EPJ Web of Conferences **257**, 00024 (2022) https://doi.org/10.1051/epjconf/202225700024
mm Universe @ NIKA2

PANCO2: A new software to measure pressure profiles from resolved thermal SZ observations

F. Kéruzoré[1,*], *E.* Artis[1], *J.-F.* Macías-Pérez[1], *F.* Mayet[1], *M.* Muñoz-Echeverría[1], *L.* Perotto[1], and *F.* Ruppin[2]

[1]Univ. Grenoble Alpes, CNRS, Grenoble INP, LPSC-IN2P3, 53, avenue des Martyrs, 38000 Grenoble, France
[2]Kavli Institute for Astrophysics and Space Research, Massachusetts Institute of Technology, Cambridge, MA 02139, USA

Abstract. We have developed a new software to perform the measurement of galaxy cluster pressure profiles from high angular resolution thermal SZ observations. The code allows the user to take into account various features of millimeter observations, such as point spread function (PSF) convolution, pipeline filtering, correlated residual noise, and point source contamination, in a forward modeling approach. A major advantage of this software is its performance, enabling the extraction of the pressure profile and associated confidence intervals via MCMC sampling in times as short as a few minutes. We present the code and its validation on various realistic synthetic maps, of ideal spherical clusters, as well as of realistic, hydrodynamically simulated objects. We plan to publicly release the software in the coming months.

1 Introduction

The thermal Sunyaev-Zeldovich (tSZ) effect is a powerful way to detect and study galaxy clusters at millimeter wavelengths, thanks to their imprint on the cosmic microwave background. The amplitude of this effect is proportional to the line of sight-integrated pressure of the intracluster medium (ICM) gas, enabling studies of the thermodynamic properties of the ICM through tSZ observations. The importance of ICM pressure distribution in cosmology (for cluster counts and tSZ power spectra analyses) and astrophysics (*e.g.* to study feedback from active galactic nuclei) therefore makes resolved SZ observations highly attractive.

The NIKA2 SZ Large Program (hereafter LPSZ, [1, 2]) is a follow-up of 50 intermediate-high redshift galaxy clusters ($0.5 < z < 0.9$, $M_{500} \in [3, 11] \times 10^{14} M_{\odot}$). SZ observations are performed using the high angular resolution NIKA2 camera, observing at 150 and 260 GHz with respective angular resolutions 18" and 11" [3]. Among its goals are the measurement of the $Y_{500} - M_{500}$ scaling relation [4] and of the mean pressure profile of the ICM. These studies require the extraction of individual pressure profiles for each cluster in the sample of the program. As a consequence, a fast and robust software able to provide an accurate pressure profile measurement must be used.

In this proceeding, we present the `PANCO2` (Pipeline for the Analysis of NIKA2 Cluster Observations) software, that has been developed in order to fill the needs of the NIKA2 SZ

*e-mail: keruzore@lpsc.in2p3.fr

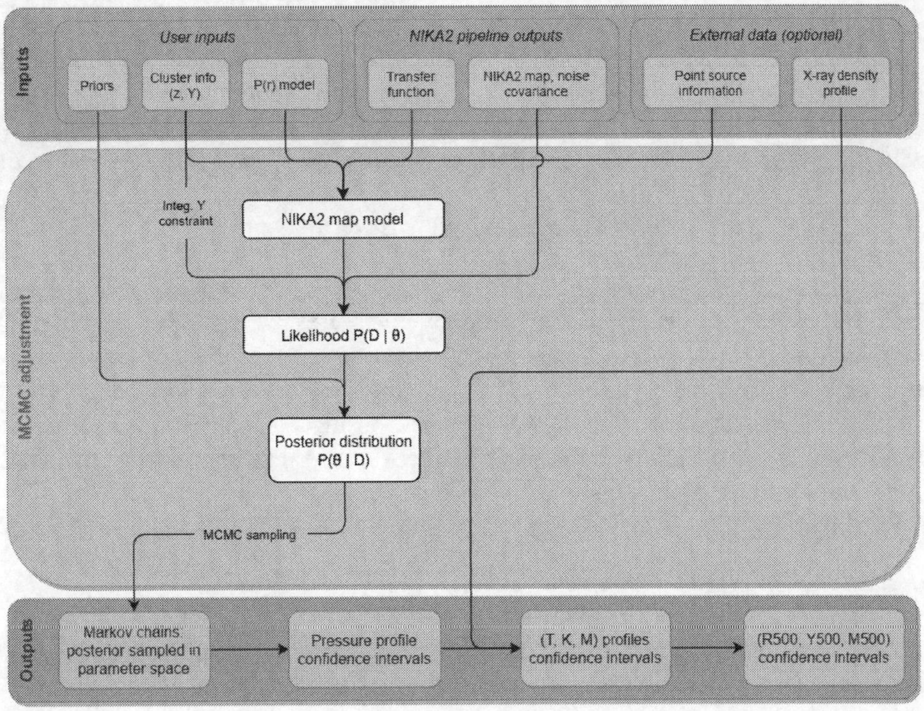

Figure 1. Schematic overview of the PANCO2 workflow. The inputs (in blue) include standard SZ data reduction products, the choice of a pressure model, as well as information on point source contamination. The MCMC analysis yields an estimation of the ICM pressure profile, which can be combined with a density profile extracted from X-ray observations to compute the temperature T, entropy K and hydrostatic mass M profiles, as well as confidence intervals.

Large Program by building upon many previous developments (see *e.g.* [5–10]). It uses a forward modeling regression approach, well-suited to the use of ground-based millimeter observations. Among the key advantages of PANCO2 are its speed – with the possibility to extract a pressure profile measurement in a few minutes – and the flexibility of the modeling, which enables the use of binned pressure profiles, accounting for correlated residual noise and data processing filtering, as well as point-source contamination. Other codes have been publicly released in recent years, offering different and complementary options (see *e.g.* [11], offering joint modeling of SZ and X-ray signal, but no treatment of point source contamination).

2 Algorithm

The main objective of PANCO2 is to extract a pressure profile estimation from SZ maps, and to combine this pressure profile with a density profile extracted form X-ray observations to infer a more detailed characterization of the ICM thermodynamics. The outline of the algorithm used by the software is presented in figure 1. In this section, we give a brief overview of the most relevant steps in the analysis.

2.1 Pressure profile modeling

Past studies of the ICM pressure distribution have relied on very constrained pressure profile models, such as the widely-used generalized Navarro-Frenk-White (gNFW) model [12]. With the advent of deep high angular resolution observations of galaxy clusters, with instruments such as NIKA2 [8], MUSTANG2 [13] or ALMA [14], the number of resolution elements in SZ maps increased, enabling a finer characterization of the intracluster medium. This has led to an increased usage of radially-binned pressure profile models, often referred to as "*non-parametric models*" [7, 9]. In such parametrizations, the model parameters are the values P_i of the pressure profile at pre-determined radii R_i, with a power-law interpolation between the bins:

$$P(R_i < r < R_{i+1}) = P_i\,(r/R_i)^{-\alpha}\,. \tag{1}$$

Using this modeling offers a larger flexibility than what can be achieved with a gNFW model, for instance by enabling the detection of localized overpressures.

`PANCO2` offers both gNFW and binned parametrizations of the pressure profile of the ICM, in order to let users decide for themselves what model is best-suited to their analyses, as well as to enable model comparison studies. When choosing a binned parametrization, the radial bins R_i are to be defined. By default, they will cover the range between the instrument angular resolution and the characteristic radius R_{500} of the studied galaxy cluster.

2.2 Forward modeling of SZ observations

`PANCO2` uses forward modeling to extract pressure profile estimates. In this approach, the physical model – *i.e.* the ICM pressure profile – is affected by the same operations that led to the observed data – *i.e.* a map of the SZ effect. The forward modeling used in `PANCO2` can be split in 4 steps, that we detail below.

1. The chosen pressure profile model is integrated along the line of sight (LoS) to create a Compton parameter y map, following:

$$y = \frac{\sigma_{\rm T}}{m_{\rm e}c^2} \int_{\rm LoS} P_{\rm e}(r)\,{\rm d}l, \tag{2}$$

   where $\sigma_{\rm T}$ is the Thompson scattering cross-section, and $m_{\rm e}$ the electron mass.
   In the case of a gNFW profile, the line of sight integration is performed numerically. In the case of a binned pressure profile, we use the analytical formulation for the integration of power laws in concentric shells detailed in [9][1].

2. The Compton y map is convolved with the instrumental beam to account for PSF smearing. If it is provided, the map is also convolved with a transfer function, to account for possible filtering in the raw data processing.

3. The resulting map is converted to surface brightness units. Since the conversion coefficient depends on the calibration of the maps, as well as on the beam size and atmosphere opacity at the time of observation, it is affected by uncertainty. It is therefore treated as a nuisance parameter in our analysis, enabling the propagation of this uncertainty to the pressure profile estimates and deriving physical properties.

4. Point source models are added to the SZ surface brightness map, with fluxes treated as model nuisance parameters. This approach, described in *e.g.* [10], allows us to account for point source contamination while also propagating uncertainties on this contamination to final physical results.

[1] A Python implementation is publicly available on GitHub: `https://github.com/CharlesERomero/MCMC_ICM_PP/blob/master/integrations/shell_pl.py`

2.3 MCMC analysis

The forward modeling procedure described in the previous section produces a surface brightness map model affected by the same filtering and contaminations as the input data. By doing so, it is possible to adjust the parameters of the pressure profile model to find the values best describing the data, through a likelihood function expressed as:

$$-2\log\mathcal{L}(\vartheta) = [D - \mathcal{M}(\vartheta)]^{\mathrm{T}}\Sigma^{-1}[D - \mathcal{M}(\vartheta)] + \left[\frac{Y^{\mathrm{meas.}} - Y(\vartheta)}{\Delta Y^{\mathrm{meas.}}}\right]^2, \tag{3}$$

where D is the observed SZ map, ϑ is a vector in the parameter space, $\mathcal{M}(\vartheta)$ the model map created from the parameters ϑ, and Σ is the noise covariance matrix. The second term of equation (3) is a constraint on the integrated Compton parameter Y, which can be obtained from lower angular resolution SZ observations (*e.g.* in SZ surveys). It serves as an effective constraint on the pressure profile at scales not probed by observations.

This likelihood function is combined with priors on the model parameters to express the posterior distribution for the parameters, $P(\vartheta|D)$. `PANCO2` offers a large flexibility for these priors, in order to be suited to many kinds of analyses. The posterior distribution is sampled using a Monte Carlo Markov Chains (MCMC) algorithm, performed using the `emcee` library [15]. Convergence of the MCMC sampling is monitored using chains autocorrelation and the $\hat{R}$ test. The final chains are clipped to remove burn-in, and are trimmed to only keep independent points.

The Markov chains produced are used to compute the probability distribution of physical quantities of interest, starting with the pressure profile of the ICM. If available, an electron density profile can be combined with the pressure to further characterize the thermodynamic properties of the ICM, by computing temperature, entropy, and hydrostatic mass profiles. The latter can then be used to compute a characteristic radius R_Δ containing a mean density Δ times greater than the Universe critical density. In turn, R_Δ is used to compute characteristic quantities integrated within the radius value, such as the integrated Compton parameter Y_Δ and the cluster hydrostatic mass M_Δ^{HSE}.

3 Validation on simulated input

The validation of `PANCO2` is performed on synthetic maps of clusters with known pressure profiles, allowing us to compare the final analysis product – *i.e.* the pressure profile reconstructed from the SZ map – to the true underlying cluster properties.

3.1 Simple spherical simulations

The first step of our validation is running `PANCO2` on a simulated NIKA2 map of spherical mock clusters. We model clusters representative of the low-mass, high redshift and high-mass, low-redshift part of the NIKA2 SZ Large Program sample [1, 2]. We generate realistic gNFW pressure profiles for both of these clusters. We then use the forward modeling algorithm described in section 2.2 to create realistic maps from their pressure profiles, including NIKA2 beam smearing and transfer function filtering, and realistic – albeit uncorrelated – residual noise. These maps are cropped to the NIKA2 field of view, corresponding to the size of the largest angular scales that are recovered from NIKA2 cluster observations. We run `PANCO2` on the resulting maps to reconstruct the clusters pressure profiles, and compare these results with the known true profile of each cluster.

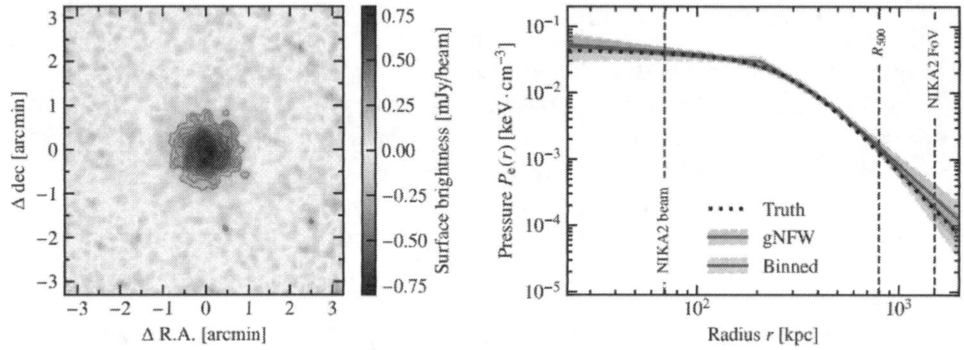

Figure 2. Left: Simulated NIKA2 2 mm surface brightness map for a low mass, high redshift galaxy cluster. Grey contours show signal-to-noise levels starting from 3σ with 1σ spacing. **Right:** Pressure profile estimates obtained by running `PANCO2` on the simulated map. Both the gNFW (red) and binned (blue) fitted pressure models agree with the input profile (black) within 1σ.

The analysis is performed on both clusters, using both a gNFW and binned pressure profile model. We present the results for both model adjustments on the low-mass, high-redshift cluster in figure 2. The results are very similar for each of the four studies, showing excellent agreement between recovered and true pressure profiles, from the angular resolution of NIKA2 to its field of view. Typical running times are around 10 minutes for a binned model fit and 30 minutes for a gNFW, using 30 walkers on 30 parallel threads. The data used for this analysis will be released with `PANCO2`.

3.2 Hydrodynamical simulations

The conclusions that can be drawn from the analysis of simple, spherical clusters are limited by the fact that real clusters are rarely such ideal objects. A more thorough validation must therefore make use of galaxy clusters with more realistic physical properties. To that end, we use simulated observations of galaxy clusters from the MUSIC[2] hydrodynamical simulation [16]. Mock NIKA2 maps of these clusters were created for a sample of 32 clusters mimicking the NIKA2 SZ Large Program in [17]. These maps include SZ signal from the line of sight projection of the ICM pressure distribution, realistic beam and transfer function filtering, as well as correlated residual noise. They therefore constitute an ideal test-case for `PANCO2`, and can help assess its ability to recover pressure profile estimates from realistic observations.

The results for one of the clusters in this sample are presented in figure 3. As for the simple simulation, the recovered pressure profile is in overall agreement with the true cluster profile, with compatibility between the two profiles within 1σ from the NIKA2 instrumental beam to beyond its field of view. The results are similar for the other objects of the 32 cluster sample. This shows that `PANCO2` can deliver unbiased estimates of the ICM pressure profile from NIKA2 SZ observations, even for a complex pressure distribution, as expected for real galaxy clusters.

4 Conclusions

We have presented `PANCO2`, a new software to perform pressure profile measurements from thermal SZ observations. The forward modeling MCMC approach used enables accounting

[2] `https://music.ft.uam.es`

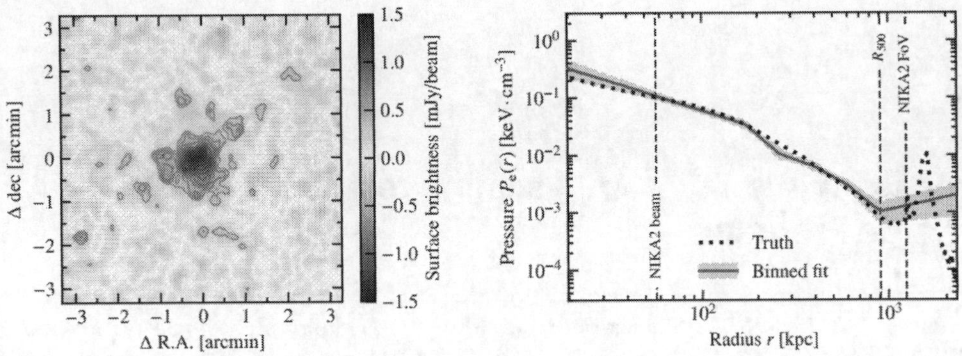

Figure 3. Left: Simulated NIKA2 2 mm map for a cluster from the MUSIC simulation at $z = 0.54$. Grey contours show signal-to-noise levels starting from 3σ with 1σ spacing. **Right:** Pressure profile estimates obtained by running PANCO2 on the simulated map. Error bars are 1σ.

for various millimeter observational artifacts to deliver unbiased pressure profile estimates and confidence intervals. This was demonstrated using simulated cluster maps, showing good agreement between underlying and recovered pressure profiles on radial ranges covered by observations. PANCO2 is currently being adapted to be able to take as input SZ data from any instrument, so that it will be usable for all current and future millimeter surveys and pointed observations. This software will be made public in the coming months.

Acknowledgements

This work is supported by the French National Research Agency in the framework of the "Investissements d'avenir" program (ANR-15-IDEX-02). 291294). F.R. acknowledges financial supports provided by NASA through SAO Award Number SV2-82023 issued by the Chandra X-Ray Observatory Center, which is operated by the Smithsonian Astrophysical Observatory for and on behalf of NASA under contract NAS8-03060.

References

[1] F. Mayet *et al.*, EPJ Web of Conferences **228**, 00017 (2020)
[2] L. Perotto *et al.*, These proceedings (2021), arXiv: 2111.01729
[3] L. Perotto *et al.*, A&A **637**, A71 (2020)
[4] F. Kéruzoré *et al.*, These proceedings (2021), arXiv: 2111.01660
[5] R. Adam *et al.*, A&A **569**, A66 (2014)
[6] R. Adam *et al.*, A&A **576**, A12 (2015)
[7] F. Ruppin *et al.*, A&A **597**, A110 (2017)
[8] F. Ruppin *et al.*, A&A **615**, A112 (2018)
[9] C. Romero *et al.*, A&A **612**, A39 (2018)
[10] F. Kéruzoré *et al.*, A&A **644**, A93 (2020)
[11] F. Castagna, S. Andreon, A&A **639**, A73 (2020)
[12] D. Nagai, A.V. Kravtsov, A. Vikhlinin, ApJ **668**, 1 (2007)
[13] C.E. Romero *et al.*, ApJ **891**, 90 (2020)
[14] L. Di Mascolo, E. Churazov, T. Mroczkowski, MNRAS **487**, 4037 (2019)
[15] D. Foreman-Mackey *et al.*, The Journal of Open Source Software **4**, 1864 (2019)
[16] F. Sembolini *et al.*, MNRAS **429**, 323 (2013)
[17] F. Ruppin *et al.*, A&A **631**, A21 (2019)

Forecasting the $Y_{500} - M_{500}$ scaling relation from the NIKA2 SZ Large Program

F. Kéruzoré[1,*], *E.* Artis[1], *J.-F.* Macías-Pérez[1], *F.* Mayet[1], *M.* Muñoz-Echeverría[1], *L.* Perotto[1], and *F.* Ruppin[2]

[1]Univ. Grenoble Alpes, CNRS, Grenoble INP, LPSC-IN2P3, 53, avenue des Martyrs, 38000 Grenoble, France

[2]Kavli Institute for Astrophysics and Space Research, Massachusetts Institute of Technology, Cambridge, MA 02139, USA

Abstract. One of the key elements needed to perform the cosmological exploitation of a cluster survey is the relation between the survey observable and the cluster masses. Among these observables, the integrated Compton parameter Y is a measurable quantity in Sunyaev-Zeldovich (SZ) surveys, which tightly correlates with cluster mass. The calibration of the relation between the Compton parameter Y_{500} and the mass M_{500} enclosed within radius R_{500} is one of the scientific goals of the NIKA2 SZ Large Program (LPSZ). We present an ongoing study to forecast the constraining power of this program, using mock simulated datasets that mimic the large program sample, selection function, and typical uncertainties on Y_{500} and M_{500}. We use a Bayesian hierarchical modelling that enables taking into account a large panel of systematic effects. Our results show that the LPSZ can yield unbiased estimates of the scaling relation parameters for realistic input parameter values. The relative uncertainties on these parameters is $\sim$ 10% for the intercept and slope of the scaling relation, and $\sim$ 34% for its intrinsic scatter, foreshadowing precise estimates to be delivered by the LPSZ.

1 Introduction

The abundance of galaxy clusters in mass and redshift is tightly linked to large scale structure formation physics. As a consequence, cluster surveys can be used to constrain cosmological parameters, provided cluster masses can be estimated (see *e.g.* [1]). Such studies therefore rely on a prior knowledge of the link between cluster masses and survey observables, often taking the form of mass-observable scaling relations (SR). In particular, catalogues of galaxy clusters detected through the Sunyaev-Zeldovich (SZ) effect may use the observable integrated Compton parameter Y as a mass proxy for cosmological studies (*e.g.* [2]).

Several approaches can be considered to estimate mass-observable scaling relations. One of them is to follow-up representative samples of galaxy clusters with dedicated observations, enabling individual mass measurements that can be used to study their correlation with survey observables. The NIKA2 SZ Large Program (hereafter LPSZ, [3, 4]) consists in a high-resolution SZ follow-up of $\sim$ 50 galaxy clusters selected from SZ surveys. The combination

*e-mail: keruzore@lpsc.in2p3.fr

of NIKA2 SZ observations and of X-ray follow-ups gives access to individual hydrostatic mass measurements for each of these clusters, and thus to studies of the $Y-M$ scaling relation.

In this paper, we present a study forecasting the ability of the NIKA2 LPSZ to constrain the $Y-M$ scaling relation. We use a Monte-Carlo approach to generate mock realistic LPSZ-like samples with a fiducial scaling relation. The estimation of the scaling relation from these samples then allows us to forecast typical biases and uncertainties in the analysis.

2 Scaling relation modelling

Self-similar scenarii of structure growth predict a power law relation between the thermal energy content of a galaxy cluster and its mass. The former can be probed by the integrated Compton parameter, defined as:

$$Y_{500} = 4\pi \frac{\sigma_{\mathrm{T}}}{m_{\mathrm{e}}c^2} \int_0^{R_{500}} r^2 P_{\mathrm{e}}(r)\,\mathrm{d}r, \tag{1}$$

where σ_{T} is the Thomson scattering cross-section, P_{e} the electron pressure in the intracluster medium (ICM), and R_{500} is a characteristic radius of the cluster, defined as the radius of a sphere around the cluster enclosing an average density 500 times greater than the critical density of the Universe at the redshift of the cluster.
The $Y-M$ scaling relation can then be written as (*e.g.* [5]):

$$E^{-2/3}(z)\left[\frac{Y_{500}}{10^{-4}\,\mathrm{Mpc}^2}\right] = 10^{\alpha}\left[\frac{M_{500}}{3\times10^{14}\,M_{\odot}}\right]^{\beta}, \tag{2}$$

where M_{500} is the mass contained within R_{500}, and $E(z) \equiv H(z)/H_0$ is the reduced Hubble parameter at the redshift of the cluster.
Defining log-scaled observable and mass values as Y and Z respectively, eq. (2) becomes a linear relation:

$$Y = \alpha_{Y|Z} + \beta_{Y|Z} Z \tag{3}$$

The complex physical processes occurring inside galaxy clusters make this linear relation a trend rather than a deterministic relation. The scaling relation can then be expressed as the probability of a cluster having an observable value given its mass:

$$P(Y|Z) = \mathcal{N}(\alpha_{Y|Z} + \beta_{Y|Z}Z, \sigma^2_{Y|Z}), \tag{4}$$

where $\sigma_{Y|Z}$ is the Gaussian intrinsic scatter around the relation.

3 Mock cluster sample generation

In order to evaluate the NIKA2 LPSZ constraining power on the $Y-M$ scaling relation, we generate realistic mock cluster samples, with the same properties as the actual LPSZ sample in the (Y_{500}, M_{500}) plane. The LPSZ is composed of 45 clusters between redshifts 0.5 and 0.9 [3, 4]. The clusters were selected from *Planck* and ACT SZ catalogues [6, 7] according to their integrated Compton parameter Y_{500}. Five bins in Y_{500} and two bins in redshift were defined, dividing the observable-redshift plane in ten boxes (see bottom right panel of figure 1). Five clusters were then selected in each of these boxes[1], creating a relatively homogeneous coverage of the mass range of interest. Assuming the $Y_{500} - M_{500}$ scaling relation from [8], the mass range covered roughly spans across $M_{500} \in [3, 11] \times 10^{14} M_{\odot}$.

[1]There were not enough high-mass, high-z clusters in the *Planck* and ACT catalogues to fill the two highest mass boxes in the LPSZ high-redshift bin. This is due to the fact that high mass, high redshift clusters are rare objects in the Universe. The resulting sample therefore only includes 45 clusters.

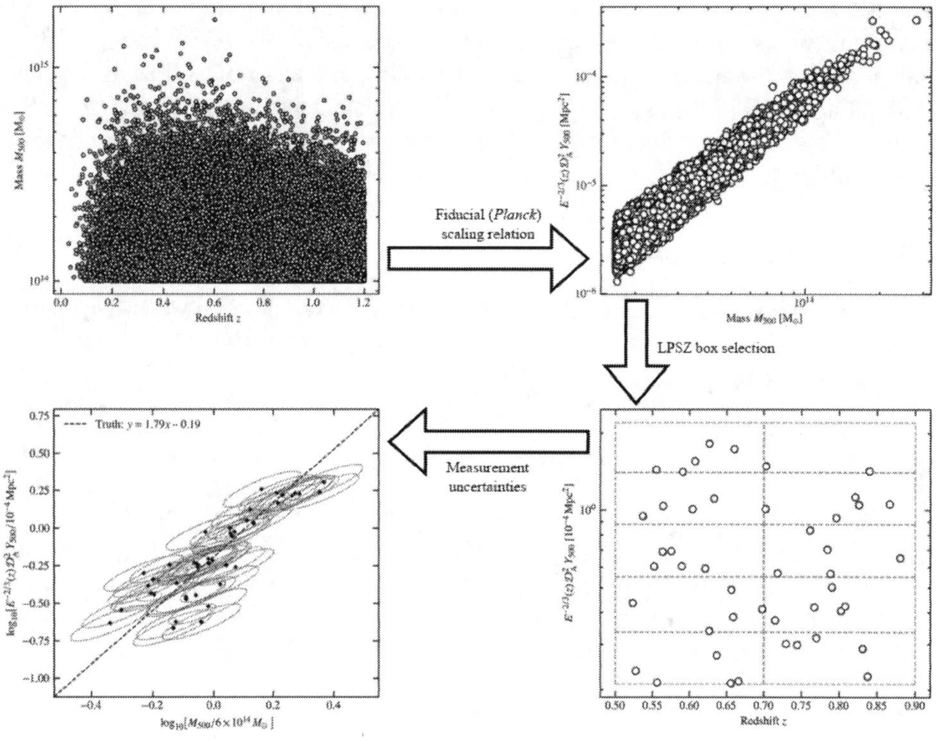

Figure 1. Schematic representation of the workflow used to generate a mock cluster sample. The produced samples mimic the NIKA2 LPSZ in terms of mass and redshift coverage, as well as data quality.

The generation of a cluster sample is performed through the following steps, summarized in figure 1:

1. A large number of points is drawn randomly in the (z, M_{500}) plane from a Tinker halo mass function [9]. Given the steepness of the mass function at low masses, we set a lower limit as $M_{500} \geqslant 10^{14} M_\odot$. This allows samples to cover a mass range larger than that of the NIKA2 LPSZ. The resulting sample is a Universe-like distribution of galaxy clusters in the mass-redshift plane.

2. A fiducial scaling relation is used to compute observable values for each of these clusters. We use eq. (4) with parameter values obtained by the *Planck* collaboration [2]:

$$\alpha_{Y|Z}^{\text{true}} = -0.19 \quad ; \quad \beta_{Y|Z}^{\text{true}} = 1.79 \quad ; \quad \sigma_{Y|Z}^{\text{true}} = 0.075. \tag{5}$$

3. Galaxy clusters are selected according to their SZ observable values and redshift similarly to the NIKA2 LPSZ selection function, *i.e.* by randomly picking 5 clusters for each box according to their redshift and Y_{500} value.

4. Correlated uncertainties on the (log scaled) mass and observable are added to each data point. We generated realistic mock NIKA2 observations of simulated clusters, with various data quality. Processing these maps with the `PANCO2` software [10] showed that realistic data quality for NIKA2 LPSZ cluster observations yielded relative uncertainties between 10% and 15% on Y_{500} and M_{500}, with a ~ 85% correlation coefficient. These results are consistent with previous NIKA2 studies of galaxy clusters (*e.g.* [11, 12]).

These steps are repeated to generate 5000 cluster samples similar to that of the NIKA2 LPSZ, with realistic values of mass and SZ observable, and a known scaling relation between the two. The estimation of the scaling relation parameters by regression on the mock observed data will then allow us to assess biases and scatter of these estimators, *i.e.* the accuracy and precision expected for the LPSZ scaling relation estimation.

It is important to note that the NIKA2 LPSZ sample creation includes one more step, that is not replicated in this study. Galaxy clusters from the LPSZ sample are selected from cluster catalogues detected by the *Planck* and ACT SZ surveys [6, 7] rather than directly from the true cluster population in the Universe. By bypassing this step, we ignore the selection function of these two surveys, which is equivalent to assuming that the *Planck* and ACT survey catalogues are a good proxy of the underlying cluster population in the Universe in the portion of the mass-redshift planed covered by the LPSZ. The validity of this hypothesis and its impact on the scaling relation are to be assessed in a future study.

4 Scaling relation regression

4.1 Regression scheme

We use the LIRA R software [13] to estimate scaling relation parameters from our samples. LIRA provides a ready-to-use regression scheme using a Bayesian hierarchical modelling of the scaling relation, with very flexible parameters and options. The Bayesian hierarchical approach enables taking into account multiple systematic effects at play in mass-observable scaling relations, such as scatter and bias in the mass estimators, correlated uncertainties on both axes of the relation, or selection effects. For more detailed information on the use of Bayesian hierarchical modelling for cluster mass-observable scaling relations, we refer the reader to the works of *e.g.* [13, 14].

LIRA uses a Gibbs sampling Monte Carlo Markov Chains (MCMC) algorithm to sample the posterior probability of the model parameters given the observed data. Its direct products are therefore Markov chains forming a sampling of the posterior distribution in the parameter space, which can be used to infer the probability distribution for the parameters of interest $(\alpha_{Y|Z}, \beta_{Y|Z}, \sigma_{Y|Z})$ from eq. (4). We run LIRA to fit the scaling relation for each of the 5000 realistic mock cluster samples generated in §3. For each mock sample, we estimate the bias ξ and the dispersion η of each parameter of interest in percent of the true parameter value from the Markov chains:

$$\xi_\vartheta \equiv 100 \times \frac{\mathrm{Med}[\vartheta_i]_i - \vartheta^{\mathrm{true}}}{|\vartheta^{\mathrm{true}}|}, \qquad \eta_\vartheta \equiv 100 \times \frac{\sqrt{\mathrm{Var}[\vartheta_i]_i}}{|\vartheta^{\mathrm{true}}|} \tag{6}$$

where ϑ_i is the i-th sample of the Markov chain for the parameter ϑ, and $\vartheta^{\mathrm{true}}$ is the fiducial value used to generate the cluster sample, given in eq. (5). $\mathrm{Med}[\dots]_i$ et $\mathrm{Var}[\dots]_i$ respectively denote the median and the variance of the Markov chains. We also define the significance of the bias of each parameter estimator ζ as:

$$\zeta_\vartheta \equiv \frac{\mathrm{Med}[\vartheta_i]_i - \vartheta^{\mathrm{true}}}{\sqrt{\mathrm{Var}[\vartheta_i]_i}} = \frac{\xi_\vartheta}{\eta_\vartheta}. \tag{7}$$

As a result, we obtain a value of ξ, ζ, η for each of the three parameters of interest for each mock LPSZ sample.

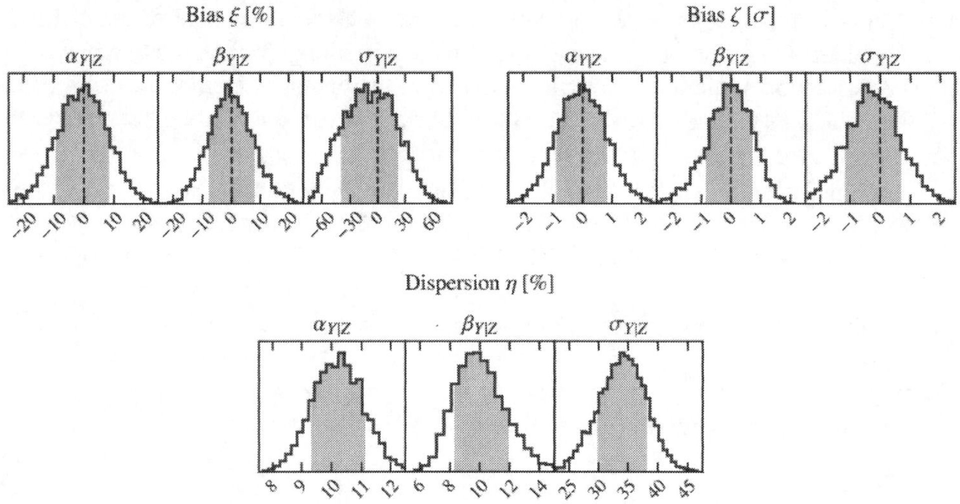

Figure 2. Distributions of the biases (*top*) and dispersions (*bottom*) of the scaling relation parameter estimators over the 5000 mock LPSZ-like samples. No significant bias is found on the parameters of interest. Relative uncertainties are on average ∼ 10% on the slope and intercept of the scaling relation, and ∼ 34% on its intrinsic scatter. Shaded regions are 1σ.

Table 1. Summary of the biases and dispersions on each parameter of interest in our scaling relation analysis. Values reported are the mean values and 1σ uncertainties.

Parameter ϑ	Bias ξ_ϑ [%]	Bias ζ_ϑ [σ]	Dispersion η_ϑ [%]
$\alpha_{Y\|Z}$	-0.5 ± 8.5	0.0 ± 0.8	10.2 ± 0.9
$\beta_{Y\|Z}$	0.0 ± 7.6	-0.1 ± 0.8	10.1 ± 1.8
$\sigma_{Y\|Z}$	-8.2 ± 28.2	-0.2 ± 0.9	34.0 ± 4.0

4.2 Results

We present the distributions of the biases and dispersions obtained for our 5000 LPSZ-like samples in figure 2. The top panel shows that the distributions of ξ and ζ are centred around zero, indicating no bias on average on the parameters of interest of the scaling relation. This shows that an unbiased estimation of the $Y_{500} - M_{500}$ scaling relation can be recovered from the NIKA2 LPSZ follow-up of galaxy clusters, within the scope of our assumptions. The bottom panel shows the distribution of the dispersions η, representing the uncertainties on the recovered scaling relation parameters from the adjustment. We see that relative uncertainties on the intercept $\alpha_{Y|Z}$ and slope $\beta_{Y|Z}$ are on average ∼ 10%, while the intrinsic scatter $\sigma_{Y|Z}$ is less constrained, with ∼ 34% uncertainty on average. These results are summarized in table 1.

5 Summary and conclusions

We presented a study forecasting the precision and accuracy expected for the estimation of the $Y_{500} - M_{500}$ scaling relation with the NIKA2 SZ Large Program. We used Monte-Carlo simulations to create mock LPSZ-like cluster samples by following a sample selection procedure similar to the one used to create the real LPSZ cluster sample, and including realistic uncertainties on the measured mass and integrated Compton parameter of each cluster. We used the LIRA software to adjust a power-law scaling relation between Y_{500} and M_{500}. The comparison of LIRA results with the true fiducial scaling relation used to generate the mock samples did not allow us to identify any significant bias in the analysis. Moreover, it allowed us to forecast the expected uncertainties on the LPSZ estimates of the scaling relation parameters. We obtain typical uncertainties of around 10% of the true values for the slope and intercept of the relation, and 34% for its intrinsic scatter. Such uncertainties are comparable with those obtained through the exploitation of earlier cluster samples using multi-wavelength data (such as *e.g.* [15]). As a combination of high angular resolution SZ and X-ray follow-ups, the NIKA2 SZ Large Program can therefore be expected to deliver quality cluster mass calibrations for SZ surveys.

The two main caveats of this study reside in its hypotheses. First, when generating mock cluster samples, we neglected selection effects in the *Planck* and ACT surveys. The inclusion of these effects will be needed to be able to state that the LPSZ can recover the $Y_{500} - M_{500}$ scaling relation of the true cluster population in the Universe. Second, we have chosen to ignore bias and scatter in the mass estimator used to weigh clusters in the LPSZ. The combination of SZ and X-rays that will be used allows us to access the hydrostatic mass of clusters, which is well known to be a biased, but low-scatter estimator. As such, this study focuses on the scaling relation between integrated Compton parameter and the hydrostatic mass of galaxy clusters. This relation needs to be coupled with external information on the value of the hydrostatic mass bias to be exploitable for cosmological purposes, enabling the propagation of the uncertainty on this bias to cosmological results. These two caveats will be addressed in a future study.

Acknowledgements

This work is supported by the French National Research Agency in the framework of the "Investissements d'avenir" program (ANR-15-IDEX-02). 291294). F.R. acknowledges financial supports provided by NASA through SAO Award Number SV2-82023 issued by the Chandra X-Ray Observatory Center, which is operated by the Smithsonian Astrophysical Observatory for and on behalf of NASA under contract NAS8-03060.

References

[1] S.W. Allen, A.E. Evrard, A.B. Mantz, Annual Review of Astronomy and Astrophysics **49**, 409 (2011)
[2] Planck Collaboration *et al.*, A&A **594**, A24 (2016)
[3] F. Mayet *et al.*, EPJ Web of Conferences **228**, 00017 (2020)
[4] L. Perotto *et al.*, These proceedings (2021), arXiv: 2111.01729
[5] A.V. Kravtsov, A. Vikhlinin, D. Nagai, ApJ **650**, 128 (2006)
[6] Planck Collaboration *et al.*, A&A **594**, A27 (2016)
[7] M. Hasselfield *et al.*, JCAP **2013**, 008 (2013)
[8] M. Arnaud *et al.*, A&A **517**, A92 (2010)
[9] J. Tinker *et al.*, ApJ **688**, 709 (2008)
[10] F. Kéruzoré *et al.*, These proceedings (2021), arXiv: 2111.06493
[11] F. Ruppin *et al.*, A&A **615**, A112 (2018)
[12] F. Kéruzoré *et al.*, A&A **644**, A93 (2020)
[13] M. Sereno, MNRAS **455**, 2149 (2016)
[14] A.B. Mantz, MNRAS **457**, 1279 (2016)
[15] M. Sereno, S. Ettori, L. Moscardini, MNRAS **450**, 3649 (2015)

Internal dark matter structure of the most massive galaxy clusters since redshift 1

Amandine M. C. Le Brun[1,2,*] and *Romain* Teyssier[3]

[1]PSL Fellow, LUTh, Observatoire de Paris, PSL Research University, CNRS, Université de Paris, 92195 Meudon, France
[2]Laboratoire AIM, IRFU/Département d'Astrophysique – CEA/DRF – CNRS – Université de Paris, Bât. 709, CEA-Saclay 91191 Gif-sur-Yvette Cedex, France
[3]Institute for Computational Science, University of Zürich, CH-8057 Zürich, Switzerland

Abstract. We investigate the evolution of the dark matter density profiles of the most massive galaxy clusters in the Universe. Using a 'zoom-in' procedure on a large suite of cosmological simulations of total comoving volume of 3 $(h^{-1}\,\mathrm{Gpc})^3$, we study the 25 most massive clusters in four redshift slices from $z \sim 1$ to the present. The minimum mass is $M_{500} > 5.5 \times 10^{14}\,\mathrm{M}_\odot$ at $z = 1$. Each system has more than two million particles within r_{500}. Once scaled to the critical density at each redshift, the dark matter profiles within r_{500} are strikingly similar from $z \sim 1$ to the present day, exhibiting a low dispersion of 0.15 dex, and showing little evolution with redshift in the radial logarithmic slope and scatter. They have the running power law shape typical of the NFW-type profiles, and their inner structure, resolved to $3.8\,h^{-1}$ comoving kpc at $z = 1$, shows no signs of converging to an asymptotic slope. Our results suggest that this type of profile is already in place at $z > 1$ in the highest-mass haloes in the Universe, and that it remains exceptionally robust to merging activity.

1 Introduction

This proceeding is entirely based upon [18] and M_Δ is always the mass within radius r_Δ, the radius within which the mean mass density is Δ times the critical density at the cluster redshift. Even though the theoretical framework for the formation of large-scale structures in a dark-matter-dominated universe was first formalised over 40 years ago [e.g. 1], observations have only recently conclusively confirmed the conceptual paradigm within which it takes place [e.g. 2–4]. The hierarchical collapse of dark matter into 'haloes' in the dark-energy dominated cold dark matter (ΛCDM) model is nowadays the basis of our understanding of hierarchical galaxy and structure formation. The most massive galaxy clusters hold a peerless place in this hierarchy. Being mostly dark matter-dominated, they have deep potential wells and thus gravity is the dominant mechanism driving their evolution. Hence, they are least affected by complex non-gravitational galaxy formation processes [e.g. 5–7]. They are detectable up to high redshift, and complementary techniques can probe both their internal structure and global properties. Hence, they are perfect objects for testing these theories. The emergence of Sunyaev-Zeldovich (SZ) surveys gave rise to a burst in the number of high-redshift clusters

*e-mail: amandine.le-brun@obspm.fr (AMCLB)

[8–11]. Many of these are high mass, and the feasibility of obtaining high-quality structural and scaling information from these objects has now been proven [e.g. 12, 13]. In particular, X-ray observations of these bright systems can probe the $[0.05-1]\,r_{500}$ radial range quite easily [e.g. 12]. Numerical simulations of structure formation in the ΛCDM cosmology make testable predictions, such as the presence of a quasi-universal, cuspy, dark matter density profile [14]. However, existing simulations are ill-suited for studying high-mass, high redshift systems, as oft their resolution is scant, and/or simulated objects are in short supply. This study extends existing works on the shape of massive dark matter haloes [15–17] by augmenting the spatial or mass resolution and the amount of systems. More essentially, we simulate the most massive galaxy clusters at $z > 0$, and not just the progenitors of the $z = 0$ ones.

2 Simulations and data processing

We moulded our simulations to assemble a fairly-big sample (~ 50) of massive ($M_{500} \gtrsim 5 \times 10^{14}$ M$_{\odot}$) objects at $z \sim 1$, both comparable to present SZ observational surveys in terms of mass and sufficient to derive robust conclusions. The deepest current observational survey follow-up data sets resolve the inner structure at the tens of kiloparsec scale [12]. A simulation reproducing these characteristics would require a $\gtrsim 1$ Gpc^3 volume to be simulated at high resolution. This being impossible given current computational resources, we adopted a strategy in which the total volume was split into three periodic boxes of 1 h^{-1} comoving Gpc on a side containing 2048^3 dark matter particles, and then used the 'zoom-in' technique to further refine individual systems from these parent large box simulations. The simulations are described in [18]. In short, the simulations were carried out with the Eulerian adaptive mesh refinement (AMR) code RAMSES [21] with cosmological parameters taken from [4]. The boxes have a spatial resolution of ~ 15 h^{-1} comoving kpc (ckpc). Haloes were found using PHEW, recentred using a shrinking sphere procedure [23] and then spherical overdensity masses M_{Δ} were computed. The 'zoom-in' simulations have an effective resolution of 8192^3 (8K) particles, resulting in a particle mass of $m_{cdm} = 1.59 \times 10^8\ h^{-1}$ M$_{\odot}$. The main haloes thus contain at least ~ 2 million particles within r_{500}. Their spatial resolutions is ~ 3.8 h^{-1} ckpc.

Here we focus on the dark matter only (DMO) 'zooms' of the 25 most massive systems selected in four different redshift slices ($z_{\rm sel}$ = 1, 0.8, 0.6 and 0[1]). The main characteristics of the subsample at each z_{sel}, including the resulting minimum mass, are summarised in Table 1 of [18]. A resolution study shows that the 8K zooms (i.e. the nominal resolution) are converged (in particular with the even-higher resolution 16K runs) over the whole resolved ($r \geq r_{\rm min}^{\rm 8K}$) radial range (see Appendix A of [18]) and that an effective resolution of 8192^3 is thus the minimum required.

3 Results

Figure 1 depicts the density profiles computed using 50 evenly-spaced logarithmic bins over the $10^{-3} \leq r/r_{500} \leq 5$ radial range with the PYMSES[2] python module. We use mass-weighted radii in this and subsequent figures. Densities were normalised by the critical density of the Universe at the cluster redshift, and the radii by the corresponding r_{500}. The resolution limits at $z = 0$ are shown as black dashed and dotted lines for the boxes and the DMO zooms, respectively. Both limits are nearly redshift-independent. The 'zoom-in' approach allows for at least a five-fold increase of spatial resolution. Being smaller than the Poisson errors,

[1]This redshift distribution has been chosen to match an observational sample at $z \gtrsim 0.5$ [24] as well as providing a 'local reference' for comparison with both observations and previous theoretical work.

[2]http://irfu.cea.fr/Projets/PYMSES/intro.html

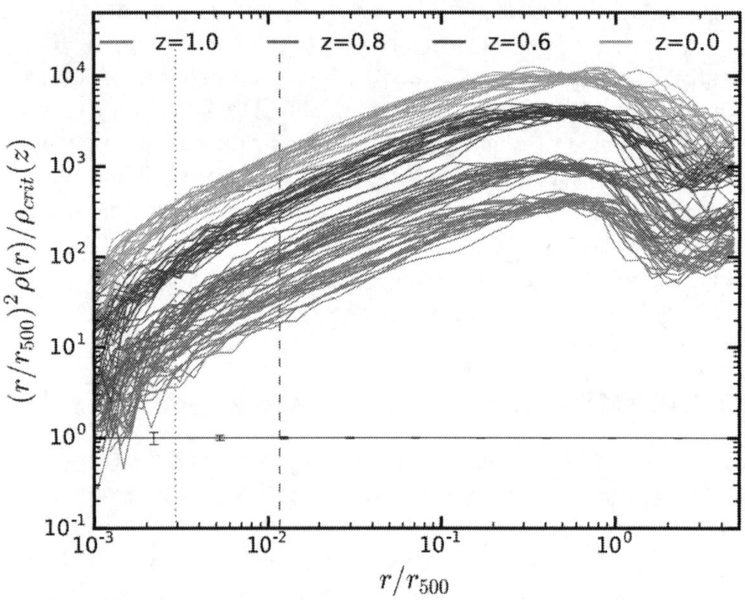

Figure 1. Spherical density profiles for each redshift slice. For clarity, the profiles of the z_{sel} = 1, 0.8, 0.6 and 0 systems have been shifted up by a factor of 2, 5, 20 and 50, respectively. The resolution limit at $z = 0$ is depicted as dashed and dotted vertical lines for the boxes and the DMO zooms, respectively. The typical Poisson errors for every fifth data point are displayed as black error bars. Figure reproduced from [18].

fluctuations are real and plausibly due to substructures and oscillations resulting from halo relaxation.

The left panels of Fig. 2 show the median scaled density profiles of each redshift slice. In the self-similar model, we expect the density profiles to be alike, regardless of cluster redshift or any other properties. Hereafter, evolution refers to the evolution of the scaled profiles, i.e. evolution in excess of the self-similar expectation. The mean density within r_{500} being proportional to $\rho_{\rm crit}(z)$ by definition, scaled profiles can solely have varying shapes in the region $r/r_{500} < 1$. Additionally, they are predicted to intersect at $r/r_{500} \sim 0.6$, the radius enclosing half of M_{500}, and their scatter is hence smallest around that radius. Beyond r_{500}, a breach of self-similarity may turn into evolution in both shape and normalisation. The scaled profiles display a barely noticeable evolution for $1 \geq z \geq 0.6$ over the whole radial range. They change by never more than a factor of ~ 1.2 with decreasing redshift (see the *middle* panel). Conversely, they evolve a bit more (by up to a factor of ~ 1.5) both in the core ($r/r_{500} \lesssim 0.2$) and the outskirts ($r/r_{500} \gtrsim 1.5$) for $z \leq 0.6$. This is consistent with the 'stable clustering' hypothesis often used for computing the non-linear matter power spectrum [e.g. 1]. The former evolution corresponds to a small increase of peakiness. The latter is mostly due to their mass growth. The evolution in the outskirts corresponds to the transition between the one- and two-halo terms moving outwards as the haloes grow. The scatter of the density profiles, as displayed in the *bottom* panel, is notably small over the redshift and radial ranges ($\lesssim 0.2$ dex). The scatter is less than 0.15 dex within r_{500}. In the outskirts, it is slightly larger than in the core and seems to increase mildly (by $\sim$ 30 per cent) with redshift. Note that these trends have to be confirmed with the whole set of simulations as each redshift slice only contains 25 systems.

The high degree of self-similarity is confirmed when studying the profile shapes. To quantify the profile shapes, we calculated the logarithmic slope $\gamma \equiv \mathrm{d}\ln\rho/\mathrm{d}\ln r$ from the smallest resolved radius ($\max(r_{min}/r_{500}) \lesssim 0.003$), using the fourth-order Savitzky-Golay algorithm over the 15 nearest bins [28]. Excluding the seven innermost and outermost bins where the method is no longer valid, the logarithmic slope profiles cover the $0.01 \leq r/r_{500} \leq 1.35$ radial range. The right panels of Fig. 2 show the median logarithmic slope profiles of each redshift slice. They display small amounts of both evolution (it never exceeds 20 per

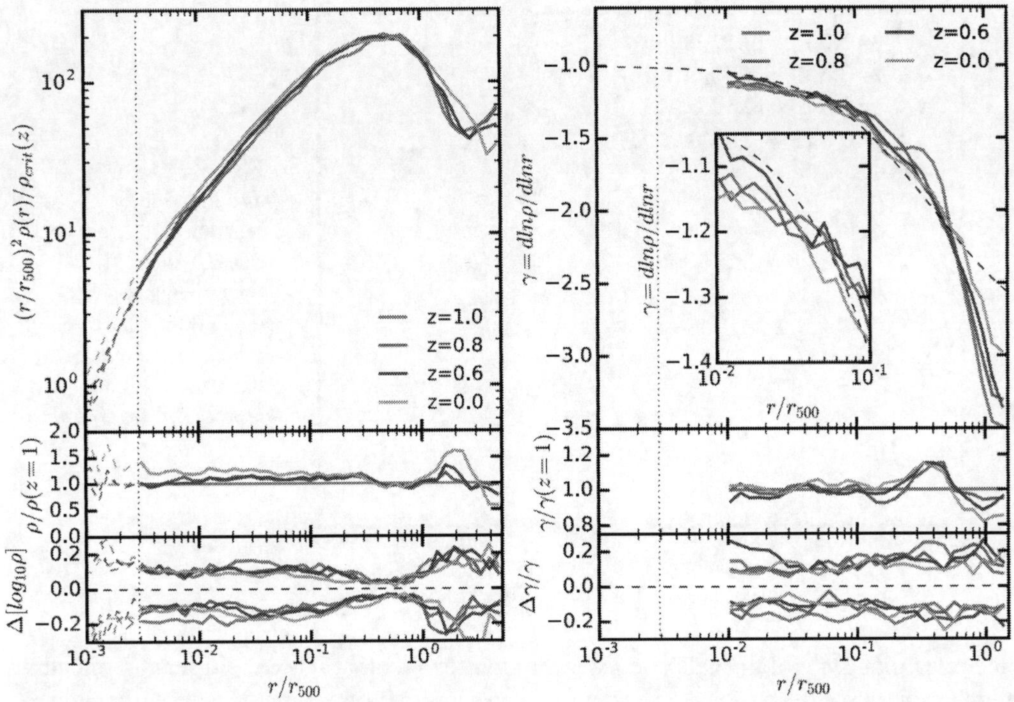

Figure 2. Median spherical density (*left*) and logarithmic slope (*right*) profiles for each redshift slice (*top*). The *middle* panels show the median profiles normalised by the median profile at $z_{sel} = 1$ whereas the *bottom* panel displays the difference between the 16/84 percentiles and the median. In all panels, the resolution limit at $z = 0$ is depicted as a dotted line and the profiles are plotted either as dashed thinner lines below that limit for the density or only from the eighth radial bin above that limit for the logarithmic slope where the Savitzky-Golay filter method is valid. The inset in the *top right* panel is a zoom of the $0.01 \leq r/r_{500} \leq 0.1$ radial range. Figure reproduced from [18].

cent; see the *middle* panel) and dispersion ($\lesssim 0.2$ dex; see the *bottom* panel) over the whole redshift and radial range. Similarly to the density profiles, the evolution is more important in the very central regions ($r/r_{500} \lesssim 0.02$) and the outer regions ($r/r_{500} \gtrsim 0.6$). More surprising, the slope of the profiles, contrarily to the profiles themselves, displays noticeable evolution for all $z \leq 1$ and not just $z \leq 0.6$. The scatter of the density slope exhibits nearly no evolution over the whole radial range and it has similar amplitude to that of the profiles (except around $r/r_{500} \sim 0.6$, where the scatter of the density is minimum by construction). Crucially, the inner slope presents no signs of converging to an asymptotic value but is still consistent with a NFW profile of typical concentration for massive galaxy clusters ($c_{200} = 3.5$) over the radial range in question (see the *top* panel and its inset; in agreement with the results of e.g. [30]).

A first investigation of the origin of the scatter in the density profiles is illustrated by Fig. 3, which depicts the M_{500}/M_{2500} ratio as a function of Δr[3]. The most relaxed clusters ($\Delta r \leq 0.04$) lean towards being more centrally concentrated ($M_{500}/M_{2500} \lesssim 4$). Yet, there is no clear correlation between both parameters in terms of Spearman's rank and null hypothesis probability, as listed in the legend of Fig. 3. In fact, the distribution is consistent with all systems having a similar M_{500}/M_{2500} ratio, no matter their relaxation state, but with a dispersion that rises with rising Δr. Hence, the scatter in the density profiles is only connected

[3] Δr is the distance between the centre of mass within the [26] virial radius r_{vir} and the centre of the shrinking sphere, in units of r_{vir}.

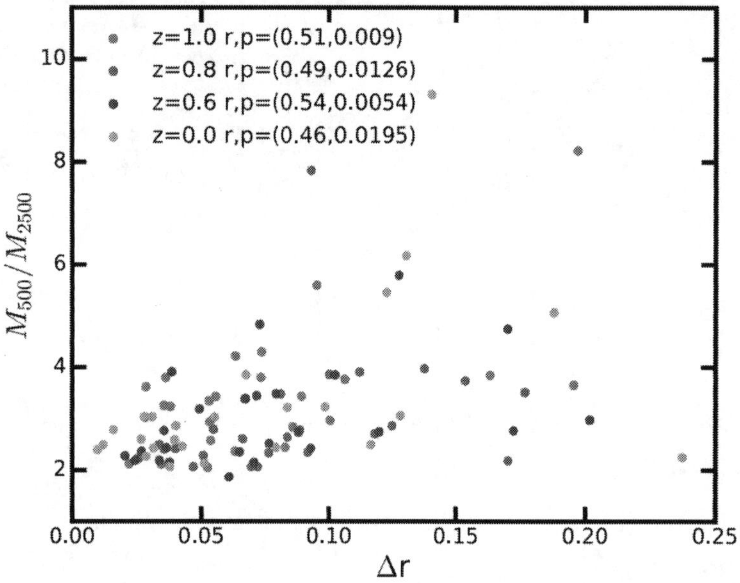

Figure 3. M_{500}/M_{2500} as a function of relaxation state for each redshift slice. The Spearman's rank and the null hypothesis probability are listed in the legend. Figure reproduced from [18].

to the relaxation state of the galaxy cluster through the fact that relaxed clusters are mostly centrally concentrated, while unrelaxed objects span a larger variety of profile shapes, including very shallow profiles. Note that the entire simulation sample is needed to reach a reliable conclusion.

4 Discussion and conclusion

We have focussed in this work on the dark matter density profiles of the 25 most massive systems selected from DMO simulations in four redshift slices (z_{sel} = 1, 0.8, 0.6 and 0). With a median mass of $M_{500} = 6.3 \times 10^{14} M_{\odot}$ at $z = 1$, the sample is composed of the rarest objects, probing for the first time the extreme limits of the cluster mass function, such as would be detectable observationally only in all-sky surveys. Surprisingly, these objects exhibit a high level of self-similarity, and their dark matter density profiles can be described with the typical NFW profile found in relaxed local systems.

From the Millenium simulations [31], high–mass systems with $M > 5 \times 10^{14}$ $M_{\odot}$ at $z = 1$ ($z = 0$) have undergone at least one major merger (mass ratio > 1 : 3) during the preceding 4 Gyr (12 Gyr). The relaxation time estimated from their crossing time ($t_{cross} \propto r_{500}/\sigma_{500}$ with $\sigma_{500} = (GM_{500}/r_{500})^{1/2}$) is close to two Hubble times ($t_H = 1/H(z)$), i.e. about 16 Gyr (29 Gyr). A similar conclusion is reached if one uses the dynamical time. We expect that these objects should still be forming and thus be highly unrelaxed. Using Δr as a dynamical indicator, or from visual inspection of the images, we find that the vast majority of the systems in question are indeed unrelaxed. Naively, one would then expect that the density profiles of such objects would exhibit large variations, linked to the wide variety of dynamical states and formation histories, and that the median profile has not yet converged to the near-universal form of relaxed systems in the local Universe. In contrast, once scaled according to the critical density at each redshift, the density profiles of the clusters in our simulations are remarkably similar, with a low dispersion of less than 0.15 dex within r_{500}. Furthermore, there is little evolution in the radial logarithmic slope or scatter with redshift. This surprising result suggests that the 'universal', 'broken/running' power-law, density profile (e.g. similar to NFW or Einasto) is already in place at $z > 1$ and that it is robust to merging activity. This conclusion is similar to that recently obtained for primordial (Earth–mass) haloes by [32] and

[33], but at scales that are 21 orders of magnitude larger. Interestingly, [34] recently found a remarkably standard self-similar evolution in the mean profile of the hot gas beyond the cooling core region in massive clusters up to $z \sim 1.9$. This would be a natural consequence of the self-similar evolution of the underlying dark matter distribution that we have shown here, since these systems are dark-matter dominated and the gas evolution, except in the very central regions, is dominated by simple gravitational physics. There is an indication of a residual link between the profile shape and dynamical state, with the most unrelaxed clusters exhibiting a larger dispersion. Ongoing work on a larger sample, covering a wider mass range, will enable us to better characterise both the evolution and the scatter. These are essential to understand the link between profile shape, dynamical state, and formation history. This new study should thus start bringing value to observational analysis in the near-future. Its results will be made available to current and future experiments and will be extremely useful to understand the impact of dynamical state (e.g. due to recent mergers) on mass measurements and profile shapes as well as for obtaining unbiased mass profiles. A precise quantification of the impact of cosmology on profile shapes is still badly needed and is the topic of a lot of ongoing studies.

Acknowledgements

This work was supported by the French Agence Nationale de la Recherche under grant ANR-11-BS56-015 and by the European Research Council under the European Union's Seventh Framework Programme (FP7/2007-2013) / ERC grant agreement number 340519. It was granted access to the HPC resources of CINES under the allocations 2015-047350, 2016-047350 and 2017-047350 made by GENCI. The authors also thank the anonymous referee for useful suggestions that helped improve the paper. AMCLB is extremely grateful to Damien Chapon for his help with PYMSES and Patrick Hennebelle for granting access to his shared memory machine. She further thanks for its hospitality the Institute for Computational Science of the University of Zürich.

References

[1] P.J.E. Peebles, *The large-scale structure of the universe* (1980)
[2] G. Hinshaw, D. Larson, E. Komatsu, D.N. Spergel, C.L. Bennett, J. Dunkley, M.R. Nolta, M. Halpern, R.S. Hill, N. Odegard et al., ApJS**208**, 19 (2013), 1212.5226
[3] Planck Collaboration XVI, A&A**571**, A16 (2014), 1303.5076
[4] Planck Collaboration XIII, A&A**594**, A13 (2016), 1502.01589
[5] W. Cui, S. Borgani, G. Murante, MNRAS**441**, 1769 (2014), 1402.1493
[6] D. Martizzi, I. Mohammed, R. Teyssier, B. Moore, MNRAS**440**, 2290 (2014), 1307.6002
[7] M. Velliscig, M.P. van Daalen, J. Schaye, I.G. McCarthy, M. Cacciato, A.M.C. Le Brun, C. Dalla Vecchia, MNRAS**442**, 2641 (2014), 1402.4461
[8] Planck Collaboration XXIX, A&A**571**, A29 (2014), 1303.5089
[9] M. Hasselfield, M. Hilton, T.A. Marriage, G.E. Addison, L.F. Barrientos, N. Battaglia, E.S. Battistelli, J.R. Bond, D. Crichton, S. Das et al., J. Cosmology Astropart. Phys.**7**, 008 (2013), 1301.0816
[10] L.E. Bleem, B. Stalder, T. de Haan, K.A. Aird, S.W. Allen, D.E. Applegate, M.L.N. Ashby, M. Bautz, M. Bayliss, B.A. Benson et al., ApJS**216**, 27 (2015), 1409.0850
[11] Planck Collaboration XXVII, A&A**594**, A27 (2016), 1502.01598
[12] I. Bartalucci, M. Arnaud, G.W. Pratt, J. Démoclès, R.F.J. van der Burg, P. Mazzotta, A&A**598**, A61 (2017), 1610.01899

[13] T. Schrabback, D. Applegate, J.P. Dietrich, H. Hoekstra, S. Bocquet, A.H. Gonzalez, A. von der Linden, M. McDonald, C.B. Morrison, S.F. Raihan et al., `ArXiv:1611.03866` (2016), `1611.03866`

[14] J.F. Navarro, C.S. Frenk, S.D.M. White, ApJ**490**, 493 (1997), `astro-ph/9611107`

[15] A. Tasitsiomi, A.V. Kravtsov, S. Gottlöber, A.A. Klypin, ApJ**607**, 125 (2004), `astro-ph/0311062`

[16] L. Gao, J.F. Navarro, C.S. Frenk, A. Jenkins, V. Springel, S.D.M. White, MNRAS**425**, 2169 (2012), `1201.1940`

[17] H.Y. Wu, O. Hahn, R.H. Wechsler, Y.Y. Mao, P.S. Behroozi, ApJ**763**, 70 (2013), `1209.3309`

[18] A.M.C. Le Brun, M. Arnaud, G.W. Pratt, R. Teyssier, MNRAS**473**, L69 (2018), `1709.07457`

[19] O. Hahn, T. Abel, MNRAS**415**, 2101 (2011), `1103.6031`

[20] A. Lewis, A. Challinor, A. Lasenby, ApJ**538**, 473 (2000), `astro-ph/9911177`

[21] R. Teyssier, A&A**385**, 337 (2002), `astro-ph/0111367`

[22] A. Bleuler, R. Teyssier, S. Carassou, D. Martizzi, Computational Astrophysics and Cosmology **2**, 5 (2015), `1412.0510`

[23] C. Power, J.F. Navarro, A. Jenkins, C.S. Frenk, S.D.M. White, V. Springel, J. Stadel, T. Quinn, MNRAS**338**, 14 (2003), `astro-ph/0201544`

[24] I. Bartalucci, M. Arnaud, G.W. Pratt, J. Démoclès, L. Lovisari, A&A**628**, A86 (2019), `1906.02339`

[25] C. Power, A. Knebe, S.R. Knollmann, MNRAS**419**, 1576 (2012), `1109.2671`

[26] G.L. Bryan, M.L. Norman, ApJ**495**, 80 (1998), `astro-ph/9710107`

[27] B. Diemer, A.V. Kravtsov, ApJ**789**, 1 (2014), `1401.1216`

[28] A. Savitzky, M.J.E. Golay, Analytical Chemistry **36**, 1627 (1964)

[29] B. Diemer, S. More, A.V. Kravtsov, ApJ**766**, 25 (2013), `1207.0816`

[30] J.F. Navarro, E. Hayashi, C. Power, A.R. Jenkins, C.S. Frenk, S.D.M. White, V. Springel, J. Stadel, T.R. Quinn, MNRAS**349**, 1039 (2004), `astro-ph/0311231`

[31] O. Fakhouri, C.P. Ma, M. Boylan-Kolchin, MNRAS**406**, 2267 (2010), `1001.2304`

[32] R.E. Angulo, O. Hahn, A.D. Ludlow, S. Bonoli, MNRAS**471**, 4687 (2017), `1604.03131`

[33] G. Ogiya, D. Nagai, T. Ishiyama, MNRAS**461**, 3385 (2016), `1604.02866`

[34] M. McDonald, S.W. Allen, M. Bayliss, B.A. Benson, L.E. Bleem, M. Brodwin, E. Bulbul, J.E. Carlstrom, W.R. Forman, J. Hlavacek-Larrondo et al., `ArXiv:1702.05094` (2017), `1702.05094`

Overdensity of SubMillimiter Galaxies in the GJ526 Field mapped with the NIKA2 Camera

J.-F. Lestrade[1,*], *R.* Adam[2], *P.* Ade[3], *H.* Ajeddig[4], *P.* André[4], *E.* Artis[5], *H.* Aussel[4], *A.* Beelen[6], *A.* Benoît[7], *S.* Berta[8], *L.* Bing[9], *O.* Bourrion[5], *M.* Calvo[7], *A.* Catalano[5], *A.* Coulais[1], *M.* De Petris[10], *F.-X.* Désert[11], *S.* Doyle[3], *E. F. C.* Driessen[8], *A.* Gomez[12], *J.* Goupy[7], *F.* Kéruzoré[5], *C.* Kramer[13], *B.* Ladjelate[13], *G.* Lagache[9], *S.* Leclercq[8], *J.-F.* Macías-Pérez[5], *A.* Maury[4], *P.* Mauskopf[3,14], *F.* Mayet[5], *A.* Monfardini[7], *M.* Muñoz-Echeverría[5], *L.* Perotto[5], *G.* Pisano[3], *N.* Ponthieu[11], *V.* Revéret[4], *A. J.* Rigby[3], *A.* Ritacco[6,15], *C.* Romero[16], *H.* Roussel[17], *F.* Ruppin[18], *K.* Schuster[8], *S.* Shu[19], *A.* Sievers[13], *C.* Tucker[3], and *R.* Zylka[8]

[1]LERMA, Observatoire de Paris, PSL Research University, CNRS, Sorbonne Université, UPMC, 75014 Paris, France
[2]LLR (Laboratoire Leprince-Ringuet), CNRS, École Polytechnique, Institut Polytechnique de Paris, Palaiseau, France
[3]School of Physics and Astronomy, Cardiff University, Queen's Buildings, The Parade, Cardiff, CF24 3AA, UK
[4]AIM, CEA, CNRS, Université Paris-Saclay, Université Paris Diderot, Sorbonne Paris Cité, 91191 Gif-sur-Yvette, France
[5]Univ. Grenoble Alpes, CNRS, Grenoble INP, LPSC-IN2P3, 53, avenue des Martyrs, 38000 Grenoble, France
[6]Institut d'Astrophysique Spatiale (IAS), CNRS, Université Paris Sud, Orsay, France
[7]Institut Néel, CNRS, Université Grenoble Alpes, France
[8]Institut de RadioAstronomie Millimétrique (IRAM), Grenoble, France
[9]Aix Marseille Univ, CNRS, CNES, LAM (Laboratoire d'Astrophysique de Marseille), Marseille, France
[10]Dipartimento di Fisica, Sapienza Università di Roma, Piazzale Aldo Moro 5, I-00185 Roma, Italy
[11]Univ. Grenoble Alpes, CNRS, IPAG, 38000 Grenoble, France
[12]Centro de Astrobiología (CSIC-INTA), Torrejón de Ardoz, 28850 Madrid, Spain
[13]Instituto de Radioastronomía Milimétrica (IRAM), Granada, Spain
[14]School of Earth and Space Exploration and Department of Physics, Arizona State University, Tempe, AZ 85287, USA
[15]Laboratoire de Physique de l'École Normale Supérieure, ENS, PSL Research University, CNRS, Sorbonne Université, Université de Paris, 75005 Paris, France
[16]Department of Physics and Astronomy, University of Pennsylvania, 209 South 33rd Street, Philadelphia, PA, 19104, USA
[17]Institut d'Astrophysique de Paris, CNRS (UMR7095), 98 bis boulevard Arago, 75014 Paris, France
[18]Kavli Institute for Astrophysics and Space Research, Massachusetts Institute of Technology, Cambridge, MA 02139, USA
[19]Caltech, Pasadena, CA 91125, USA

Abstract. Using the NIKA2 dual band millimeter camera installed on the IRAM30m telescope, we have mapped a relatively large field (~ 70 arcmin^2) in the direction of the star GJ526 to investigate the nature of the sources found

*e-mail: jean-francois.lestrade@obspm.fr

with the MAMBO camera at 1.2 mm ten years earlier. We have found that they must be dust-obscured galaxies (SMGs) in the background beyond the star. The new NIKA2 map at 1.15 mm reveals additional sources and, in fact, an overdensity of SMGs predominantly distributed along a filament-like structure in projection on the sky across the whole observed field. We speculate this might be a cosmic filament at high redshift as revealed in cosmological hydrodynamical simulations. Measurement of spectroscopic redshifts of the SMGs in the candidate filament is required now for a definitive confirmation of the nature of the structure.

1 Introduction

Surveys at submillimeter and millimeter-wavelengths have revealed a dust-obscured population of galaxies at high redshift [1, 2]. These massive SubMillimeter Galaxies (SMGs) are forming stars at prodigious rates, sometimes exceeding 1000 $M_{\odot}$/yr, with correspondingly bright far infrared luminosities larger than $10^{12}\, L_{\odot}$ [3]. SMGs are enshrouded in dust and so have only faint or undetectable optical counterparts but contribute to the far infrared cosmic background emitting approximately half of all of the light in the universe [4, 5]. Theories predict that galaxy formation preferentially occurs along large-scale dark matter filamentary or sheetlike overdense regions that are related to initial fluctuations of the density field in the primordial universe [6] and formed the *cosmic web* as depicted by cosmological hydrodynamical simulations. It is thought that star formation and galaxy growth at an early stage of cosmic evolution are critically tied to this network of dark matter filaments. A large fraction of the neutral hydrogen from the inter galactic medium falls into them and constantly replenishes the gas reservoirs of galaxies through cold gas accretion ; the cold gas streams along the filaments, survives the shocks at the virial radii of the dark matter halos and infalls onto the galaxies to fuel their star formation and growth [7, 8]. However, definitive observational confirmation of cold gas accretion along filaments of the *cosmic web* is actively sought.

A region of the sky several hundreds of arcseconds in size corresponds to an area of several cMpc at high reshift (scaling factor cMpc/$''$ does not depends strongly on z at high z in standard cosmology) that can be mapped efficiently at millimeter wavelengths using a modern camera such as NIKA2 on the IRAM30m radiotelescope [9–14]. SMGs detected in such a field must be embedded in filaments according to the current paradigm and should be signposts of the *cosmic web* in the distant universe. We present NIKA2 data that are a contribution to this endeavour due to serendipity.

In 2007, the field in the direction of the nearby star GJ526 was mapped in 1.2 mm continuum waveband using the MAMBO camera installed on the IRAM 30-meter radiotelescope at the time and revealed a quasi-alignment of the star and five compact sources in projection on the sky. It was not possible to decisively conclude whether these sources were background SMGs or a large, clumpy circumstellar disk seen edge-on around this star [15]. As cautiously stated in this paper, only future observations could definitely decide since these sources would stay fixed if in the background or move at the same rate and in the same direction as the star if clumps of a circumstellar disk. So a possible displacement as large as 22.6$''$ in the south-west direction according to the star proper motion after ten years in 2017 motivated us to observe this field with the NIKA2 camera to determine the origin of these sources.

In this contribution, we report our finding that the five MAMBO sources did not move over this 10 year period and so they must be SMGs in the distant background beyond the star. It was argued in [15] that this was statistically unlikely in assuming a mean number surface density of SMGs but we are now led to believe that the GJ526 field presents instead an overdensity of SMGs. In this paper, we discuss the remarkable spatial distribution of these sources in the 1.15 mm NIKA2 map along a filament-like structure reminiscent of the *cosmic web* revealed in the cosmological hydrodynamical simulations.

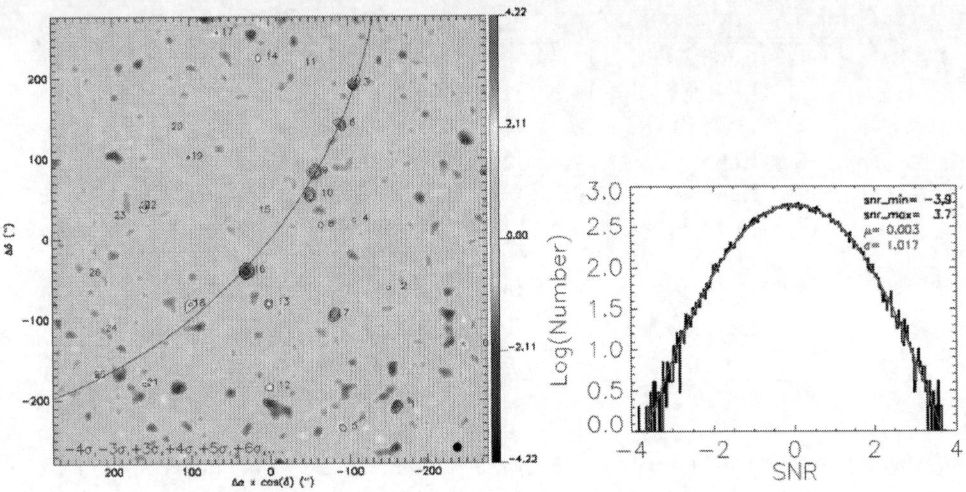

Figure 1. *Left :* GJ526 NIKA2 map at 1.15mm (color scale : mJy/beam, contours : signal-to-noise ratio). Curve traces the filament-like structure identified across the map. Beam size is 11.2″ (black filled circle at brc). Map center (J2000) : 13h45m44.52s and +14°53′20.6″. *Right :* Signal-to-noise ratio histogramm of the jack-knifed map at 1.15 mm (in red, Gaussian distribution with zero mean and σ=1 for comparison).

2 NIKA2 observations and the 1.15 mm map

The NIKA2 observations were obtained during the science verification observations of the NIKA2 camera on 2017 april 24, 2018 february 14, 17, 18, 19, 20 and 2018 march 14, in average weather conditions (atmospheric opacity between 0.20 and 0.45 at 225 GHz) and at elevation between 33° and 67°. The observing strategy consisted of a series of 9′ × 4.5′ on-the-fly scans with a scanning speed of 40″/sec along two orthogonal orientations (+60°, +150°) in the AZ-EL coordinate system at the telescope to minimise residual stripping patterns in the maps. These observations are split into 167 scans that correspond to a total of 9.3 hours of observations on source. The time ordered information (TOI) of each detector was processed in order to remove the contribution of atmospheric and electronic correlated noises and to obtain the maps at 1.15 mm and 2.0 mm following the procedure described in [16] and Ponthieu et al. (in preparation). Point sources are detected by matching the maps with a Gaussian Point-Spread-Function. Position accuracy is better than 3 arcseconds (for each coordinate) for sources above 4σ and the absolute photometric accuracy is ~15 % [14].

In the limited space of this presentation, we shall discuss only the NIKA2 map at 1.15 mm and plan a complete presentation including the NIKA2 map at 2.0 mm in a forthcoming paper. We present the resulting 1.15 mm NIKA2 map in Fig. 1 and astrometry and photometry of the ten unambiguously detected sources with a signal-to-noise ratio (SNR) larger than 4 in Table 1. In order to determine this detection threshold, we have analysed the map noise in producing a jack-knifed map to compute SNRs in all pixels and found that their distribution are Gaussian, centered on zero and have a standard deviation of unity as expected for normal noise (see Fig. 1 right). In these conditions, no spurious sources with $SNR > 4$ are statistically

 https://doi.org/10.1051/epjconf/202225700027

Table 1: NIKA2 astrometry and photometry of unambiguously detected sources at 1.15 mm.

ID H3	α_{2000} (h m s)	δ_{2000} (° ′ ″)	S_{260GHz} (mJy)	SNR 260GHz
1	13:45:33.48	14:49:50.5	3.96± 0.95	5.3
3	13:45:37.18	14:56:31.9	3.67± 0.81	6.1
6	13:45:38.36	14:55:42.2	2.95± 0.68	5.7
7	13:45:38.97	14:51:45.4	2.87± 0.63	6.1
9	13:45:40.61	14:54:42.8	3.13± 0.65	6.9
10	13:45:41.09	14:54:14.8	2.97± 0.62	6.8
13	13:45:44.69	14:51:59.4	2.08± 0.53	4.7
16	13:45:46.63	14:52:39.7	4.11± 0.75	9.6
18	13:45:51.51	14:51:59.0	2.15± 0.57	4.5
22	13:45:55.51	14:54: 1.1	2.40± 0.63	4.6

expected in the 500″ × 500″ map owing to the number of independent beams (11.2″ at 1.15 mm) in it. There are also sixteen additional sources with 3 < SNR < 4 but three or four spurious positive sources are statistically expected at this significance level. Source IDs in Fig. 1 and Table 1 are assigned to all SNR > 3 sources.

We found that four of the five MAMBO sources (see section GJ526 in Table 2 of [15]) have been detected anew in the NIKA2 map at 1.15 mm and are identified as ID 7, 9, 10, 16 in Table 1 herein. The fifth MAMBO source – MM134543+145317, a 4.3σ detection in [15] – has been searched in the 1.15 mm NIKA2 map and identified but only at a level of 2.8σ. Nonetheless, such a difference in SNR between the NIKA2 and MAMBO observations is statistically expected. As already mentionned in the introduction, our main finding is that all five MAMBO sources did not move between 2007 and 2017 and so they must be SMGs in the distant background.

The MAMBO flux densities given at 250 GHz in Table 2 of [15] are comparable with the NIKA2 flux densities in Table 1 herein to better than 2.4 times the quadratically combined uncertainties of the two measurements. In Table 1, the flux density uncertainties include both statistical and systematic photometric uncertainties (this latter one is ~15% [14]).

3 Filament-like structure in the GJ526 field

As stated above, there are ten sources unambiguously detected (SNR> 4) above 2 mJy at 1.15 mm in the whole NIKA2 map. This can be usefully compared to the cumulative source number counts in other fields at a similar wavelength. To make this comparison meaningful, we shall restrict the map to its central region within a radial distance of 150″ where the noise rms is uniform at the 10% level. In this region, there are six sources (ID 7, 9, 10, 13, 16, 18) that are unambiguously detected above 2 mJy at 1.15 mm. Combination of several single dish studies of the COSMOS field, the Lockman Hole North field and the GOODS field, provides the cumulative source number count $N(S_{1.2mm} > 2.0\text{mJy}) = 500$ sources/deg^2 [17]. The Poisson probability that six sources be found within $r < 150''$ given this mean surface number density is as low as 3.7%. However, the GJ526 field we picked for our NIKA2 observations is one out of 50 similar-sized fields targeted in the initial MAMBO survey to search for circumstellar disks [15]. So this overdensity in the GJ526 field might very well be a statistical fluctuation in the whole MAMBO survey (note that no other MAMBO field shows such an overdensity). We turn now to the more intriguing spatial distribution of these SMGs along a filament-like structure.

It is remarkable that seven SMGs (IDs 3, 6, 9, 10, 16, 18, 25) are spatially distributed along a slightly curved filament-like structure in projection on the sky as highlighted in Fig. 1 with an arc (solid line). The first six sources are unambiguously detected and reported in Table 1 while the last one (ID 25) is only a possible 3.3σ detection with a 1.15 mm flux density of 2.61 ± 0.79 mJy. We can estimate the probability that such a quasi-alignment occurs in using the average source number surface density $N(S_{1.2mm} > 2.0\text{mJy})$=500 sources/degree^2, noted μ below, in small, consecutive sectors with a deviation angle δ from strict alignment of $\pm 10°$ (see sketch in Fig. 2). Then, for a triplet of sources, the Poisson probability for the third source not to deviate more than δ from alignment with the first two sources is $\mu \times \pi r^2 \times 2\delta/360°$, where r is the distance between the second and the third sources.

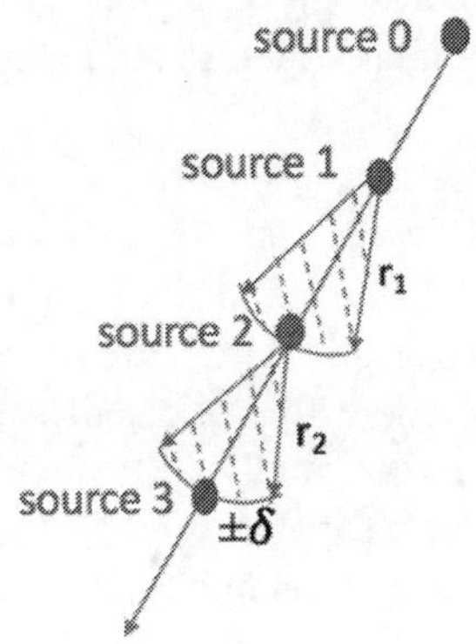

Figure 2. Estimation of chance alignment of sources.

As sketched in Fig. 2, applying this formula for the first triplet in the NIKA2 map (source ID's 3, 6, 9), the probability of occurence of quasi-alignement within $\pm 10°$ is 500 sources/deg^2 $\times$ $\pi(70''/3600'')^2$ $\times$ $20°/360°$ = 3.3% where r_1 = 70″ is the distance between source IDs 6 and 9. Similarly, the probability is 0.6% for the second triplet (ID's 6, 9, 10) with r_2 = 30″, it is 11% for the third triplet (IDs 9, 10, 16) with r_3 = 130″, it is 4.3% for the fourth triplet (ID's 10, 16, 18) with r_4 = 80″, and it is 15% for the fifth triplet (ID's 16, 18, 25) with r_5 = 150″. The combined probability ($\prod_{i=1}^{5} \mu\pi r_i^2 2\delta/360°$) is then as low as $1.4\ 10^{-7}$ and so it is very unlikely that this filament-like structure has occured by chance, even in accounting for the fact that 50 similar-sized fields have been observed in the initial MAMBO survey. So this apparent structure in the NIKA2 map is statistically significant.

This filament-like structure stretches across the whole map which is 500″ in size. So, it extends at least over ~4 cMpc in using the scaling factor 8.3kpc/″ that does not depend strongly on redshift comprised between 1 and 6 in standard cosmology [1]. In addition, the angular separations r_i specified above between these seven SMGs correspond to linear separations comprised between ~0.25 cMpc and ~1.25 cMpc in using the same redshift and scaling factor. This extent and these separations are comparable with the spatial distribution of dark matter haloes hosting galaxies in the high resolution cosmological simulation *NewHorizon* [19] for instance. This match between observations and simulations suggests that the structure in the NIKA2 map may be a cosmic filament. A complete presentation of this hypothesis and additional evidences, noticeably high luminosities ($L_{FIR} > 10^{12}\ L_\odot$) and high star formation rates (SFR > 300 $M_\odot$/yr) of the seven SMGs estimated with the 2.0 mm NIKA2 map and supplemental data from Herschel/SPIRE, will be published elsewhere.

4 Conclusion

In the 1.15 mm NIKA2 map of the GJ526 field presented here, there is an overdensity of SMGs characterized by an apparent filamentary structure in projection on the sky that is un-

[1] https://irsa.ipac.caltech.edu/Missions/herschel.html

likely chance occurence and is visually reminiscent of a cosmic filament as cosmological simulations have shown (*Illustris* [18], *NewHorizon* [19], *Uchuu* [20]). However, confirmation of such a filament requires now measurements of spectroscopic redshifts of the milli-Jansky unlensed SMGs in using molecular and atomic lines with the most sensitive arrays such as NOEMA and ALMA. Also, comparison of our observed structure to the most recent cosmological simulations listed above is presently hampered by the fact they do not track specifically the rapidly star forming galaxies (main properties of SMGs). In fact, the less recent GADGET-2 simulation [21] was designed specifically to track massive galaxies with high star formation rates (> 280 $M_{\odot}$/yr) but the study does not present their spatial distribution. Specific tools to search cosmological simulations to better isolate SMGs are needed and complementary to the high-speed and deep mapping capability attained with NIKA2.

While galaxy surveys in the optical domain have already revealed the distinctive pattern of voids, sheets, filaments and nodes characterising the structure of the universe at low redshift, spatial distribution of SMGs have the potential to reveal similar structure but at high redshift.

Acknowledgements

This work is based on observations carried out during the science verification program of NIKA2 on the IRAM 30m telescope. IRAM is supported by INSU/CNRS (France), MPG (Germany) and IGN (Spain). We are grateful to the IRAM staff for their dedicated support. This research has made use of the SIMBAD database, operated at CDS, Strasbourg, France. This research has made use of the NASA/IPAC Extragalactic Database (NED), which is operated by the Jet Propulsion Laboratory, California Institute of Technology, under contract with the National Aeronautics and Space Administration.

References

[1] I. Smail, R.J. Ivison, A.W. Blain, ApJ **490**, L5 (1997)
[2] D.H. Hughes, S. Serjeant, J. Dunlop *et al.*, Nature **394**, 241 (1998)
[3] M.L. Strandet, A. Weiss, J.D. Vieira *et al.*, ApJ **822**, 80 (2016)
[4] J.L. Puget, A. Abergel, J.P. Bernard *et al.*, Astron. Astrophys. **308**, L5 (1996)
[5] H. Dole, G. Lagache, J.L. Puget *et al.*, Astron. Astrophys. **451**, 417 (2006)
[6] J.R. Bond, L. Kofman, D. Pogosyan, Nature **380**, 603 (1996)
[7] D. Kereš, N. Katz, D.H. Weinberg, R. Davé, MNRAS **363**, 2 (2005)
[8] A. Dekel, Y. Birnboim, G. Engel *et al.*, Nature **457**, 451 (2009)
[9] A. Monfardini, A. Benoit, A. Bideaud *et al.*, ApJS **194**, 24 (2011)
[10] A. Catalano *et al.*, Astron. Astrophys. **569**, A9 (2014)
[11] M. Calvo *et al.*, Journal of Low Temperature Physics, **184**, 816 (2016)
[12] O. Bourrion *et al.*, Journal of Instrumentation, **11**, 11001 (2016)
[13] R. Adam *et al.*, Astron. Astrophys. **609**, A115 (2018)
[14] L. Perotto, N. Ponthieu, J.-F. Macías-Pérez *et al.*, Astron. Astrophys. **637**, A71 (2020)
[15] J.F. Lestrade, M.C. Wyatt, F. Bertoldi *et al.*, Astron. Astrophys. **506**, 1455 (2009)
[16] R. Adam, B. Comis, J.F. Macías-Pérez *et al.*, Astron. Astrophys. **569**, A66 (2014)
[17] R..R. Lindner, A.J. Baker, A. Omont *et al.*, ApJ **737**, 83 (2011)
[18] V. Springel, R. Pakmor, A. Pillepich *et al.*, MNRAS, **475**, 676 (2018)
[19] Y. Dubois, R. Beckmann, F. Bournaud *et al.*, Astron. Astrophys. **651**, A109 (2021)
[20] T. Ishiyama, F. Prada, A. Klypin *et al.*, MNRAS **506**, 4210 (2021)
[21] R. Davé, B.D. Oppenheimer, N. Katz *et al.*, MNRAS, **408**, 2051 (2010)

Cosmology with the SZ spectrum: Measuring the Universe's temperature with galaxy clusters

Gemma Luzzi[1,*], *Emanuele* D'Angelo[2], *Herve* Bourdin[3], *Federico* De Luca[3], *Pasquale* Mazzotta[3], *Filippo* Oppizzi[3], and *Gianluca* Polenta[1]

[1]Space Science Data Center, Italian Space Agency, via del Politecnico, 00133, Roma, Italy
[2]Dipartimento di Matematica, Università degli Studi di Roma "Tor Vergata," via della Ricerca Scientifica, 1, I-00133 Roma, Italy
[3]Dipartimento di Fisica, Università degli Studi di Roma "Tor Vergata," via della Ricerca Scientifica, 1, I-00133 Roma, Italy

Abstract. The hot gas in clusters of galaxies creates a distinctive spectral distortion in the cosmic microwave background (CMB) via the Sunyaev-Zel'dovich (SZ) effect. The spectral signature of the SZ can be used to measure the CMB temperature at cluster redshift ($T_{\rm CMB}(z)$) and to constrain the monopole of the y-type spectral distortion of the CMB spectrum. In this work, we start showing the measurements of $T_{\rm CMB}(z)$ for a sample extracted from the Second Catalog of galaxy clusters produced by Planck (PSZ2) and containing 75 clusters selected from CHEX-MATE. Then we show the forecasts for future CMB experiments about the constraints on the monopole of the y-type spectral distortion of the CMB spectrum via the spectrum of the SZ effect.

1 Introduction

The COBE-FIRAS experiment revealed that the average CMB spectrum is extremely close to a perfect black-body with present-day temperature $T_0 = 2.7260 \pm 0.0013$ K [1] and with possible distortions $< 10^{-5}$. Departures of the CMB frequency spectrum from a pure black body encode information about the thermal history of the early Universe [2]. According to the Big Bang model, under the assumptions of adiabatic expansion and photon number conservation, the CMB temperature evolves with redshift z as $T_{\rm CMB}(z) = T_0(1+z)$ which is one of the core predictions of standard big bang cosmology that may be violated under nonstandard scenarios [3]. Testing its validity is important for cosmology and fundamental physics. The possibility of determining $T_{\rm CMB}(z)$, T_0 and the monopole of y-type distortion of the CMB spectrum from measurements of the Sunyaev–Zeldovich effect had been suggested long ago [4, 5]. Building upon these two works, in Sect. 2) we present constraints on the CMB temperature evolution with redshift and independent estimate of T_0. In Sect 3 we show forecasts for future CMB experiments about constraints on the y-type distortion of the CMB spectrum.

*e-mail: gemma.luzzi@ssdc.asi.it

2 Constraining the evolution of the CMB temperature for the CHEX-MATE sample of Planck SZ clusters

Deviations of the standard CMB temperature scaling with redshift are usually described using the parametrization proposed by [6], $T_{\rm CMB}(z) = T_0(1+z)^{(1-\beta)}$, where β is a constant parameter ($\beta = 0$ in the standard scenario). The spectral signature of the SZ depends on the frequency ν through the non-dimensional frequency $x = \frac{h\nu(z)}{kT(z)} = \frac{h\nu_0(1+z)}{kT_0(1+z)} = \frac{h\nu_0}{kT_0}$, it is redshift-invariant only for the standard scaling of $T_{\rm CMB}(z)$, in all other scenarios becomes z-dependent, resulting in a small dilation/contraction of the SZ spectrum on the frequency axis. There have been various attempts to constrain the CMB temperature evolution via the SZ method. Starting from the initial constraints using a few clusters of [7, 8], notable improvements in constraining the β value are obtained by using large galaxy cluster catalogues together with very precise CMB data from the Planck satellite [9–11], from the South Pole Telescope [12] and from The Atacama Cosmology Telescope [13]. The core of the present analysis rests upon a sample of the second Planck SZ cluster catalog (PSZ2) [14] selected by CHEX-MATE [15] and Planck PR2 temperature maps[1] at frequencies from 100 to 857 GHz [16]. For this sample we derive information about the R_{500} and about the electronic temperature T_e, which will be included as prior information in the subsequent analysis for the extraction of the CMB temperature at the redshift of each cluster. Here we show results only for 77 clusters in the CHEX-MATE sample. The analysis for the full sample, consisting of 118 clusters, will be presented in D'Angelo et al in preparation.

2.1 SZ maps at Planck frequencies and modelling of the SZ signal

It is widely known that the *Planck* maps include radiation from a variety of astrophysical sources, chiefly the CMB, the SZ signal, Galactic foregrounds, CO emission, zodiacal light and point sources. In order to separate these components from the tSZ signal we apply to the *Planck* PR2 data the same recipe used by [17] on PR1 data. In particular CMB and dust anisotropies are spectrally modelled using all HFI frequency maps outside each galaxy cluster ($7R_{500} \le R \le 12R_{500}$), and spatially modelled using a linear combination of wavelet filtered frequency maps at 857 and 217 GHz. SZ signal maps at 100, 143, 217 and 353GHz are used to build the likelihood for the $T_{\rm CMB}(z)$ extraction. For each cluster, we assume that the pressure structure follows a spherically symmetric distribution, $P(r)$, first proposed by [18]. Apart from the amplitude, which is left free to vary, and the size, which is determined by the analysis of the CHEX-MATE data, all the other parameters of the pressure profile are fixed at the values of [19]. We can thus factorize the SZ signal of a cluster into two terms: the SZ template and the SZ amplitude $\Delta T_{\rm SZ}(\nu, \Theta_n)$ which depends on the unknown parameters we want to extract; we account for the relativistic correctyions to the SZ effect and we use the formulas provided in [20]. Assuming that the SZ data are gaussianly distributed allows us to construct the single cluster likelihood function. The likelihood for the n-th cluster is:

$$P(\mathrm{map_{obs}}|\Theta_n) \propto \exp\left\{-\sum_{\nu}\sum_{ij}\frac{[\Delta T_{\rm SZ}(\nu,\Theta_n)\times \mathrm{template}(\nu,i,j) - \mathrm{map_{obs}}(\nu,i,j)]^2}{\mathrm{varmap}(\nu,i,j) + \delta CMB^2}\right\} \quad (1)$$

where Θ_n are the cluster parameters (Comptonization parameter y_n, peculiar velocity v_{pec}, electron temperature T_e) and the CMB temperature at the redshift of the cluster, while

[1] http://www.sciops.esa.int/index.php?project=planck\&page=Planck_Legacy_Archive

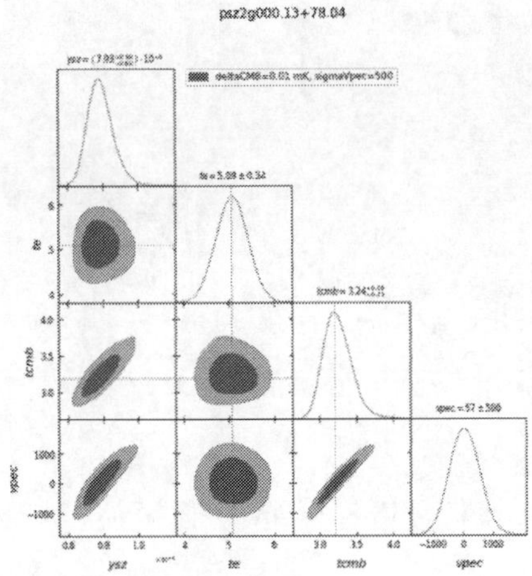

Figure 1. 1D and 2D posteriors for one cluster of the sample. We adopt a gaussian prior for the electronic temperature, while keeping free the CMB temperature parameter and the comptonization parameter. Vertical grey lines denote respectively: for T_e the central value from X-ray data, for T_{CMB} the value expected by the standard scaling. The posterior for T_e coincides with the prior.

varmap(ν, i, j) is the variance in thermodynamic temperature for each pixel and for each frequency channel, as provided by the Planck collaboration. Kinematic SZ and the CMB primary anisotropies have the same spectral shape in the non-relativistic limit, thus in cleaning for CMB we also remove the main kinematic component of the SZ. We account for possible CMB and KSZ residuals in our maps after the cleaning procedure. We model the CMB residual signal as in [21], accounting for a systematic effect which is a not exactly known offset, the uncertainty about which is described by a Gaussian pdf around zero and standard deviation δCMB. To estimate δCMB we rely on mock maps at Planck frequencies. We generate SZ signal maps using data from X-ray. CMB maps and dust emission signals are added to the clusters maps. We apply the cleaning procedure as in [17], getting as output both the SZ signal maps and the CMB map. We evaluate the amplitude of the δCMB by taking the mean of the RMS of the CMB residual maps (difference between CMB in input and restored CMB map). While to account for possible KSZ residuals, we model them as a kinematic SZ component and adopt as a prior a Gaussian with a universal vanishing mean and with a 500 km/s standard deviation. In order to extract $T_{\rm CMB}(z)$ we adopt the publicly available MCMC sampler Cobaya[2] [22, 23], able to reconstruct the posterior distribution for the parameters of interest. The MCMC chains were analyzed by using GetDist[3]. We explore the space of the cluster parameters and the CMB temperature at the redshift of the cluster. We adopt a gaussian prior on T_e, derived by the analysis of the CHEX-MATE data. Not including the informative prior on T_e results in a bias on $T_{\rm CMB}$, while not including the $v_{\rm pec}$ term results in an underestimation of the error on $T_{\rm CMB}$, the two effects have been tested with simulated clusters. Examples of parameter extraction performed on SZ maps are provided in figure 1.

[2]https://cobaya.readthedocs.io/en/latest/
[3]https://github.com/cmbant/getdist

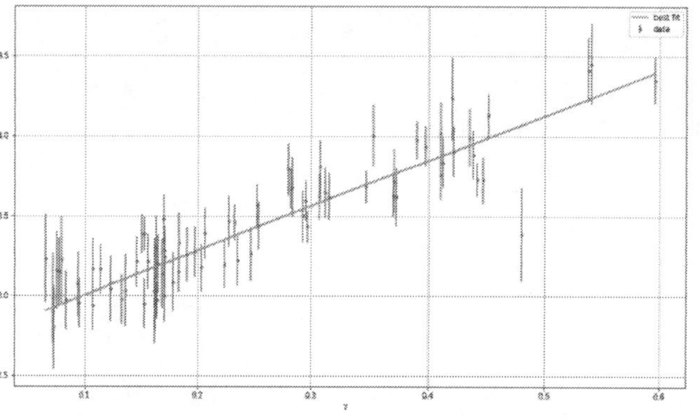

Figure 2. Measurements of the CMB temperature as a function of redshift for 77 *Planck* selected clusters. The orange line is the best fit (with $\beta = -0.02$) to the scaling $T_{\rm CMB}(z)$=$T_0(1+z)^{1-\beta}$.

2.2 CMB temperature measurements, constraints on the β parameter and independent estimate of T_0

In figure 2 we plot the measurements of the CMB temperature as a function of redshift. We report expected values and standard deviation of each single $T_{\rm CMB}(z)$ distribution. Our measurements reach a precision of up to 3%. By fitting the $T_{\rm CMB}(z)$ data with the parametrization $T_{CMB}(z)$=$T_0(1+z)^{1-\beta}$ we get for 77 clusters $\beta = -0.02 \pm 0.02$, at 1σ uncertainty, in agreement with the predictions of the standard model. Lately there has been a renewed interest in independent estimate of T_0. Motivated by the works of [24–28], assuming the standard evolution of the CMB temperature with redshift, we can use our measurements of $T_{\rm CMB}(z)$ to put constraints on the value of $T_{\rm CMB}$ at redshift zero. By considering T_0 as a free parameter when fitting the data with the standard relation $T_{\rm CMB}(z) = T_0(1+z)$, we get an independent estimate of the CMB temperature at redshift zero: $T_0 = 2.729 \pm 0.014$ K, at 1σ uncertainty, a value consistent with the COBE-FIRAS measurement, $T_0 = 2.7260 \pm 0.0013$ K.

3 Constraining the monopole y distortion via the SZ spectrum

Departures of the CMB frequency spectrum from a pure black body encode information about the thermal history of the early Universe. Notably, spectral distortions of the CMB can be created, after thermalization becomes inefficient at redshifts $z \lesssim 10^6$, by energy-releasing processes that can drive matter and radiation out of thermal equilibrium. Examples include injection of photons or other particles that interact electromagnetically. The effects associated with such processes may be described as y-type and μ-type distortions, characterising the exchange of energy between photons and electrons through Compton scattering. In particular, as Compton scattering becomes inefficient for $z \lesssim 10^4$, y-type distortions may be induced in these redshifts, and so the detection of a primordial y-distortion signal, y_p, would probe the thermal history of our Universe during the recombination and reionization eras. Here we apply the method proposed by [4, 5] to constrain the CMB y_p distortion via the SZ spectrum. The expected residual distortion due to a CMB $y_p = 10^{-6}$ for a typical cluster with $y_{SZ} = 10^{-4}$, is shown in figure 3. Given the amplitude of the signal (about 100 Jy / sr at 300 GHz) it is clear that only future CMB experiments could be able to improve the constraints on the CMB y_p distortion via the SZ spectrum.

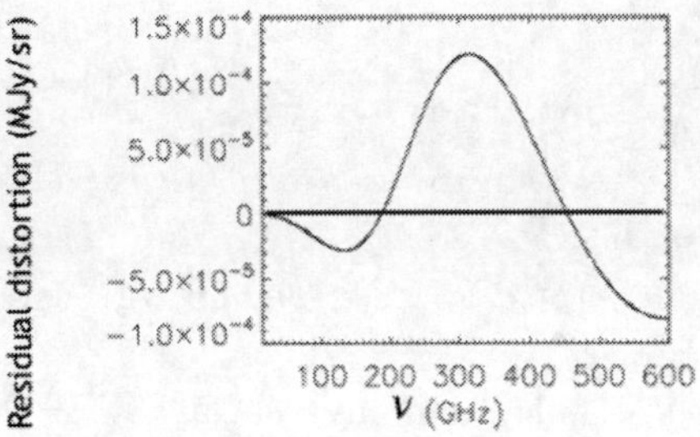

Figure 3. Departures from an SZ spectrum (for a cluster with $y_{SZ} = 10^{-4}$) due to a CMB spectrum with distortion $y_p = 10^{-6}$.

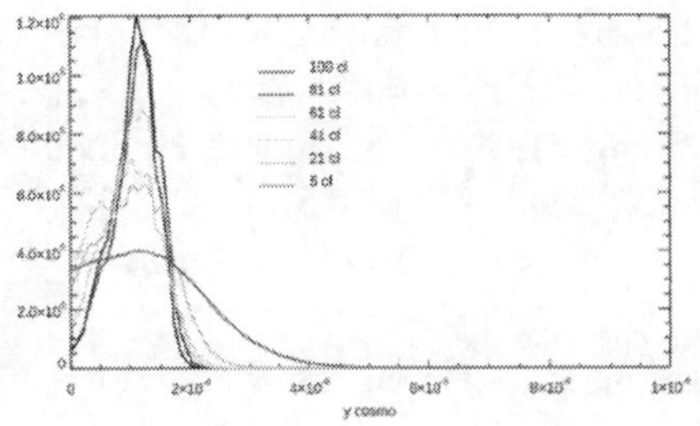

Figure 4. Joint pdf on the parameter y_p from simulated Millimetron SZ spectra (5 to 100 clusters considered).

3.1 Forecasts for Millimetron

The Millimetron Space Observatory (MSO) [4] is a large, cryogenic space observatory covering the spectral range from $70\mu m$ to 10mm. The antenna has an opening of 10m and is cooled below 10K. The observatory will work from deep space, both as a single telescope and as part of a very long baseline interferometer (VLBI). In 2011, ASI funded a phase A study for the Long wave-Array Camera Spectrometer (LACS) (PI P. de Bernardis), a focal plane instrument comprising a Fourier transform spectrometer with non-coherent detector mosaics. The main characteristics of the telescope-LACS combination are the following: spectral coverage: 100 GHz - 1000 GHz (in 4 sub-bands); maximum spectral resolution: 1 GHz; angular resolution: from 10 arc seconds to 1 arc minute; field of view about 5 arc minutes; sensitivity: about 100 Jy / sr in a 10 GHz band in 1 hour of integration. With LACS, both differential and absolute spectral measurements are possible. The SZ effect in galaxy clusters constitutes the main scientific goal for LACS. By exploiting the very accurate spectroscopic capabilities of the instrument, as well as its high angular resolution, it will be possible to perform a reliable separation of all spectral components. In figure 4 we show forecasts for the extraction of the CMB y_p distortion via the SZ spectrum for a number of Millimetron clusters ranging from 5 to 100. Even if the simulation adopted are quite simplistic, since the SZ spectra in the range 100 - 600 GHz are recovered assuming perfect cleaning of foregrounds, it seems that this differential approach can be complementary to absolute spectroscopic measurements of the CMB y distortions.

[4] http://millimetron.ru/en/

References

[1] D.J. Fixsen, Ap. J. **707**, 916 (2009), 0911.1955
[2] J. Chluba, *Future steps in cosmology using spectral distortions of the cosmic microwave background* (2018), 1806.02915
[3] A. Avgoustidis, G. Luzzi, C.J.A.P. Martins, A.M.R.V.L. Monteiro, JCAP **1202**, 013 (2012), 1112.1862
[4] R. Fabbri, F. Melchiorri, V. Natale, Astrophys. & Sp. Sci. **59**, 223 (1978)
[5] Y. Rephaeli, Ap. J. **241**, 858 (1980)
[6] J.A.S. Lima, A.I. Silva, S.M. Viegas, M.N.R.A.S. **312**, 747 (2000)
[7] E.S. Battistelli, M. De Petris, L. Lamagna, F. Melchiorri, E. Palladino, G. Savini, A. Cooray, A. Melchiorri, Y. Rephaeli, M. Shimon, Ap. J. Lett. **580**, L101 (2002), astro-ph/0208027
[8] G. Luzzi, M. Shimon, L. Lamagna, Y. Rephaeli, M. De Petris, A. Conte, S. De Gregori, E.S. Battistelli, Ap. J. **705**, 1122 (2009), 0909.2815
[9] I. de Martino, R. Génova-Santos, F. Atrio-Barandela, H. Ebeling, A. Kashlinsky, D. Kocevski, C.J.A.P. Martins, Ap. J. **808**, 128 (2015), 1502.06707
[10] G. Hurier, N. Aghanim, M. Douspis, E. Pointecouteau, A. & A. **561**, A143 (2014), 1311.4694
[11] G. Luzzi, R.T. Génova-Santos, C.J.A.P. Martins, M. De Petris, L. Lamagna, JCAP **1509**, 011 (2015), 1502.07858
[12] A. Saro et al. (SPT), Mon. Not. Roy. Astron. Soc. **440**, 2610 (2014), 1312.2462
[13] Y. Li et al. (2021), 2106.12467
[14] Planck Collaboration, P.A.R. Ade, N. Aghanim, M. Arnaud, M. Ashdown, J. Aumont, C. Baccigalupi, A.J. Banday, R.B. Barreiro, R. Barrena et al., A & A**594**, A27 (2016), 1502.01598
[15] The CHEX-MATE Collaboration, :, M. Arnaud, S. Ettori, G.W. Pratt, M. Rossetti, D. Eckert, F. Gastaldello, R. Gavazzi, S.T. Kay et al., arXiv e-prints arXiv:2010.11972 (2020), 2010.11972
[16] Planck Collaboration, R. Adam, P.A.R. Ade, N. Aghanim, Y. Akrami, M.I.R. Alves, F. Argüeso, M. Arnaud, F. Arroja, M. Ashdown et al., A & A**594**, A1 (2016), 1502.01582
[17] H. Bourdin, P. Mazzotta, A. Kozmanyan, C. Jones, A. Vikhlinin, ApJ**843**, 72 (2017), 1707.02248
[18] D. Nagai, A.V. Kravtsov, A. Vikhlinin, ApJ**668**, 1 (2007), astro-ph/0703661
[19] M. Arnaud, G.W. Pratt, R. Piffaretti, H. Böhringer, J.H. Croston, E. Pointecouteau, A & A**517**, A92 (2010), 0910.1234
[20] S. Nozawa, N. Itoh, Y. Kohyama, A & A**440**, 39 (2005), astro-ph/0501114
[21] G. D'Agostini, M. Raso (2000), hep-ex/0002056
[22] J. Torrado, A. Lewis, Journal of Cosmology and Astroparticle Physics **2021**, 057 (2021)
[23] J. Torrado, A. Lewis, *Cobaya: Bayesian analysis in cosmology* (2019), 1910.019
[24] M.M. Ivanov, Y. Ali-Haïmoud, J. Lesgourgues, Phys. Rev. D **102**, 063515 (2020)
[25] B. Bose, L. Lombriser, Phys. Rev. D **103**, L081304 (2021)
[26] Y. Wen, D. Scott, R. Sullivan, J.P. Zibin, Phys. Rev. D **104**, 043516 (2021)
[27] V.V. Klimenko, A.V. Ivanchik, P. Petitjean, P. Noterdaeme, R. Srianand, Astronomy Letters **46**, 715–725 (2020)
[28] Bengaly, Carlos A. P., Gonzalez, Javier E., Alcaniz, Jailson S., Eur. Phys. J. C **80**, 936 (2020)

HI and H_2 gas evolution over cosmic times: ColdSIM

Umberto Maio[1,2,*]

[1]INAF, Italian National Institute of Astrophysics, via G. Tiepolo 11, 34143 Trieste, Italy
[2]Max Planck Institute for Astrophysics, Karl-Schwarzschild-Str. 1, 85748 Garching b. M., Germany

Abstract. We present first results of cold cosmic gas evolution obtained through a set of state-of-the-art numerical simulations (ColdSIM). We model time-dependent atomic and molecular non-equilibrium chemistry coupled to hydrodynamics, star formation, feedback effects, various UV backgrounds as suggested by the recent literature, HI and H_2 self-shielding, H_2 dust grain catalysis, photoelectric heating and cosmic-ray heating. By means of such non-equilibrium calculations we are finally able to reproduce the latest HI and H_2 observational data. Consistently with available determinations, neutral-gas mass density parameter results around $\Omega_{\rm neutral} \sim 10^{-3}$ and increases from lower to higher redshift (z). The molecular-gas mass density parameter shows peak values of $\Omega_{\rm H_2} \sim 10^{-4}$, while expected H_2 fractions can be as high as 50% of the cold gas mass at $z \sim$ 4-8, in line with the latest measurements from high-z galaxies. These values agree with observations up to $z \sim 7$ and both HI and H_2 trends are well reproduced by our non-equilibrium H_2-based star formation modelling. Corresponding H_2 depletion times remain below the Hubble time and comparable to the dynamical time at all epochs. This implies that non-equilibrium molecular cooling is efficient at driving cold-gas collapse in a variety of environments and since the first half Gyr. Our findings suggest that, besides HI, non-equilibrium H_2 analyses are key probes for assessing cold gas and the role of UV background radiation.

1 Introduction

Atomic and molecular gas are fundamental phases of the cosmic medium since they constitute the bulk of neutral cosmic gas and lead to star formation and cosmic structure evolution. Neutral gas is constituted by gas with temperatures below 10^4 K (i.e. cold gas), while its chemical composition is characterised by neutral H atoms (HI), as well as by large amounts of H_2 molecules, the most abundant ones in the Universe. These latter drive gas collapse and its conversion into stars. Feedback effects from newly born structures, then, act on existing gas via a number of additional processes, such as supernova (SN) explosions, winds, metal enrichment, UV photoionization or photodissociation. These can affect HI and H_2 evolution dramatically. Observationally, HI is currently well determined up to high redshift, $z \simeq 5$, instead H_2 is being constrained up to $z \simeq 7$ by the latest determination based on e.g. ALMA, VLA, NOEMA, UKIRT, etc.. From a theoretical point of view, chemistry abundances in the cosmic space evolve off equilibrium conditions, over an expanding background, and for this reason species number densities must be computed considering explicitly the related time

*e-mail: umberto.maio@inaf.it

variations (non-equilibrium chemistry). HI is expected to dominate cold material, because recombination processes are efficient around or below 10^4 K in UV shielded gas. H_2 is more troublesome to study, since its formation in the cosmic space takes place through several channels. In presence of free electrons and protons, H_2 can be formed via H^- and H_2^+ catalysis. In dense cold neutral gas, three-H processes are also effective for H_2 production. After metal spreading during stellar evolution and SN events, heavy elements can condensate into dust grains and enhance H_2 abundances. H_2 dust grain catalysis is metal-dependent and usually accompanied by photoelectric heating. Besides metals, SNe can produce cosmic rays, too, and these further influence H_2 evolution. Albeit studied and observed in the local Universe, these processes are poorly known in cosmological environments and deserve careful investigations. Moreover, the establishment of a cosmological UV background at different z can alter chemical species in a way that is not trivial to predict. In the following, we outline the physical and chemical modelings required to address cosmic atomic and molecular gas in the most generic non-equilibrium conditions (Sect. 2). We discuss the main results obtained for gas mass density parameters according to different assumptions and their comparisons to observational HI and H_2 determinations (Sect. 3). Then, we draw our conclusions (Sect. 4). We adopt a flat ΛCDM cosmological model with present-day expansion parameter normalised to 100 km/s/Mpc of $h = 0.7$, and baryon, matter and cosmological-constant density parameters of $\Omega_{0,b} = 0.045$, $\Omega_{0,m} = 0.27$ and $\Omega_{0,\Lambda} = 0.73$, respectively. The $z = 0$ cosmological critical density is $\rho_{0,crit} \simeq 277.4\, h^2\, M_\odot/kpc^3$.

2 Method

Numerical treatment for cosmic gas and structure formation is implemented in an updated version of the P-Gadget3 code (based on [1]) that solves the equations for gravity and smoothed particles hydrodynamics. Non-equilibrium number densities are followed for e^-, H, H^+, H^-, He, He^+, He^{++}, H_2, H_2^+, D, D^+, HD, HeH^+ by solving first-order differential equations that provide the temporal variations of each species, according to the relevant creation and destruction processes [2–5]. Besides atomic ionizations and recombinations, H_2 formation is followed according to the H^- and H_2^+ channels, as well as three-body processes. To account for density-dependent HI self-shielding we fit available tabulated data [6], while H_2 self-shielding is based on [7], but we also check that more recent shielding formulations lead to similar results. The role of different UV backgrounds at different epochs is assessed employing photoheating and photoionisation rates suggested by [8] (HM), [9] (P19) and [10] (FG20). These implementations are needed to follow molecule evolution during cosmic gas cooling and star formation in pristine regimes. Star formation in dense shielded regions, stellar feedback and heavy-element production (C, N, O, Ne, Mg, Si, S, Ca, Fe, etc.) from stars with different masses and metallicities during SNII, AGB and SNIa phases is accounted for, as well [3, 11]. To include H_2 evolution in enriched environments, we additionally consider chemical rates and energy transfer of dust grain catalysis and photoelectric heating [12–15]. Dust grain temperature, $T_{gr}(z)$, is estimated assuming a power-law emission at each z [16], although adopting either a constant value in the range T_{gr} = 40-120 K or z-dependent values equal to the CMB temperature – i.e., $T_{gr}(z) = T_{CMB}(z)$ – does not affect significantly the final results. Dust chemistry and cooling/heating rates scale linearly with gas metallicity, as expected from local-Universe studies. Cosmic-ray heating in star forming regions is modelled following [17] and scaling the transferred energy by the local star formation rate. By means of such modelings, we run cosmological boxes of 10 Mpc/h side length in which the initial gas and dark-matter density fields are sampled by 512^3 particles for each species. Stellar particles are spawned when gas densities reach at least 10 cm^{-3} and the resulting stellar evolution is based on a reference Salpeter initial mass function (IMF), although we

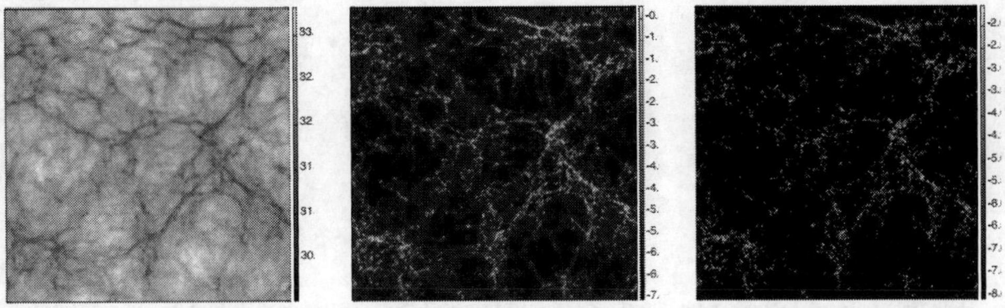

Figure 1. Simulation maps in $\log_{10}$ scale of the gas entropy in cgs units (left), HI (center) and H_2 (right) fractions, as derived by projecting a thin slice of 1/10th the box length along the third direction.

checked that adopting an alternative Chabrier IMF results are affected by less than a factor of 2. SN feedback takes place with a reference efficiency of 0.1, while wind feedback is modelled by a constant wind velocity of 350 km/s. Spreading of heavy elements is mimicked by smoothing the above-mentioned individual metallicities over the hydrodynamic kernel. Let us note that for a detailed picture one should be able to resolve both rare structures on large scales and small objects on tiny scales. Because of numerical feasibility, a fair trade-off to assess non-equilibrium chemistry, halo statistics and luminosity functions is often achieved by employing boxes of the kind considered here. Larger boxes could be used to track better massive rare structures, but this would come at detriment of gas chemistry features, as also demonstrated in the literature. Reducing the box size to obtain better resolution would be a viable option to follow more precisely atomic and molecular species, but, besides statistical issues, this would imply very long computing times. With our set-up we can already reach H_2 mass fractions of $\sim 50\%$ in dense collapsing regions, thus any issue related to box size or resolution would impact our results by less than a factor of 2. This is better than the uncertainties on particular chemical rates, IMF and physical processes at low metallicities (e.g. gas grain processes). Unknowns in stellar evolution, such as exact yield determinations, stellar rotation, impacts of the dredge-ups, AGB mass loss, etc. can have some effects. Here, we follow all the relevant stellar phases consistently with mass-dependent yields and lifetimes, hence our findings should be robust.

3 Results

A pictorial view of the simulations performed can be seen in figure 1, where maps of gas entropy, HI and H_2 fractions are shown at $z = 4.9$, after the HM UV background sets in. It is easy to recognise the filamentary shielded structures in the cosmic web as the *loci* where HI gas is found and H_2 molecules form. The knots represent cold collapsing regions where gas is driven by non-equilibrium cooling and star formation takes place.
From a quantitative point of view, given HI and H_2 masses, cold gas is often quantified in terms of the comoving neutral and H_2 mass density parameters, $\Omega_{\rm neutral}$ and $\Omega_{\rm H_2}$, defined as the corresponding mass densities divided by $\rho_{0,crit}$. Observational determinations (see e.g. [18] and references therein) suggest that neutral gas at $z \lesssim 5$ features values around $\Omega_{\rm neutral} \sim 10^{-3}$, increasing smoothly towards the epoch of reionization. On the contrary, $\Omega_{\rm H_2}$ observations in the (sub-)mm and IR [18–20] give a more complex picture of molecular gas, with peak values of $\Omega_{\rm H_2} \sim 10^{-4}$ at $z \sim 2-4$ and a drop of at least one dex at earlier times. In this respect, impressive progresses have been made by recent ALMA data in constraining H_2 gas up to $z \sim 7$ and in raising previously derived indirect lower limits at $z \simeq$ 2-3 [21].

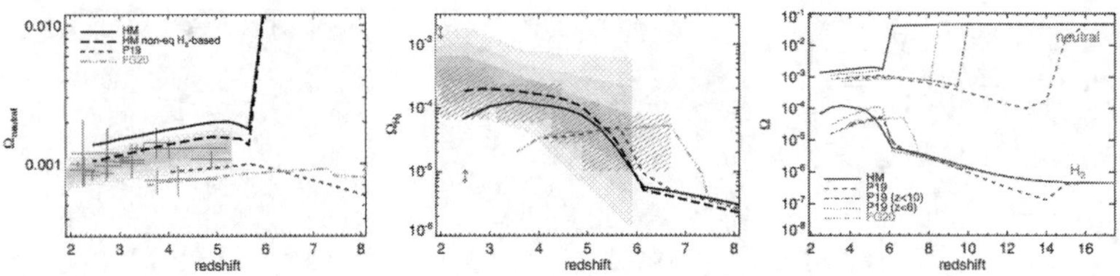

Figure 2. $\Omega_{neutral}$ (left) and Ω_{H_2} (center) redshift evolution from simulations including HM (black solid lines), P19 (blue short-dashed lines) and FG20 (orange dot-dot-dot-dashed lines) UV backgrounds. For the HM case, results from the simulation run including non-equilibrium H_2-density dependent star formation (non-equilibrium H_2-based star formation) are shown, as well (black long-dashed lines). Data points and shaded areas correspond to recent observational determinations [18–20, 22], while $z < 3$ lower and upper limits for Ω_{H_2} are the estimates by [21] and [23] , respectively. On the right, both mass density parameters are shown on a larger redshift range. Two additional cases adopting non-null P19 UV rates at $z < 10$ (blue dot-dashed lines) and $z < 6$ (blue dotted lines), respectively, are also displayed.

Figure 2 displays the trends expected for $\Omega_{neutral}$ and Ω_{H_2} from simulations including HM, P19 and FG20 UV backgrounds. They all include HI and H_2 self-shielding, the various H_2 formation channels, as well as density-dependent star formation and feedback effects. In the HM case, we also check the implications of H_2-density dependent star formation based on the non-equilibrium species abundance calculations (non-equilibrium H_2-based star formation). In the left and central panels, model predictions are compared to our collection of recent HI and H_2 observational determinations. From the numerical results, it is clear that the various UV background models are roughly consistent with $\Omega_{neutral}$ observational data. The HM non-equilibrium H_2-based case leads to slightly lower $\Omega_{neutral}$ values, in better agreement with observations. This is a consequence of the smaller star formation feedback and the consequently larger amounts of neutral H mass converted into H_2. The behaviour of the Ω_{H_2} evolution is more strongly affected by modelling assumptions, although it is evident that P19 and FG20 backgrounds are in tension with the trend inferred by observations. The HM non-equilibrium H_2-based case features Ω_{H_2} values that are larger by up to a factor of a few with respect to the HM scenario. Consistently with what just said about $\Omega_{neutral}$, this is a result of the less efficient star formation feedback in destroying molecules. We warn the reader that all the several physical and chemical processes involved here play a role in shaping HI and H_2 gas. UV radiation and local chemical heating or cooling cumulate to provide the desired neutral and ionised H fractions. Molecule formation paths are usually led by the H^- channel in pristine gas and by dust grain catalysis in enriched one. Gas self-shielding is crucial to obtain reasonable abundances at all z, instead, the individual effects of photoelectric or cosmic-ray heating induce only small variations on mass density parameters. These channels, that are effective in quite heterogeneous conditions, can explain the observationally inferred lack of environmental dependence in molecular-mass build-up at $z \lesssim 3.5$ [24] and the large molecular fractions of 50% or higher, recently inferred at $z \sim$ 4-6 [25, 26], can be easily justified by local H_2 formation both in pristine and in enriched cold material. The depletion times implied by the H_2 gas mass densities result of the order of 1/10th the Hubble time at all z. This is a very interesting piece of information, since it suggests that cosmic gas can collapse and structures can form even in the early epoch of reionization. We note that the sharp increase for both HM models at $z > 6$ is due to the UV rates proposed by [8]. Since they are equal to zero at those early cosmic epochs, early cosmic gas is not photoheated above 10^4 K and remains mostly neutral with $\Omega_{neutral}$ reaching the baryon density parameter at $z \simeq 6$, as neatly visible in the right panel. In the P19 and FG20 cases the UV background sets on at $z \simeq 15$ and $z \simeq 8$, respectively, and, indeed, an analogous convergence is found at those redshifts. Of course, the effects of the UV background onset depend on the strength of the rates at different z. This is confirmed by the two alternative P19 scenarios displayed, where the UV background is assumed to set on at $z = 10$ or $z = 6$ with null rates at higher z. As said, H_2 evolution rely on the available charges in the cosmic medium. At high z, molecular fractions are always low, close to typical early cosmic values of $\sim 10^{-6}$. While cosmic evolution progresses, the injection of UV radiation can either enhance Ω_{H_2} through ionization of neutral particles (as is the case for HM at $z < 6$) or inhibit it when UV rates are too large (as it happens for P19 and FG20 at $z < 8$, in which regime, despite partially boosted, Ω_{H_2} hardly reaches values of 10^{-4} due to dominant H_2 destruction). In fact, by a direct comparison of the different UV rates, one can see that HM photoionization rates are much smaller than P19 ones at $z \lesssim 6$ and FG20 ones at $z \lesssim 8$. The typically larger P19 and FG20 UV rates explain why corresponding $\Omega_{\rm neutral}$ values, albeit compatible with observational limits, are slightly lower than HM ones at $z < 6$. The high-z tail of the P19 model is additionally responsible for possible H ionization and H_2 destruction at $z \gtrsim 10$.

4 Conclusions

We have quantified the evolution of cold atomic and molecular cosmic gas and interpreted state-of-the-art observations in the (sub-)mm and IR ranges by means of three-dimensional cosmological N-body hydrodynamic chemistry simulations following: time-dependent non-equilibrium abundance calculations, the impact of different UV backgrounds, HI and H_2 self-shielding, dust grain catalysis, photoelectric effect, cosmic-ray heating, star formation, stellar feedback and production of heavy-elements from stars with different masses (SNII, AGB, SNIa) and metallicities. This is among the first studies addressing cosmic HI and H_2 evolution by three-dimensional cosmological hydrodynamic simulations that include non-equilibrium chemistry networks and such levels of detail in the implementation of cosmic-gas processes. We find that:

– the evolution of cold cosmic gas ($\Omega_{\rm neutral}$ and Ω_{H_2}) obtained with basic time-dependent non-equilibrium chemistry is broadly consistent with the latest observational determinations;

– details about the UV background are modest for $\Omega_{\rm neutral}$ (within the z range where observations are available), but can be relevant for Ω_{H_2}, in particular early ($z \gtrsim 8$) UV onsets could require either lower UV rates or larger gas self-shielding;

– a non-equilibrium H_2-based star formation prescription coupled to the UV background leads to results that are in slightly better agreement with HI and H_2 determinations;

– H_2 evolution is globally driven by the H^- channel, with a relevant contribution from H_2 grain catalysis at high metallicities and on local scales: these essential paths can explain the observationally inferred lack of environmental dependence in molecular-mass build-up at $z \lesssim 3.5$ and the large H_2 fractions detected at $z \simeq$ 4-6;

– a precise determination of dust grain temperature has little relevance and is generally sub-dominant to metallicity effects;

– cosmic-ray heating in star forming regions, quantified in accordance to different models for the cosmic-ray ionization rate, affects the H_2 abundance at ten-per-cent level, globally;

– wind feedback impacts results mostly at lower redshift, while SN explosions affects early star forming regimes, with effects from IMF variations remaining usually small at all times;

– despite the insurgence of UV radiation, H_2 depletion times reach values lower than the Hubble time within the first half Gyr.

These results stress the relevance of non-equilibrium chemistry treatments for reproducing the cosmic abundances of atomic and molecular species. They also suggest the need to perform dedicated numerical implementations to understand in depth the physics of cosmic gas and to improve upon the works largely available in the literature. We finish by noting that the ability to form significant amounts of H_2 and to reach short depletion times during the epoch of reionization means that future discoveries of molecular-rich star forming galaxies at early times will be possible in the next years. Therefore, new data from upcoming international facilities will be decisive to shed light on the still unanswered questions about the origin of chemical species and the whole cosmic baryon cycle.

References

[1] V. Springel, MNRAS **364**, 1105 (2005), 0505010
[2] U. Maio, K. Dolag, B. Ciardi, L. Tornatore, MNRAS **379**, 963 (2007), 0704.2182
[3] U. Maio, B. Ciardi, K. Dolag, L. Tornatore, S. Khochfar, MNRAS **407**, 1003 (2010), 1003.4992
[4] U. Maio, S. Khochfar, J.L. Johnson, B. Ciardi, MNRAS **414**, 1145 (2011), 1011.3999
[5] U. Maio, E. Tescari, MNRAS **453**, 3798 (2015), 1509.03637
[6] A. Rahmati, A.H. Pawlik, M. Raičević, J. Schaye, MNRAS **430**, 2427 (2013), 1210.7808
[7] B.T. Draine, F. Bertoldi, ApJ **468**, 269 (1996), astro-ph/9603032
[8] F. Haardt, P. Madau, ApJ **461**, 20 (1996), astro-ph/9509093
[9] E. Puchwein, F. Haardt, M.G. Haehnelt, P. Madau, MNRAS **485**, 47 (2019), 1801.04931
[10] C.A. Faucher-Giguère, MNRAS **493**, 1614 (2020), 1903.08657
[11] L. Tornatore, S. Borgani, K. Dolag, F. Matteucci, MNRAS **382**, 1050 (2007), 0705.1921
[12] D. Hollenbach, C.F. McKee, ApJS **41**, 555 (1979)
[13] E.L.O. Bakes, A.G.G.M. Tielens, ApJ **427**, 822 (1994)
[14] S. Cazaux, A.G.G.M. Tielens, ApJ **604**, 222 (2004)
[15] K. Omukai, T. Tsuribe, R. Schneider, A. Ferrara, ApJ **626**, 627 (2005)
[16] B.T. Draine, H.M. Lee, ApJ **285**, 89 (1984)
[17] M. Padovani, A.V. Ivlev, D. Galli, P. Caselli, A&A **614**, A111 (2018), 1803.09348
[18] C. Péroux, J.C. Howk, ARA&A **58**, 363 (2020), 2011.01935
[19] D.A. Riechers, L.A. Boogaard, R. Decarli, J. González-López, I. Smail, F. Walter, M. Aravena, C.L. Carilli, P.C. Cortes, P. Cox et al., ApJ **896**, L21 (2020), 2005.09653
[20] T.K. Garratt, K.E.K. Coppin, J.E. Geach, O. Almaini, W.G. Hartley, D.T. Maltby, C.J. Simpson, A. Wilkinson, C.J. Conselice, M. Franco et al., ApJ **912**, 62 (2021), 2103.08613
[21] S. Berta, D. Lutz, R. Nordon, R. Genzel, B. Magnelli, P. Popesso, D. Rosario, A. Saintonge, S. Wuyts, L.J. Tacconi, A&A **555**, L8 (2013), 1304.7771
[22] R. Decarli, M. Aravena, L. Boogaard, C. Carilli, J. González-López, F. Walter, P.C. Cortes, P. Cox, E. da Cunha, E. Daddi et al., ApJ **902**, 110 (2020), 2009.10744
[23] A. Klitsch, C. Péroux, M.A. Zwaan, I. Smail, D. Nelson, G. Popping, C.C. Chen, B. Diemer, R.J. Ivison, J.R. Allison et al., MNRAS **490**, 1220 (2019), 1909.08624
[24] B. Darvish, N.Z. Scoville, C. Martin, B. Mobasher, T. Diaz-Santos, L. Shen, ApJ **860**, 111 (2018), 1805.10291
[25] L.J. Tacconi, R. Genzel, A. Sternberg, ARA&A **58**, 157 (2020), 2003.06245
[26] M. Dessauges-Zavadsky, M. Ginolfi, F. Pozzi, M. Béthermin, O. Le Fèvre, S. Fujimoto, J.D. Silverman, G.C. Jones, L. Vallini, D. Schaerer et al., A&A **643**, A5 (2020), 2004.10771

Sunyaev Zel'dovich effect in galaxy clusters cavities: Thermal or non-thermal origin?

Paolo Marchegiani[1,2,*]

[1]Dipartimento di Fisica, Università Sapienza, P.le Aldo Moro 2, 00185, Roma, Italy
[2]Wits Centre for Astrophysics, School of Physics, University of the Witwatersrand, 1 Jan Smuts Avenue, Johannesburg 2050, South Africa

Abstract. Several galaxy clusters host X-ray cavities, often filled with relativistic electrons emitting in the radio band. In the cluster MS 0735.6+7421 the cavities have been detected through the Sunyaev Zel'dovich (SZ) effect, but it has not been possible to determine if this effect is thermal (produced by a very high temperature gas filling the cavity) or non-thermal (produced by the relativistic electrons that produce the diffuse radio emission detected in the cavity). In this paper we discuss the role of the density of the high temperature gas inside the cavities in determining whether the dominant SZ effect is the thermal or the non-thermal one, and how it can be possible to distinguish between the two possibilities, discussing the role of observations at higher energy bands.

1 Introduction

Several relaxed galaxy clusters show the presence inside their Intra Cluster Medium (ICM) of couples of cavities in the X-ray emission, usually located in opposite directions compared to the central Brightest Cluster Galaxy [1, 2]. The origin of these cavities is probably due to the mechanical power produced by AGN jets that is converted to heat during the expansion of the lobe in the ICM [3]. Many X-ray cavities host a diffuse radio emission [4], showing that relativistic electrons and magnetic fields are present in them. However, it has been found that the energy stored in these relativistic electrons, determined under the minimum energy assumption, is well below the energy necessary to inflate the cavity [5], indicating that another component inside the cavity should be dominant in energy. This dominant component can be a high temperature thermal gas with a temperature of the order of $kT \sim 100$ keV or higher according to hydrodynamic simulation [6, 7], or a population of relativistic protons.

While radio observations are suitable to constrain the high energy part of the relativistic electrons spectrum and the magnetic field, they are not able to provide information on the low energy part of the electrons spectrum, and on the thermal gas inside the cavity. The Sunyaev-Zel'dovich effect (SZE) produced inside the cavity, i.e. the distortion of the Cosmic Microwave Background (CMB) spectrum by inverse Compton scattering (ICS) [8], has instead been proposed as a probe to constrain the low energy electrons or the thermal gas properties [9–11]. With this kind of information it is possible to constrain the energetic content of the different components inside the cavity, because the low energy part of the electrons spectrum is the dominant one in determining the total energy stored in them [12].

*e-mail: paolo.marchegiani@uniroma1.it

The galaxy cluster MS 0735.6+7421 hosts a couple of X-ray cavities [13–15], where a diffuse radio emission filling the cavities is also observed [4, 16]. These cavities have also been observed through the SZE at 30 GHz using the Combined Array for Research in Millimeter-wave Astronomy (CARMA) interferometer [17], which detected a clear deficit in the SZ signal in the direction of the cavities. From these observations, it has not been possible to determine if the dominant component inside the cavities is thermal or non-thermal, because the data were compatible with both a non-thermal electrons population having a low value of the minimum moment ($p_1 \sim 1 - 10$), and with a thermal population with very high temperature of $kT \sim 1000$ keV, or even higher (see fig.8 in [17]).

In a recent paper [18] we discussed this problem, pointing up that these two different populations, i.e. the non-thermal electrons and the high temperature gas, are linked each other because of the effect of the Coulomb interactions. In particular, a higher density of the thermal gas implies a higher rate of the Coulomb losses, which have the effect of reducing the density of the low energy non-thermal electrons and therefore of flattening their spectrum, and as a consequence of reducing the intensity of the non-thermal SZE. Therefore, this effect should be taken into account when interpreting the results of observations of the SZE inside the cluster cavities in terms of thermal or non-thermal populations.

In this paper, we summarize the results obtained in [18], and expand further the discussion about how it can be possible to discriminate between the thermal and the non-thermal origin of the SZE inside the cavities, also considering the expected SZE in spectral bands with higher energies than the microwave one.

2 Modeling the spectrum of non-thermal electrons

The time evolution of the non-thermal electrons spectrum $N_e(p,t)$, written as a function of the time t and the normalized momentum $p = \beta\gamma$, is given the following equation [19]:

$$\frac{\partial N_e(p,t)}{\partial t} = \frac{\partial}{\partial p}\left[\left(-\frac{2}{p}D_{pp} + \sum_i b_i(p)\right)N_e(p,t) + D_{pp}\frac{\partial N_e(p,t)}{\partial p}\right], \qquad (1)$$

where D_{pp} is the diffusion term in the momentum space, associated to Fermi-II acceleration processes, which we model as $D_{pp} = \chi p^2/4$ so that the characteristic acceleration time is $\tau_{acc} = p^2/4D_{pp} = \chi^{-1}$ [20]. The energy loss term $b_i(p)$ is given by the sum of the Coulomb losses, $b_{Coul} \propto n_{th}$ [21], where n_{th} is the density of the thermal gas, of the radiative losses (synchrotron and inverse Compton scattering), $b_{rad} \propto (B^2 + B^2_{ICS})p^2$ [22], and of the adiabatic expansion losses, $b_{ad} \propto (dV/dt)p$, where we assume that the volume is changing with a Sedov-like expansion $V(t) \propto (t/t_0)^{6/5}$ [23].

We assume an initial electrons spectrum with a power law shape, $N_e(p,t_0) = k_0 p^{-s}$ with $s = 2.7$, and initial values of the magnetic field and the thermal density so that at the present time they are equal respectively to the equipartion value $B = 4.7\ \mu$G [16] and to a reference value chosen in the range $10^{-6} - 10^{-3}$ cm^{-3}. We therefore calculate the electrons spectrum after a time of the order of the age of the bubble, estimated to be $\sim$ 160 Myr from X-ray observations [15] and MHD simulations [24], and choose the electrons normalization k_0 and the Fermi-II acceleration parameter χ in order to match the observed spectrum of the radio emission, with the best parameters values resulting to be $k_0 = 1.5 \times 10^{-3}$ cm^{-3} and $\chi = 3.5 \times 10^{-16}$ s^{-1} [18].

In Fig.1, left panel, the resulting radio spectrum, compared with the observed data [4, 16], is shown for four values of the thermal density inside the cavity, while the electrons spectra for the same values of the thermal density are shown in the right panel. It is interesting to see that, while the value of the thermal density does not impact too much on the shape of the radio

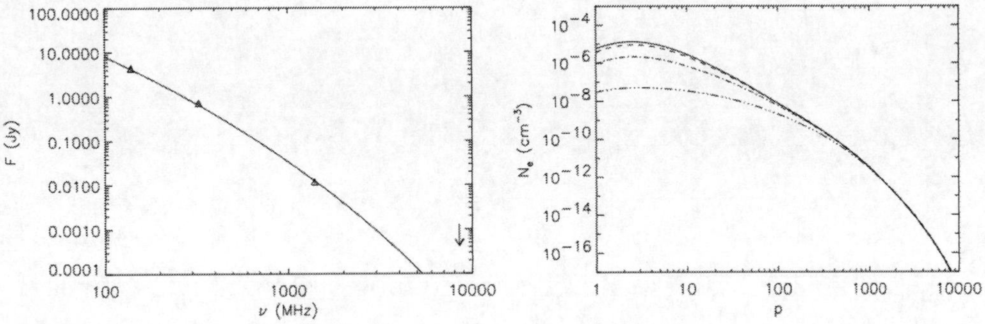

Figure 1. Radio spectrum after 160 Myr from the injection (left panel) and the corresponding non-thermal electrons spectra (right panel) for four values of the thermal density: 10^{-6} (solid line), 10^{-5} (dashed line), 10^{-4} (dot-dashed line), and 10^{-3} (three dots-dashed line) cm^{-3}. Figures are from [18].

spectrum (in fact the relative differences between different models are smaller than 10% in the whole frequency range between 100 MHz and 10 GHz), it heavily affects the spectrum of the electrons at low energies. As shown in the next Section, this fact has strong consequences on the SZE produced by these electrons.

3 The non-thermal Sunyaev-Zel'dovich effect

Using the electrons spectra calculated in the previous Section, it is possible to calculate the non-thermal SZE, using the full relativistic formalism [25, 26], where the intensity of the SZE is given by $\Delta I(x) = \tau[J_1(x) - I_0(x)]$. Here $x = h\nu/(k_B T_0)$ is the frequency normalized at the CMB temperature T_0, $I_0(x)$ is the unperturbed CMB spectrum, τ is the electrons optical depth, proportional to the electrons density integrated along the line of sight, and $J_1(x)$ is calculated by convolving the CMB spectrum with the single scattering redistribution function, which is given by the convolution of the electrons spectrum with the Compton scattering redistribution function for an electron with a given energy [27].

The resulting SZE spectra for the four values of the thermal density that were used in the previous Section are shown in the first panel of Fig.2. As it is possible to see, smaller values of the thermal density imply reduced Coulomb losses, and therefore a higher density of the non-thermal electrons at low energies and as a consequence a more intense SZE.

The SZE produced by the non-thermal electrons inside the cavity can be compared with the thermal SZE produced by the high temperature gas inside the same cavity. Since we don't have observational information about the properties of this gas, we can assume several values for its temperature, and assume for its density the values for which its pressure is equal to the one of the external ICM at the location of the cavity, $P_{cav} = 6 \times 10^{-11}$ erg cm^{-3} [14], i.e. $n_{th} = P_{cav}/(kT)$. Using the full relativistic formalism, we calculate the thermal SZE for these temperatures and the optical depths derived from these values of the density, and the non-thermal SZE for the Coulomb losses rates provided by the same values of the thermal density. In the last three panels of Fig.2 the results for three values of the temperature are shown.

As it is possible to see, values of the temperature until $\sim$ 1500 keV allow to have quite high values of the thermal density, and this fact increases the intensity of the thermal SZE and at the same time decreases the one of the non-thermal SZE because of the Coulomb losses.

https://doi.org/10.1051/epjconf/202225700030

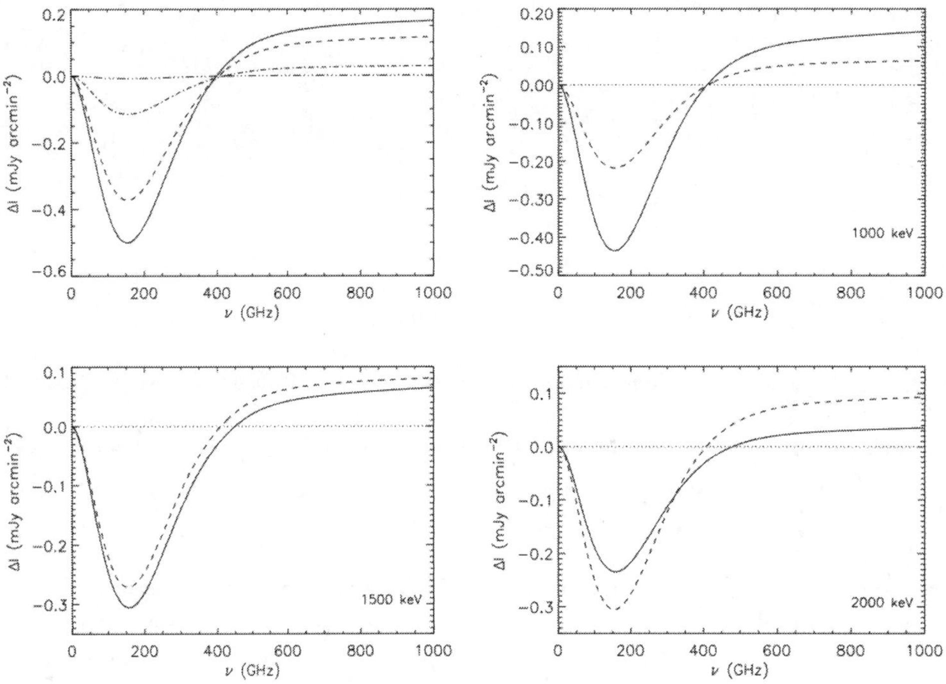

Figure 2. First panel: non-thermal SZE produced by the electrons with the spectra shown in Fig.1, with the same meaning of the line styles. Other panels: thermal SZE inside the cavities (solid lines) compared with the non-thermal SZE calculated for the corresponding values of the thermal density (dashed lines) for values of the temperature of 1000, 1500, and 2000 keV. The dotted line in last three panels indicates the zero level. Figures are from [18].

For higher temperatures, we can expect that the non-thermal SZE should be dominant, while for lower temperatures the thermal effect can be dominant, even if it is possible that the thermal density is lower than the value assumed here, so it is not possible to state whether in this case the dominant effect is definitely the thermal one.

4 Thermal or non-thermal SZE?

In this Section we discuss how it can be possible to determine if the dominant SZE inside the cavity is the thermal or the non-thermal one, adding some other considerations as compared to the previous paper [18], to which we refer instead for other details on the models, on the properties of the electrons populations used in the calculations, on the comparison between the SZE inside the cavity and the one of the surrounding ICM, and on the observational challenges to be faced in order to derive this kind of information.

As discussed in the previous paper, the angular size of the cavities in MS 0735.6+7421 is of the order of 1 arcmin; therefore instruments with angular resolution of the order of 10 arcsec like NIKA2 are well suited to resolve the SZ signal inside the cavities, and to distinguish it from the one coming from the rest of cluster. While NIKA2 can provide information at the frequencies of 150 and 260 GHz, a wider spectral coverage could be useful to better distinguish between a thermal and a non-thermal origin of the SZE; this can be done with instruments operating at 90 GHz, like Mustang-2 at the Green Bank Telescope (GBT) and

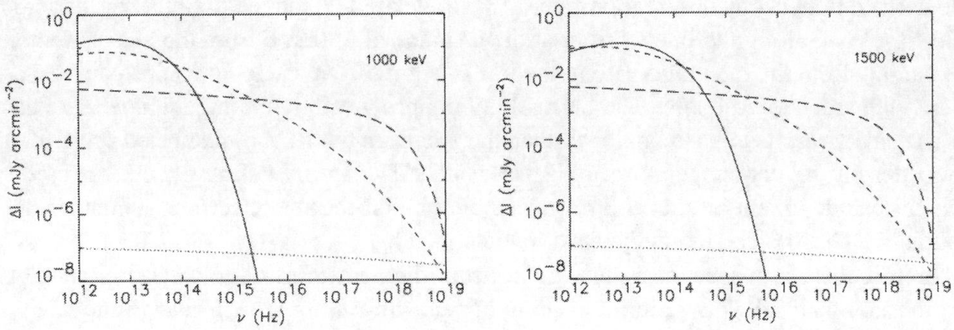

Figure 3. SZE spectra at high frequencies for the thermal gas inside the cavity (solid lines) with temperatures of 1000 (left panel) and 1500 keV (right panel), compared with the SZE spectra of the non-thermal electrons (dashed lines), and with the bremsstrahlung of the high temperature gas inside the cavity (dotted line) and of the external ICM along the line of sight of the cavity center (long dashed lines).

Mistral at the Sardinia Radio Telescope (SRT), and at higher frequencies, like the planned space telescope Millimetron [28], with its planned angular resolution of a few arcsec and a spectral coverage until a few THz.

However, also a full spectral frequency coverage of the SZE inside the cavity might not be sufficient to determine if its origin is thermal or non-thermal. In fact, in Fig.2 we can see that for temperatures in the range 1000 – 1500 keV the thermal and the non-thermal SZE are expected to have very similar spectral shapes (note for example that the crossover frequency in the top right panel of Fig.2 is basically the same for the two effects).

A possible solution might be to look at the SZE emission at frequencies higher than the THz, where it should reach its maximum, depending on the properties of the electrons population. For example, it has been suggested that the Hard X-ray band can be good to distinguish between the ICS emission of the non-thermal electrons and the bremsstrahlung of the high temperature thermal gas, because the efficiency of the ICS process is much higher than the bremsstrahlung one [7]. In order to check this possibility, in Fig.3 we compare the thermal and non-thermal SZE spectra calculated at high frequencies, from the THz until 10^{19} Hz (corresponding to a photon energy of $\sim$ 40 keV), for values of the gas temperature inside the cavity of 1000 and 1500 keV.

As it is possible to see, the spectra of the thermal ad the non-thermal SZE are quite similar until frequencies of $\nu \sim 10^{14}$ Hz, while they start to diverge at higher frequencies, with the thermal SZE falling down quite abruptly, and the non-thermal SZE continuing instead with a power-law shape until the Hard X-ray band, with a progressive steepening due to the corresponding high energy steepening of the electrons spectrum, visible in the right panel of Fig.1. Therefore, a hypothetical detection of the SZE inside the cavity at frequencies $> 10^{14}$ Hz, i.e. from the near infrared band to higher frequencies, should be attributed to the non-thermal electrons rather than to the high temperature gas. Such an effect should appear as a positive diffuse emission spatially coincident with the cavity.

In Fig.3 we also show an estimate of the surface brightness in the direction of the cavity center due to the thermal bremsstrahlung of the high temperature gas inside the cavity and of the external ICM present along the line of sight. This last quantity has been calculated assuming the presence of a gas with constant temperature of 5.5 keV and with the double beta radial profile derived from X-ray observations (see [18] for details), extended until the virial radius of 2.23 Mpc [14], along the line of sight at the distance of 150 kpc from the cluster center, and excluding from the integration along the line of sight the cavity region, approximated as a sphere with radius of 100 kpc. The bremsstrahlung produced by the high temperature gas inside the cavity results to be much lower than the one from the ICM, as expected because of its lower density, higher temperature, and smaller size. The bremsstrahlung produced by the ICM is instead dominant on the SZE produced by the non-thermal electrons at frequencies higher than $\sim 2 \times 10^{15}$ Hz, i.e. in the far ultraviolet band.

Therefore there is a spectral region, going from the near infrared to the far ultraviolet, where the non-thermal SZE should be the dominant diffuse component inside the cavity. Should a diffuse emission spatially coincident with the cavity in these spectral bands be detected, it should therefore be attributed to the non-thermal SZE. In particular, the most favorable spectral region results to be at a wavelength around 400 nm, i.e. between the near ultraviolet and the visible spectral regions. Instruments that can be suitable to detect such an emission might be UVOT aboard the *Swift* satellite, or the WFC3 camera aboard the Hubble Space Telescope; in order to establish the effective possibility to detect this signal, a dedicated study would be required.

References

[1] Fabian A.C., 2012, ARA&A, 50, 455
[2] Gitti M., Brighenti F., McNamara B.R. 2012, AdAst, 2012, 950641
[3] McNamara B. R., Nulsen, P. E. J., 2007, ARA&A, 45, 117
[4] Birzan L. et al., 2020, MNRAS, 496, 2613
[5] Ito H., Kino M., Kawakatu N., et al., 2008, ApJ, 685, 828
[6] Sternberg A., Soker N., 2009, MNRAS, 398, 422
[7] Prokhorov D.A., Moraghan A., Antonuccio-Delogu V., Silk J., 2012, MNRAS, 425, 1753
[8] Sunyaev R.A., Zel'dovich Ya.B., 1972, CoASP, 4, 173
[9] Colafrancesco S., 2005, A&A, 435, L9
[10] Pfrommer C., Enßlin, T.A., Sarazin C.L., 2005, A&A, 430, 799
[11] Prokhorov D.A., Antonuccio-Delogu V., Silk J., 2010, A&A, 520, A106
[12] Colafrancesco S., Marchegiani P., 2011, A&A, 535, A108
[13] McNamara B.R., Nulsen P.E.J., Wise M.W., et al., 2005, Nature, 433, 45
[14] Gitti M., McNamara B.R., Nulsen P.E.J., Wise M.W., 2007, ApJ, 660, 1118
[15] Vantyghem A.N., McNamara B.R., Russell H.R., et al., 2014, MNRAS, 442, 3192
[16] Birzan L., McNamara B.R., Nulsen P.E.J., et al., 2008, ApJ, 686, 859
[17] Abdulla Z. et al., 2019, ApJ, 871, 195
[18] Marchegiani P., 2021, MNRAS, 503, 4183
[19] Schlickeiser R., 2002, *Cosmic Ray Astrophysics*, Springer-Verlag, Berlin
[20] Brunetti G., Lazarian A., 2007, MNRAS, 378, 245
[21] Gould R.J., 1972, Physica, 60, 145
[22] Winner G., Pfrommer C., Girichidis P., Pakmor R., 2019, MNRAS, 488, 2235
[23] Enßlin T.A., Gopal-Krishna, 2001, A&A, 366, 26
[24] Ehlert K., Pfrommer C., Weinberger R., Pakmor R., Springel V., 2019, ApJ, 872, L8
[25] Wright E.L., 1979, ApJ, 232, 348
[26] Colafrancesco S., Marchegiani P., Palladino E., 2003, A&A, 397, 27
[27] Enßlin T.A., Kaiser C.R., 2000, A&A, 360, 417
[28] Kardashev N.S. et al., 2014, Physics-Uspekhi, 57, 12

EPJ Web of Conferences **257**, 00031 (2022) https://doi.org/10.1051/epjconf/202225700031
mm Universe @ NIKA2

Joint constraints on the galaxy cluster pressure profile from *Planck* and SPT-SZ

Jean-Baptiste Melin[1,*], *Gabriel W.* Pratt[1,2], and *Monique* Arnaud[1,2]

[1]IRFU, CEA, Université Paris-Saclay, F-91191 Gif-sur-Yvette, France
[2]Université Paris Diderot, AIM, Sorbonne Paris Cité, CEA, CNRS, F-91191, Gif-sur-Yvette, France

Abstract. We present the average pressure profile of 461 galaxy clusters from the 2500d South Pole Telescope catalogue measured in the *Planck* and SPT-SZ data jointly. The joint measurement is performed in Fourier space for the first time, and takes advantage of the complementarity between the high resolution of SPT (1.75 arcmin beam) and the coverage of large angular scales by *Planck* (> 30 arcmin). The redshift leverage of the catalogue allows the study of the evolution of the profile. We also constrain the profile shape in the outskirts.

1 Introduction

The thermal Sunyaev-Zeldovich (SZ) effect [1, 2] is the inverse Compton scattering of the Cosmic Microwave Background photons by hot electrons in galaxy clusters. It depends on the Compton y-parameter

$$y = \frac{\sigma_{\rm T}}{m_{\rm e}c^2} \int_{los} P_e dl \tag{1}$$

where $\sigma_{\rm T}$ is the Thomson cross-section, $m_{\rm e}$ is the electron rest mass, c is the speed of light light and $P_e = n_e \times T_e$ is the pressure (the product of the electronic density and temperature). The integral is performed along the line-of-sight (los). Observing the SZ effect in galaxy clusters allows one to probe directly their electronic pressure profiles. Structure formation in the Universe being dominated by the gravitational interaction on the largest scales, the pressure profiles are expected to be self-similar: they can be rescaled to a unique profile ('the universal pressure profile') according to the cluster mass and redshift. The pressure profile can provide a wealth of information on structure formation, in particular via its evolution with redshift, or in studying the assembly process in the cluster outskirts.
The South Pole Telescope (SPT) collaboration released recently the data of the SPT-SZ survey [3]. This dataset covers approximately 2,500 deg^2 and is complementary to the *Planck* all-sky dataset in terms of resolution and frequency coverage. The goal of this work is to combine the two data sets and average the profiles of the clusters detected in the SPT-SZ survey to improve the signal-to-noise and radial resolution of the measured average pressure profile. We also study the evolution of the average pressure profile and examine the outskirt region. The datasets are presented in Sect. 2, the analysis method in Sect. 3 and the results in Sect. 4. We conclude in Sect. 5.

*e-mail: jean-baptiste.melin@cea.fr

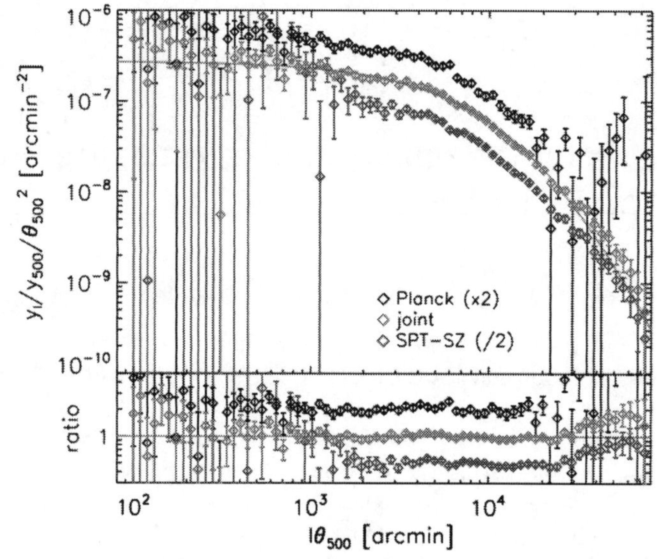

Figure 1. Average pressure profile of the 461 SPT-SZ clusters in harmonic space. The black, red and blue diamonds correspond to the *Planck* only, *Planck*+SPT-SZ and SPT-SZ only analysis respectively. The black (resp. blue) diamonds have been multiplied (resp. divided) by two for clarity. The solid green line is the PIPV profile, which is in very good agreement with the average pressure profile.

2 Datasets

We work with the fiducial SPT-SZ 2500d cluster catalogue published by the SPT collaboration [4], which contains 677 candidates. We keep only the candidates with assigned redshift, and we remove the objects located at the boundary or outside the SPT-SZ public footprint, or having a SPT-SZ flux $Y^{0.75} \geqslant 2 \times 10^{-4}$ arcmin^2, for which the filter transfer function is not accurate [Fig. B.3. of 5]. After this selection, our sample contains 461 clusters. The cluster mass M_{500}[1] is provided with the catalogue; these values were renormalised by a factor of 0.8 to make them compatible with the mass definition adopted by Arnaud et al. [6], and the corresponding radius R_{500} and angular size θ_{500} were then obtained.

We use the six *Planck* public all sky maps of the High Frequency Instrument (100, 143, 217, 353, 545, 857 GHz). We upgraded the pixel size from HEALPix [7] $N_{side} = 2048$ to $N_{side} = 8192$ (0.43 arcmin) and rotated the coordinate system from galactic to equatorial to make the format compatible with the SPT-SZ data [details given in 5]. We also used the three SPT-SZ only public maps at 90, 150, 200 GHz [3].

3 Method

The observed clusters were convolved by the instrumental point spread functions (PSF) of the experiments and also by the filter transfer function (FTF) for SPT. Averaging clusters requires radial rescaling of the profiles by θ_{500}. To perform the cleanest possible PSF and FTF deconvolution across cluster sizes, we performed our analysis in Fourier space. We also use the flat sky approximation to directly map the Fourier (wavenumber k) to the harmonic

[1]mass enclosed in a sphere of radius R_{500} whose mean density is 500 times the critical density at the cluster redshift.

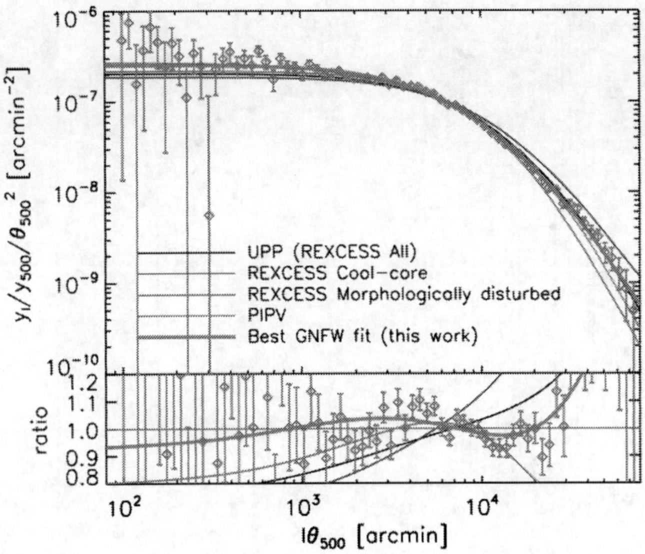

Figure 2. Best fit (thick solid pink line) of the average pressure profile of the SPT-SZ clusters in harmonic space compared to published pressure profiles. The red diamonds are the measurements from the joint analysis (identical to Figure 1). The solid green line is the PIPV profile and the black solid line the UPP. The cool-core (solid blue line) and morphologically disturbed (solid red line) profiles based on subsamples of the UPP are also displayed.

(multipole l) space. Our measurements are thus presented as $y_l/y_{500}/\theta_{500}^2$ vs. $l\theta_{500}$ (harmonic space) instead of the common $y(\theta)/y_{500}$ vs. θ/θ_{500} (real space), where θ is the angular radius. The correspondance between $l\theta_{500}$ and θ/θ_{500} can be established using the relation $l \sim \frac{180}{\theta[\mathrm{deg}]}$, so $l\theta_{500}[\mathrm{arcmin}] \sim 10800\,\frac{\theta_{500}}{\theta}$.

We first cut 10×10 deg tangential maps centered on each of the 461 candidates for the six+three *Planck* and SPT-SZ maps. We applied an internal linear combination (ILC) [8] to produce a Compton y-map for each candidate. The ILC works in Fourier space and allows one to take into account the point spread function (PSF) of each map and the FTF of SPT. We corrected each map for the pixel window function and divided by $\theta_{500}^2 y_{500}$. We then bin averaged in $||\vec{k}||\theta_{500}$ bins using inverse variance weights. This bin averaging step performs simultaneously the azimutal averaging and the spatial rescaling of the profiles. We note that before applying the ILC to the maps, we rescaled each SPT-SZ map by a factor depending on the mass and redshift of the cluster to take into account the Malmquist bias of the clusters in the SPT-SZ maps. The detailed description of the computation of this rescaling factor, together with a full explanation of the method, is given in Melin et al. [9].

4 Results

4.1 Joint *Planck* and SPT-SZ pressure profile

Results for the full sample are presented in Figure 1. The average profile extracted from *Planck* only, SPT-SZ only and *Planck*+SPT-SZ jointly are in very good agreement. They are also in very good agreement with the profile determined from *Planck* and *XMM-Newton* data on 62 nearby massive clusters [PIPV, 10]. *Planck* data provides information at large angular

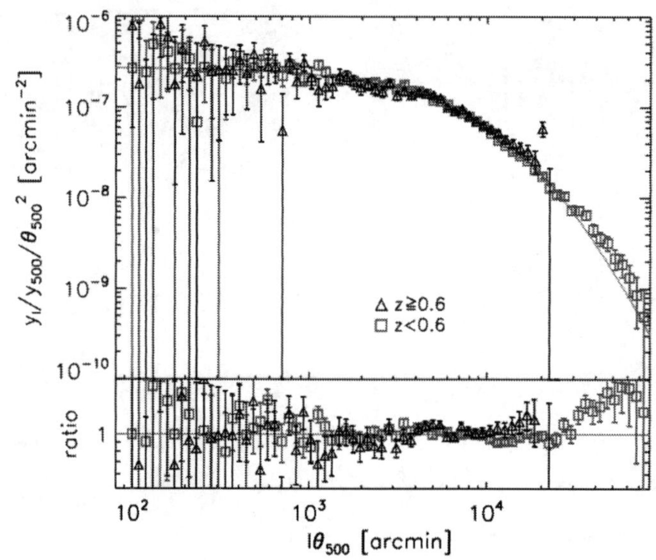

Figure 3. Evolution of the pressure profile of the SPT-SZ clusters. Blue squares (resp. black triangles) show measurements for the low (resp. high) redshift ($z < 0.6$) sample. The solid green line is the PIPV profile as in Figure 1 and 2.

scales (low multipole l) while the SPT-SZ data provides information at small angular scales (high multipoles). In particular, *Planck* outperforms SPT-SZ for scales $l\theta_{500} < 2500$ arcmin and SPT-SZ outperforms *Planck* for $l\theta_{500} > 6500$ arcmin.

In Figure 2, we present the best fit of the joint pressure profile (solid pink line). It is close to the PIPV profile (solid green line) at low $l\theta_{500}$ (outer part of the profile in real space) and close to the Universal Pressure Profile [UPP, 6] (black) at high $l\theta_{500}$ (the inner part of the profile in real space). The differences between our best-fitting profile, the UPP, and the PIPV at high $l\theta_{500}$ (inner part of the profile in real space) are smaller than the difference between the cool-core (solid blue line) and morphologically disturbed (solid red line) profiles.

4.2 Evolution with redshift

We split our sample in two at $z = 0.6$ (close to the median redshift $z = 0.56$) and studied the pressure profile of the two subsamples. Results are shown in Figure 3. The profile of the low redshift sample (blue squares) is in good agreement with the profile of the high redshift sample (black triangles). They are both in agreement with the PIPV profile. We also note that the high redshift sample lacks data points at $l\theta_{500} > 2 \times 10^4$ arcmin due to the limited resolution of the SPT-SZ survey for high redshift ($z \geqslant 0.6$) cluster profile.

4.3 The cluster outskirts

We now study the cluster outskirts. We projected the data points onto a grid and used the inverse fast Fourier transform to obtain a two dimensional Compton y-parameter profile. We obtained the envelope error by Monte Carlo sampling. The two dimensional profile is in good agreement with the PIPV and the best fit profiles. Aung et al. [11], Baxter at al. [12] and Zhang et al. [13] predicted that an accretion shock should be visible in the stacked profiles of relaxed clusters at a characteristic scale of $\theta/\theta_{200\mathrm{m}} \sim 2$ (equivalent to $\theta/\theta_{500} \sim 3$). We do not detect any obvious characteristic scale in this first study but we do not have dynamical state indicators for the cluster sample so we cannot select relaxed clusters. Moreover, we performed the average in θ_{500} and not in $\theta_{200\mathrm{m}}$. It is also possible that the sensitivities of the experiments are not sufficient to measure the effect. We will perform additional tests for the cluster outskirts in Melin et al. [9].

5 Conclusion

We presented the average pressure profile based on a joint analysis of *Planck* and SPT-SZ data for a sample of 461 clusters, a number much larger than previous analyses which includes tens of clusters. For the first time, we proposed to work and constrain the pressure profile in harmonic space, which is better suited for averaging profiles spanning a large range of mass and redshift. The measured profile is close to the pressure profile determined by the *Planck* collaboration on a sample of 62 clusters [10] in the outer regions. It confirms the excess of pressure at large radius ($\theta > \theta_{500}$) with respect to the Universal Pressure Profile [6]. In the inner regions, our measured profile is in very good agreement with the Universal Pressure Profile. Dividing our sample into low and high redshift bins, we do not find a strong evolution. However, we lack some information in the inner part of the profile for the high redshift ($z \geqslant 0.6$) subsample due to the limited resolution of the experiments. This fact reinforces the importance of high resolution ground based SZ experiments such as NIKA2, which complements the missing information of space-based and ground-based SZ surveys. We also studied the outskirts of the average profile. The average two dimensional Compton y-profile features multiple pressure drops at scales $\theta/\theta_{500} > 2$ but we do not detect an obvious characteristic scale due to the accretion shock in this first study. We will perform additional tests in a forthcoming publication [9].

Acknowledgements

We would like to thank the anonymous referee for his/her useful comments and suggestions which helped to improve this proceeding.

References

[1] R.A. Sunyaev, Y.B. Zeldovich, Comments on Astrophysics and Space Physics **2**, 66 (1970)

[2] R.A. Sunyaev, Y.B. Zeldovich, Comments on Astrophysics and Space Physics **4**, 173 (1972)

[3] R. Chown, Y. Omori, K. Aylor, B.A. Benson, L.E. Bleem, J.E. Carlstrom, C.L. Chang, H.M. Cho, T.M. Crawford, A.T. Crites et al., ApJS**239**, 10 (2018), 1803.10682

[4] L.E. Bleem, B. Stalder, T. de Haan, K.A. Aird, S.W. Allen, D.E. Applegate, M.L.N. Ashby, M. Bautz, M. Bayliss, B.A. Benson et al., ApJS**216**, 27 (2015), 1409.0850

[5] J.B. Melin, J.G. Bartlett, P. Tarrío, G.W. Pratt, A&A**647**, A106 (2021), 2009.08822

[6] M. Arnaud, G.W. Pratt, R. Piffaretti, H. Böhringer, J.H. Croston, E. Pointecouteau, A&A**517**, A92 (2010), 0910.1234

[7] K.M. Górski, E. Hivon, A.J. Banday, B.D. Wand elt, F.K. Hansen, M. Reinecke, M. Bartelmann, ApJ**622**, 759 (2005), astro-ph/0409513

[8] M. Remazeilles, J. Delabrouille, J.F. Cardoso, MNRAS**410**, 2481 (2011), 1006.5599

[9] J.B. Melin, G.W. Pratt, M. Arnaud, in prep. (2022)
[10] Planck Collaboration, P.A.R. Ade, N. Aghanim, M. Arnaud, M. Ashdown, F. Atrio-Barandela, J. Aumont, C. Baccigalupi, A. Balbi, A.J. Banday et al., A&A**550**, A131 (2013), 1207.4061
[11] H. Aung, D. Nagai, E.T. Lau, arXiv e-prints arXiv:2012.00977 (2020), 2012.00977
[12] E.J. Baxter, S. Adhikari, J. Vega-Ferrero, W. Cui, C. Chang, B. Jain, A. Knebe, arXiv e-prints arXiv:2101.04179 (2021), 2101.04179
[13] C. Zhang, I. Zhuravleva, A. Kravtsov, E. Churazov, arXiv e-prints arXiv:2103.03850 (2021), 2103.03850

Multi-probe analysis of the galaxy cluster CL J1226.9+3332: Hydrostatic mass and hydrostatic-to-lensing bias

M. Muñoz-Echeverría[1,*], *R.* Adam[2], *P.* Ade[3], *H.* Ajeddig[4], *P.* André[4], *M.* Arnaud[4], *E.* Artis[1], *H.* Aussel[4], *I.* Bartalucci[5], *A.* Beelen[6], *A.* Benoît[7], *S.* Berta[8], *L.* Bing[9], *O.* Bourrion[1], *M.* Calvo[7], *A.* Catalano[1], *M.* De Petris[10], *F.-X.* Désert[11], *S.* Doyle[3], *E. F. C.* Driessen[8], *A.* Ferragamo[10], *A.* Gomez[12], *J.* Goupy[7], *F.* Kéruzoré[1], *C.* Kramer[13], *B.* Ladjelate[13], *G.* Lagache[9], *S.* Leclercq[8], *J.-F.* Lestrade[14], *J.-F.* Macías-Pérez[1], *A.* Maury[4], *P.* Mauskopf[3,20], *F.* Mayet[1], *A.* Monfardini[7], *A.* Paliwal[10], *L.* Perotto[1], *G.* Pisano[3], *E.* Pointecouteau[15], *N.* Ponthieu[11], *G. W.* Pratt[4], *V.* Revéret[4], *A. J.* Rigby[3], *A.* Ritacco[16,6], *C.* Romero[17], *H.* Roussel[18], *F.* Ruppin[19], *K.* Schuster[8], *S.* Shu[21], *A.* Sievers[13], *C.* Tucker[3], and *G.* Yepes[22]

[1]Univ. Grenoble Alpes, CNRS, LPSC-IN2P3, 53, avenue des Martyrs, 38000 Grenoble, France
[2]LLR, CNRS, École Polytechnique, Institut Polytechnique de Paris, Palaiseau, France
[3]School of Physics and Astronomy, Cardiff University, Queen's Buildings, The Parade, Cardiff, CF24 3AA, UK
[4]AIM, CEA, CNRS, Université Paris-Saclay, Université Paris Diderot, Sorbonne Paris Cité, 91191 Gif-sur-Yvette, France
[5]INAF, IASF-Milano, Via A. Corti 12, 20133 Milano, Italy
[6]Institut d'Astrophysique Spatiale (IAS), CNRS, Université Paris Sud, Orsay, France
[7]Institut Néel, CNRS, Université Grenoble Alpes, France
[8]Institut de RadioAstronomie Millimétrique (IRAM), Grenoble, France
[9]Aix Marseille Univ, CNRS, CNES, LAM, Marseille, France
[10]Dipartimento di Fisica, Sapienza Università di Roma, Piazzale Aldo Moro 5, I-00185 Roma, Italy
[11]Univ. Grenoble Alpes, CNRS, IPAG, 38000 Grenoble, France
[12]Centro de Astrobiología (CSIC-INTA), Torrejón de Ardoz, 28850 Madrid, Spain
[13]Instituto de Radioastronomía Milimétrica (IRAM), Granada, Spain
[14]LERMA, Observatoire de Paris, PSL Research University, CNRS, Sorbonne Université, UPMC, 75014 Paris, France
[15]Univ. de Toulouse, UPS-OMP, CNRS, IRAP, 31028 Toulouse, France
[16]Laboratoire de Physique de l'École Normale Supérieure, ENS, PSL Research University, CNRS, Sorbonne Université, Université de Paris, 75005 Paris, France
[17]Department of Physics and Astronomy, University of Pennsylvania, 209 South 33rd Street, Philadelphia, PA, 19104, USA
[18]Institut d'Astrophysique de Paris, CNRS (UMR7095), 98 bis boulevard Arago, 75014 Paris, France
[19]Kavli Institute for Astrophysics and Space Research, Massachusetts Institute of Technology, Cambridge, MA 02139, USA
[20]School of Earth and Space Exploration and Department of Physics, Arizona State University, Tempe, AZ 85287, USA
[21]Caltech, Pasadena, CA 91125, USA
[22]Departamento de Física Teórica and CIAFF, Facultad de Ciencias, Modulo 8, Universidad Autánoma de Madrid, 28049 Madrid, Spain

Abstract. We present a multi-probe analysis of the well-known galaxy cluster CL J1226.9+3332 as a proof of concept for multi-wavelength studies within

*e-mail: miren.munoz@lpsc.in2p3.fr[1]

the framework of the NIKA2 Sunyaev-Zel'dovich Large Program (LPSZ). CL J1226.9+3332 is a massive and high redshift (z = 0.888) cluster that has already been observed at several wavelengths. A joint analysis of the thermal SZ (tSZ) effect at millimeter wavelength with the NIKA2 camera and in X-ray with the *XMM-Newton* satellite permits the reconstruction of the cluster's thermodynamical properties and mass assuming hydrostatic equilibrium. We test the robustness of our mass estimates against different definitions of the data analysis transfer function. Using convergence maps reconstructed from the data of the CLASH program we obtain estimates of the lensing mass, which we compare to the estimated hydrostatic mass. This allows us to measure the hydrostatic-to-lensing mass bias and the associated systematic effects related to the NIKA2 measurement. We obtain $M_{500}^{\mathrm{HSE}} = (7.65 \pm 1.03) \times 10^{14} M_{\odot}$ and $M_{500}^{\mathrm{lens}} = (7.35 \pm 0.65) \times 10^{14} M_{\odot}$, which implies a HSE-to-lensing bias consistent with 0 within 20%.

1 Introduction

Galaxy clusters are excellent probes for cosmology, in particular for the understanding of large-scale structure formation processes [1, 2]. For this purpose, a precise knowledge of cluster masses is needed. The NIKA2 SZ Large Program (LPSZ) [3] consists in 300 hours of Guaranteed Time dedicated to the observation and detailed analysis of 45 high redshift (z in 0.5-0.9) galaxy clusters taking advantage of the capabilites of the NIKA2 camera: a large field of view combined with high angular resolution at 150 and 260 GHz [4, 6–8]. These characteristics allow us to accurately map galaxy clusters via the thermal Sunyaev-Zel'dovich effect [12] and, together with high quality X-ray observations from *XMM-Newton*, will enable the investigation of the SZ–mass scaling relation at high redshift for cosmological exploitation.

After a first analysis of a LPSZ cluster for a science verification study [9] and a second one for proving the quality of NIKA2 in the most challenging case [10], we present here a study on the robustness of the hydrostatic mass reconstruction combining SZ and X-ray data. We test the impact of systematic effects from the estimation of the filtering induced by the data processing pipeline and from the pressure model fitting. Systematic effects that may arise from other effects (e.g. density estimates at large radii) are not studied in this work.

2 The galaxy cluster CL J1226.9+3332

The object of this analysis is the well-known galaxy cluster CL J1226.9+3332, also known as PSZ2-G160.83+81.66. It is located at (R.A., Dec.)$_{J2000}$ = (12h26m58.37s, +33d32m47.4s) according to the X-ray peak from [13] with mass $M_{500}^{\mathrm{HSE}} \sim 5.70 \times 10^{14} M_{\odot}$ and redshift 0.888 [11], which makes it the highest-redshift cluster of the LPSZ sample. Discovered by ROSAT [14], it has already been observed at several wavelengths [15–24], appearing, from the first SZ observations with BIMA [19] as a very spherical cluster.

X-ray analyses using *XMM-Newton* and Chandra data [16] found that the temperature of the cluster in a region at ~ 40" to the south-west of the X-ray peak is much higher than the average temperature of the intra-cluster medium (ICM). Posterior SZ analyses with MUSTANG [25] and NIKA [5, 15], as well as lensing [23] and galaxy distribution studies [18], showed also the presence of such a substructure. Various explanations have been proposed [17, 18], suggesting mainly that the cluster consists of a two-halo system, which is being observed after the less massive cluster passed through the central one and the gas followed it. In short, CL J1226.9+3332 shows a relaxed morphology at large scales and evidence of disturbance in the core.

3 Map-making and data processing filtering

The cluster was observed for 3.6 hours during the 15th NIKA2 science-purpose observation campaign (13-20 February 2018) as part of the NIKA2 Guaranteed Time at the IRAM 30-m telescope. The data were calibrated and reduced using the standard NIKA2 collaboration LPSZ pipeline described in [8, 10, 26]. We present in Figure 1 the obtained maps at 150 and 260 GHz. The map at 150 GHz shows the cluster as a decrement of flux, contaminated by the positive signal from point sources. These point sources are also detected in the 260 GHz map.

The filtering induced by the data processing in the 150 GHz map is computed using simulations via a transfer function (TF) that relates, in the Fourier space, the real astrophysical signal (the cluster) with the filtered one. The standard LPSZ analysis [9, 10] uses one-dimensional transfer functions which are obtained by averaging the filtering in Fourier-domain annuli at a fixed angular scale, supposing therefore isotropy. The black line in the left panel in Figure 2 shows the 1D TF for the 150 GHz map in Figure 1. In this work we test a new approach: a two-dimensional transfer function (2D TF) presented in Figure 2. That allows us to deal with anisotropy effects coming for example from the effect of scanning angles and the deviation from sphericity of clusters. Color lines in the left panel in Figure 2 show the one-dimensional TFs for the different directions represented in the 2D TF and illustrate the impact of anisotropies in the angular frequency space.

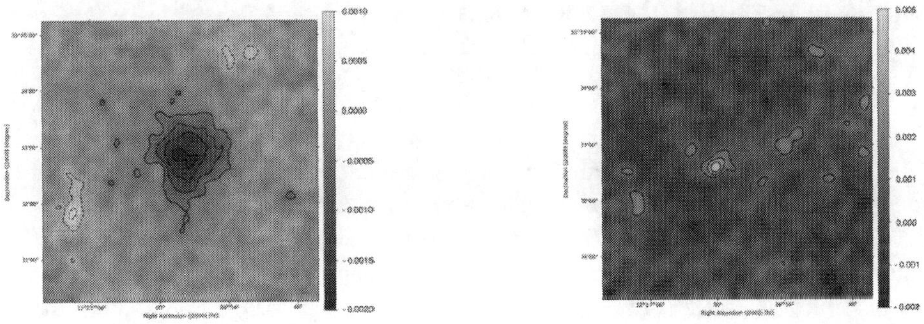

Figure 1. NIKA2 maps obtained with the LPSZ pipeline at 150 GHz (left) and 260 GHz (right) in Jy/beam units. Contours show S/N levels multiples of $\pm3\sigma$. Both maps have been smoothed with a 10" FWHM Gaussian kernel.

4 SZ analysis: pressure profile determination

The CL J1226.9+3332 surface brightness map is proportional to the integrated electron pressure along the line of sight. To recover the latter, we perform a Monte Carlo Markov chain (MCMC) fit of a model to the 150 GHz map considering the Compton parameter signal associated with the cluster, the point sources and a zero level. The contamination from point sources at 150 GHz is estimated to account for their positive flux that may compensate the tSZ signal of the cluster. The details are given in [10, 31]. In this work we performed the fit by adjusting two different pressure profile models: 1) a generalized Navarro-Frenk-White

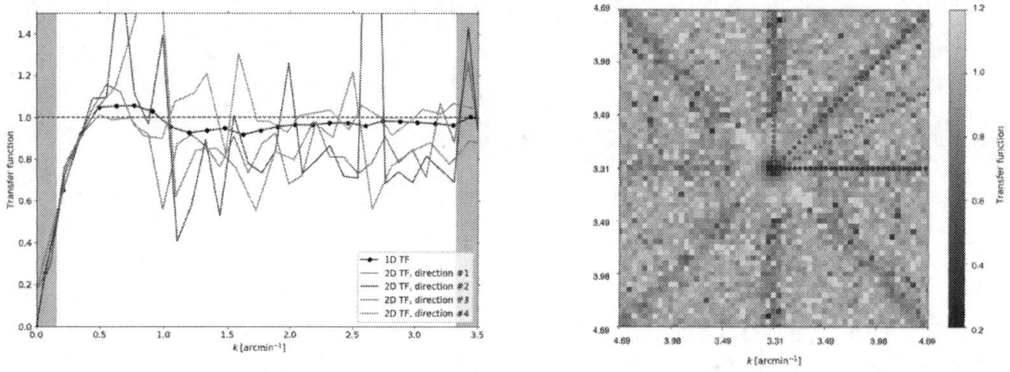

Figure 2. 1D (left) and 2D (right) transfer functions representing the data reduction-induced filtering for the 150 GHz map in Figure 1. Color lines in the left panel show the value of the transfer function for different directions in the 2D TF, represented with the same colors in the right panel.

(gNFW, [27]) and 2) a radially binned model. For the first one we fit for the 5 parameters of the model. The radially binned profile is defined by $P_e(r) = P_i\left(\frac{r}{r_i}\right)^{-\alpha_i}$, where for each radial bin r_i a value of the pressure P_i and the slope α_i are fitted. Moreover, we repeated the fit considering either 1D or 2D transfer functions. In the following we take as reference the radially binned pressure model with the 1D transfer function.

5 Hydrostatic mass and robustness tests

Combining the spherical pressure profile resulting from the fit with the electron density profile obtained from X-ray data, we can compute, under the hydrostatic equilibrium (HSE) hypothesis [9, 10, 15], the hydrostatic mass profile for CL J1226.9+3332 and with it the probability distribution for R_{500}^{HSE}[1]. We note here that before deriving the pressure, we fitted a gNFW model to the radially binned profiles. This allowed us to have smooth pressure profiles to calculate the derivative of the profile for the mass estimate, but still testing the use of a different model when reconstructing the pressure from the maps.

We show in Figure 3 (left) the obtained probability distributions for R_{500}^{HSE}, as well as the tSZ flux Y_{500} and M_{500}^{HSE}. These results are compared to previous HSE mass estimates obtained from NIKA maps [15]. Our results give higher R_{500}^{HSE} and M_{500}^{HSE} values than expected from previous studies. Nevertheless, we observe a correlation between R_{500}^{HSE} and M_{500}^{HSE}. This is expected since R_{500}^{HSE} and M_{500}^{HSE} are simultaneously obtained from the over-density definition giving directly proportional $(R_{500}^{\mathrm{HSE}})^3$ and M_{500}^{HSE}. Therefore, the comparison of probability distributions seems more representative when comparing different hydrostatic mass estimates. In the left panel in Figure 3 we also compare the results of the robustness tests we have performed starting from the 150 GHz map in Figure 1. We compare the probability distributions obtained with the reference model to the results using 2D transfer function instead. In the latter case, slightly lower values of R_{500}^{HSE} and M_{500}^{HSE} are favored. We also show results for the two pressure profile models. We find tighter constraints for the gNFW model when it is directly fitted to the map than when the radially binned profile is fitted and then gNFWs are adjusted to the resulting radially binned profiles. However, computing the pressure derivative in a consistent way (here always deriving from a gNFW), our mass estimate

[1] Radius at which the mean mass (HSE mass in this case) density of the cluster is 500 times the critical density of the universe at its redshift.

is robust. From the combination of the posterior probability distributions obtained for the four cases that we have considered, we measure an hydrostatic mass for CL J1226.9+3332 of $M_{500}^{\rm HSE} = (7.65 \pm 1.03) \times 10^{14} M_{\odot}$, where ~ 5-10 % of the uncertainty comes from the contribution of the tested systematic effects.

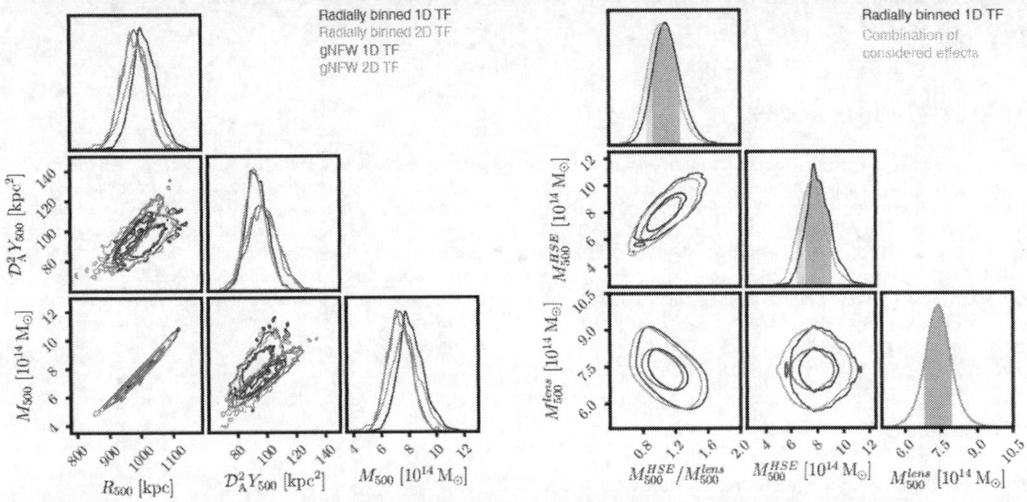

Figure 3. *Left:* $R_{500}^{\rm HSE}$, Y_{500} and $M_{500}^{\rm HSE}$ probability distributions for the radially binned pressure model with 1D TF in blue (reference), radially binned 2D TF in purple, gNFW 1D TF in red and gNFW 2D TF in orange. The grey star, cyan circle and green diamond correspond to the results from [15] for the tested PPC, FPC and NNN models, respectively. *Right:* Hydrostatic mass with respect to lensing mass at R_{500}. In dark shades of blue, 1σ and 2σ contours for $M_{500}^{\rm HSE}$ from the reference model and in light shades of blue $M_{500}^{\rm HSE}$ as the combination of the four probability distributions in the left panel.

6 Hydrostatic-to-lensing mass bias

We have performed an analysis of the convergence maps reconstructed from the CLASH data [23] to obtain lensing mass estimates for CL J1226.9+3332. Combining them we obtain $M_{500}^{\rm lens} = (7.35 \pm 0.65) \times 10^{14} M_{\odot}$. Details on this analysis will be given in [28]. The hydrostatic-to-lensing mass bias $b_{\rm HSE/lens} = 1 - M_{500}^{\rm HSE}/M_{500}^{\rm lens}$, can be an interesting indicator of effects coming from the non-thermal contributions to the pressure of the cluster and systematic effects in mass estimations [29, 30]. Supposing that they are uncorrelated estimates, we have combined the probability distributions obtained for $M_{500}^{\rm HSE}$ and $M_{500}^{\rm lens}$ to get a distribution of the ratio, which can be translated into a probability distribution of the hydrostatic-to-lensing mass bias for CL J1226.9+3332. In the right panel in Figure 3 we present the probability distributions for $M_{500}^{\rm HSE}/M_{500}^{\rm lens}$, $M_{500}^{\rm HSE}$ and $M_{500}^{\rm lens}$. We have computed the ratio using the $M_{500}^{\rm HSE}$ from the reference model, as well as using the combination of the four probability distributions derived for $M_{500}^{\rm HSE}$, therefore accounting for all the considered effects. The figure shows that for both cases the bias is consistent with 0 within ~ 20%.

7 Conclusions

Combined tSZ and X-ray analysis allow us to estimate the hydrostatic mass of this high redshift galaxy cluster, CL J1226.9+3332. We have demonstrated that our hydrostatic mass estimates are robust against the application of the 1D or 2D transfer function. The reconstruction of the pressure profile derivative is a key element for the determination of the hydrostatic mass profiles, and needs to be dealt with carefully. This analysis has also been an important step towards developing a standard pipeline to compare and combine hydrostatic and lensing mass estimates. For CL J1226.9+3332 we find a hydrostatic-to-lensing mass bias consistent with 0.

Acknowledgements

We would like to thank the IRAM staff for their support during the campaigns. The NIKA2 dilution cryostat has been designed and built at the Institut Néel. In particular, we acknowledge the crucial contribution of the Cryogenics Group, and in particular Gregory Garde, Henri Rodenas, Jean Paul Leggeri, Philippe Camus. This work has been partially funded by the Foundation Nanoscience Grenoble and the LabEx FOCUS ANR-11-LABX-0013. This work is supported by the French National Research Agency under the contracts "MKIDS", "NIKA" and ANR-15-CE31-0017 and in the framework of the "Investissements d'avenir" program (ANR-15-IDEX-02). This work has benefited from the support of the European Research Council Advanced Grant ORISTARS under the European Union's Seventh Framework Programme (Grant Agreement no. 291294). F.R. acknowledges financial supports provided by NASA through SAO Award Number SV2-82023 issued by the Chandra X-Ray Observatory Center, which is operated by the Smithsonian Astrophysical Observatory for and on behalf of NASA under contract NAS8-03060.

References

[1] S. Bocquet *et al.*, The Astrophysical Journal **878**, 55 (2019)
[2] Planck Collaboration *et al.*, Astronomy & Astrophysics **594**, A24 (2016)
[3] F. Mayet *et al.*, EPJ Web of Conferences **228**, 00017 (2020)
[4] R. Adam *et al.*, Astronomy & Astrophysics **609**, A115 (2018)
[5] R. Adam *et al.*, Astronomy & Astrophysics **614**, A118 (2018)
[6] M. Calvo *et al.*, Journal of Low Temperature Physics **184**, 816 (2016)
[7] O. Bourrion *et al.*, Journal of Instrumentation **11**, 11001 (2016)
[8] L. Perotto *et al.*, Astronomy & Astrophysics **637**, A71 (2020)
[9] F. Ruppin *et al.*, Astronomy & Astrophysics **615**, A112 (2018)
[10] F. Kéruzoré *et al.*, Astronomy & Astrophysics **644**, A93 (2020)
[11] Planck Collaboration *et al.*, Astronomy & Astrophysics **594**, A27 (2016)
[12] R. A. Sunyaev, Y. B. Zeldovich, Comments on Astro. and Space Physics **4**, 173 (1972)
[13] K. W. Cavagnolo *et al.*, The Astrophysical Journal Supplement Series **182**, 12–32 (2009)
[14] H. Ebeling *et al.*, The Astrophysical Journal Letters **548**, L23 (2008)
[15] R. Adam *et al.*, Astronomy & Astrophysics **576**, A12 (2015)
[16] B. J. Maughan *et al.*, The Astrophysical Journal **659**, 1125-1137 (2007)
[17] B. J. Maughan *et al.*, MNRAS **351**, 1193-1203 (2004)
[18] M. J. Jee, J. A. Tyson, The Astrophysical Journal **691**, 1337-1347 (2009)
[19] M. Joy *et al.*, The Astrophysical Journal **551**, L1-L4 (2001)
[20] M. Bonamente *et al.*, The Astrophysical Journal **647**, 25-54 (2006)
[21] S. W. Allen *et al.*, MNRAS **383**, 879–896 (2007)
[22] M. Postman *et al.*, The Astrophysical Journal Supplement Series **199**, 25 (2012)
[23] A. Zitrin *et al.*, The Astrophysical Journal **806**, 4 (2015)
[24] C. Romero *et al.*, Astronomy & Astrophysics **612**, A39 (2018)
[25] P. M. Korngut *et al.*, The Astrophysical Journal **734**, 10 (2011)
[26] A. Rigby *et al.*, MNRAS **000**, 1–21 (2020)
[27] D. Nagai *et al.*, The Astrophysical Journal **668**, 1-14 (2007)
[28] M. Muñoz-Echeverría *et al.*, in prep. (2021)
[29] L. Lovisari *et al.*, Astronomy & Astrophysics **644**, A78 (2020)
[30] G. W. Pratt *et al.*, Space Sci.Rev. **215**, 25 (2019)
[31] F. Kéruzoré *et al.*, in prep. (2021)

The LPSZ-CLASH galaxy cluster sample: Combining lensing and hydrostatic mass estimates

M. Muñoz-Echeverría[1,*], *R.* Adam[2], *P.* Ade[3], *H.* Ajeddig[4], *P.* André[4], *M.* Arnaud[4], *E.* Artis[1], *H.* Aussel[4], *I.* Bartalucci[5], *A.* Beelen[6], *A.* Benoît[7], *S.* Berta[8], *L.* Bing[9], *O.* Bourrion[1], *M.* Calvo[7], *A.* Catalano[1], *M.* De Petris[10], *F.-X.* Désert[11], *S.* Doyle[3], *E. F. C.* Driessen[8], *A.* Ferragamo[10], *A.* Gomez[12], *J.* Goupy[7], *F.* Kéruzoré[1], *C.* Kramer[13], *B.* Ladjelate[13], *G.* Lagache[9], *S.* Leclercq[8], *J.-F.* Lestrade[14], *J.-F.* Macías-Pérez[1], *A.* Maury[4], *P.* Mauskopf[3,20], *F.* Mayet[1], *A.* Monfardini[7], *A.* Paliwal[10], *L.* Perotto[1], *G.* Pisano[3], *E.* Pointecouteau[15], *N.* Ponthieu[11], *G. W.* Pratt[4], *V.* Revéret[4], *A. J.* Rigby[3], *A.* Ritacco[16,6], *C.* Romero[17], *H.* Roussel[18], *F.* Ruppin[19], *K.* Schuster[8], *S.* Shu[21], *A.* Sievers[13], *C.* Tucker[3], and *G.* Yepes[22]

[1]Univ. Grenoble Alpes, CNRS, LPSC-IN2P3, 53, avenue des Martyrs, 38000 Grenoble, France
[2]LLR, CNRS, École Polytechnique, Institut Polytechnique de Paris, Palaiseau, France
[3]School of Physics and Astronomy, Cardiff University, Queen's Buildings, The Parade, Cardiff, CF24 3AA, UK
[4]AIM, CEA, CNRS, Université Paris-Saclay, Université Paris Diderot, Sorbonne Paris Cité, 91191 Gif-sur-Yvette, France
[5]INAF, IASF-Milano, Via A. Corti 12, 20133 Milano, Italy
[6]Institut d'Astrophysique Spatiale (IAS), CNRS, Université Paris Sud, Orsay, France
[7]Institut Néel, CNRS, Université Grenoble Alpes, France
[8]Institut de RadioAstronomie Millimétrique (IRAM), Grenoble, France
[9]Aix Marseille Univ, CNRS, CNES, LAM, Marseille, France
[10]Dipartimento di Fisica, Sapienza Università di Roma, Piazzale Aldo Moro 5, I-00185 Roma, Italy
[11]Univ. Grenoble Alpes, CNRS, IPAG, 38000 Grenoble, France
[12]Centro de Astrobiología (CSIC-INTA), Torrejón de Ardoz, 28850 Madrid, Spain
[13]Instituto de Radioastronomía Milimétrica (IRAM), Granada, Spain
[14]LERMA, Observatoire de Paris, PSL Research University, CNRS, Sorbonne Université, UPMC, 75014 Paris, France
[15]Univ. de Toulouse, UPS-OMP, CNRS, IRAP, 31028 Toulouse, France
[16]Laboratoire de Physique de l'École Normale Supérieure, ENS, PSL Research University, CNRS, Sorbonne Université, Université de Paris, 75005 Paris, France
[17]Department of Physics and Astronomy, University of Pennsylvania, 209 South 33rd Street, Philadelphia, PA, 19104, USA
[18]Institut d'Astrophysique de Paris, CNRS (UMR7095), 98 bis boulevard Arago, 75014 Paris, France
[19]Kavli Institute for Astrophysics and Space Research, Massachusetts Institute of Technology, Cambridge, MA 02139, USA
[20]School of Earth and Space Exploration and Department of Physics, Arizona State University, Tempe, AZ 85287, USA
[21]Caltech, Pasadena, CA 91125, USA
[22]Departamento de Física Teórica and CIAFF, Facultad de Ciencias, Modulo 8, Universidad Autánoma de Madrid, 28049 Madrid, Spain

Abstract. Starting from the clusters included in the NIKA sample and in the NIKA2 Sunyaev-Zel'dovich Large Program (LPSZ) we have selected a sam-

*e-mail: miren.munoz@lpsc.in2p3.fr[1]

ple of six common objects with the Cluster Lensing And Supernova survey with Hubble (CLASH) lensing data. For the LPSZ clusters we have at our disposal both high-angular resolution observations of the thermal SZ with NIKA and NIKA2 and X-ray observations with *XMM-Newton* from which hydrostatic mass estimates can be derived. In addition, the CLASH dataset includes lensing convergence maps that can be converted into lensing estimates of the total mass of the cluster. One-dimensional mass profiles are used to derive integrated mass estimates accounting for systematic effects (data processing, modeling, etc.). Two-dimensional analysis of the maps can reveal substructures in the cluster and, therefore, inform us about the dynamical state of each system. Moreover, we are able to study the hydrostatic mass to lensing mass bias, across different morphology and a range of redshift clusters to give more insight on the hydrostatic mass bias. The analysis presented in this proceeding follows the study discussed in [20].

1 Introduction

Clusters of galaxies have been shown to be excellent cosmological probes [1–3]. Nevertheless, cluster-based cosmological results show a trend towards lower σ_8 than CMB-based results [4]. This may be due to a lack of knowledge on cluster physics and its impact on the reliable evaluation of cluster masses. Different mass estimates are affected by different systematic uncertainties [4], so combining observables may help building a consistent and more accurate mass estimate. In this work we will combine hydrostatic equilibrium and lensing mass estimates using the common sample between NIKA, the LPSZ and CLASH.

The NIKA2 Sunyaev-Zel'dovich Large Program (LPSZ) [5] studies 45 high redshift (z in 0.5-0.9) galaxy clusters selected in SZ. One of its objectives is to reconstruct the hydrostatic equilibrium (HSE) mass of the clusters by combining the electron pressure reconstruction using the thermal SZ (tSZ) effect with the NIKA2 camera [6–9] and the electron density and temperature from X-rays with *XMM-Newton* satellite. The CLASH collaboration [10] observed 25 massive galaxy clusters of redshift 0.2 to 0.9. From the combined strong and weak lensing analysis, they reconstructed convergence maps (κ-maps) that are publicly available [11] and are the starting point for the lensing study in this work.

2 The LPSZ-CLASH sample

We present in Figure 1 five of the six clusters in our sample. The study of the LPSZ-CLASH cluster PSZ2-G045.32-38.46 is ongoing and it will not be presented in this work. PSZ2-G160.83+81.66 (z=0.89), PSZ2-G144.83+25.11 (z=0.58) and PSZ2-G228.16+75.20 (z=0.55) are part of the LPSZ sample and we will study their HSE and lensing mass profiles. For PSZ2-G144.83+25.11 we will use the results from the thermal SZ and X-ray analysis done in [15]. Furthermore, MACS J1423.8+2404 (z=0.55) and MACS J0717.5+3745 (z=0.55) were part of the NIKA sample [16]. For the HSE mass analysis of MACS J1423.8+2404 we will use the results from [17] and the tSZ-map for the study of MACS J0717.5+3745 will be taken from [18].

3 Mass profiles reconstruction

3.1 Hydrostatic mass profiles

The HSE mass reconstruction was done, on the one side, by combining the electron pressure profile obtained from SZ data and the electron density from X-rays as explained in

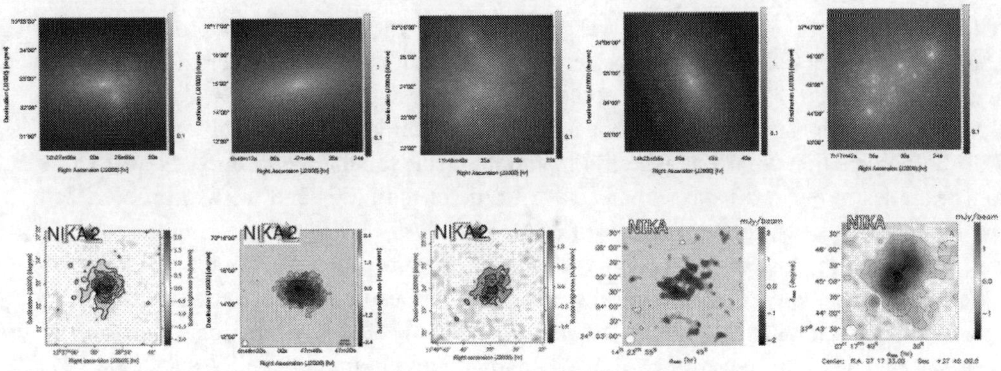

Figure 1. The LPSZ-CLASH sample, from left to right: PSZ2-G160.83+81.66, PSZ2-G144.83+25.11, PSZ2-G228.16+75.20, MACS J1423.8+2404, MACS J0717.5+3745. CLASH κ-maps on top. NIKA2 and NIKA SZ-maps at the bottom.

[15, 17, 19]. On the other side, we combine *XMM-Newton* X-ray spectroscopic temperature and electron density profiles [14]. For PSZ2-G160.83+81.66 and PSZ2-G228.16+75.20, the contamination of point sources was estimated as in [19] and taken into account for the electron pressure fit on the SZ map. For PSZ2-G144.83+25.11 the southwestern over-pressure region was masked to fit the pressure profile as well as the electron density [15]. The tSZ signal in MACS J1423.8+2404 is very contaminated by point sources and they were subtracted before the pressure profile fit (M2 method in [17]).

3.2 Lensing mass profiles

For the lensing mass estimation we followed the procedure described in [20]. Convergence maps (κ-maps) are surface mass density maps (Σ) in critical density units (Σ_{crit}), where $\Sigma_{crit} = \frac{c^2}{4\pi G}\frac{D_s}{D_l D_{ls}}$ and D_s, D_l and D_{ls} correspond to the angular diameter distance between the observer and the source, the observer and the lens (the cluster) and the lens and source, respectively. CLASH convergence maps have been scaled to $D_s/D_{ls} = 1$ [11], therefore, the κ-map for each cluster can be easily converted into a Σ-map.

In order to reconstruct the three dimensional density profiles of the clusters, we assume spherical symmetry and fit the analytical projected Navarro-Frenk-White (NFW, [21, 22]) density profile to the radially averaged profiles of the Σ-maps. We perform a Markov Chain Monte Carlo (MCMC) fit to get the best-fit values of the free parameters r_s and c_{200}, the characteristic cluster radius and the concentration, respectively. Finally, we can reconstruct the lensing mass profiles by spherically integrating the best-fit NFW density profile for each cluster. For some clusters in our sample, two different models of κ-maps were reconstructed, Light-Traces-Mass and Pseudo-Isothermal Elliptical Mass Distribution with elliptical NFW, hereafter LTM and PIEMD+eNFW. Details can be found in [11–13, 21]. For a coherent comparison with the HSE mass profiles, surface mass density profiles are centered in the same positions as the pressure and electron density profiles in Section 3.1. As a consequence, the centers of some projected density profiles may not coincide with the density peaks in the maps.

4 Hydrostatic and lensing mass comparison

The reconstructed HSE and lensing mass profiles for PSZ2-G160.83+81.66, PSZ2-G228.16+75.20, PSZ2-G144.83+25.11 and MACS J1423.8+2404 are presented in Figure 2, with 1σ and 2σ confidence envelopes. The envelopes of the lensing mass profiles are very thin compared to the HSE ones. This is due to the small amount of parameters (2 in the NFW, instead of more in [30, 31]) that we fit compared to the case of the HSE mass, as well as the fact that we consider the convergence maps as true, meaning that the radially averaged profiles of the Σ-maps that we use for the fit only contain statistical uncertainties. Nevertheless, the uncertainties in our results are consistent with the ones in [32] and we will proceed with them. Due to the complex morphology of MACS J0717.5+3745 and the simplicity of our model, we choose to analyse this cluster with a different approach in Section 5.

We concentrate here in comparing the different mass estimates obtained for the cluster sample at the characteristic R_{500} radius. In Figure 3 we show the probability distribution in the plane $M_{500}^{\rm HSE}$ and $M_{500}^{\rm lens}$ for the four clusters. The $M_{500}^{\rm HSE}$ are obtained from SZ and X-ray. The lensing masses are derived from the combination of the two lensing mass estimates (except for PSZ2-G228.16+75.20 for which we only have one lensing mass), to account for modeling effects. For PSZ2-G160.83+81.66 both $M_{500}^{\rm lens}$ are in agreement and consistent with the $M_{500}^{\rm HSE}$. The complex morphology of PSZ2-G228.16+75.20 [29] makes the spherical modeling results difficult to interpret. Even if our $M_{500}^{\rm HSE}$ estimate is consistent with [25], a more thorough study of this cluster may be needed. For PSZ2-G144.83+25.11 the SZ data (see [15]) is of very good quality and the spherical electronic pressure profile, having masked the overpressure region, fits very nicely the data. This gives a very well defined HSE mass estimate. The two $M_{500}^{\rm lens}$ estimates for PSZ2-G144.83+25.11 are in agreement within 1σ uncertainties, but the differences between the κ-map models enlarges the error bar in the final $M_{500}^{\rm lens}$ distribution. An extreme case of this effect is seen for MACS J1423.8+2404: the $M_{500}^{\rm lens}$ estimation for each of the lensing models is so different that the combined distribution reveals two distinct peaks. These results highlight the impact that systematics can have in our mass estimate. We see that the data quality, the cluster morphology and dynamical state, as well as the chosen observables and the modeling influence the mass estimate.

4.1 Hydrostatic-to-lensing mass bias

The hydrostatic-to-lensing mass bias, $b_{\rm HSE/lens} = 1 - M_{500}^{\rm HSE}/M_{500}^{\rm lens}$, is computed from the distributions in the left panel in Figure 3 and presented in the center as the ratio between the HSE and the lensing mass. The results vary largely from cluster to cluster and systematic effects in mass estimates propagate to the bias. The mean hydrostatic-to-lensing mass ratio of this small sample is $M_{500}^{\rm HSE}/M_{500}^{\rm lens} = 0.87 \pm 0.47$. This is in agreement with hydrodynamic simulations [23], but we notice that the dispersion in our result is very large. This is also consistent with the value needed to reconcile the cosmological constraints obtained from Planck CMB power spectra to the cluster counts, $M_{500}^{\rm HSE}/M_{500}^{\rm tot} = 0.58 \pm 0.04$ [2, 24]. For such comparisons we have to consider the lensing mass as an accurate estimate of the total mass of clusters.

We have also compared the hydrostatic-to-lensing mass bias in a common inner region (R_{1000}) to avoid any extrapolation of the lensing mass profile. We show in the right panel in Fig. 3 the ratios at the R_{1000} fixed by the SZ+X-ray HSE mass profile. In this case from the combination of the four clusters, we obtain a mean value of $M^{\rm HSE}/M^{\rm lens}(R_{1000}^{\rm HSE}) = 0.97 \pm 0.29$. Notice that this ratio can not be directly compared to the R_{500} one as we expect some variations with radius.

5 Identification of structures by comparing SZ and lensing maps

When comparing HSE and lensing mass results with spherical profiles we mix up different effects that can hardly be disentangled and may bias the final results. This motivates a map to

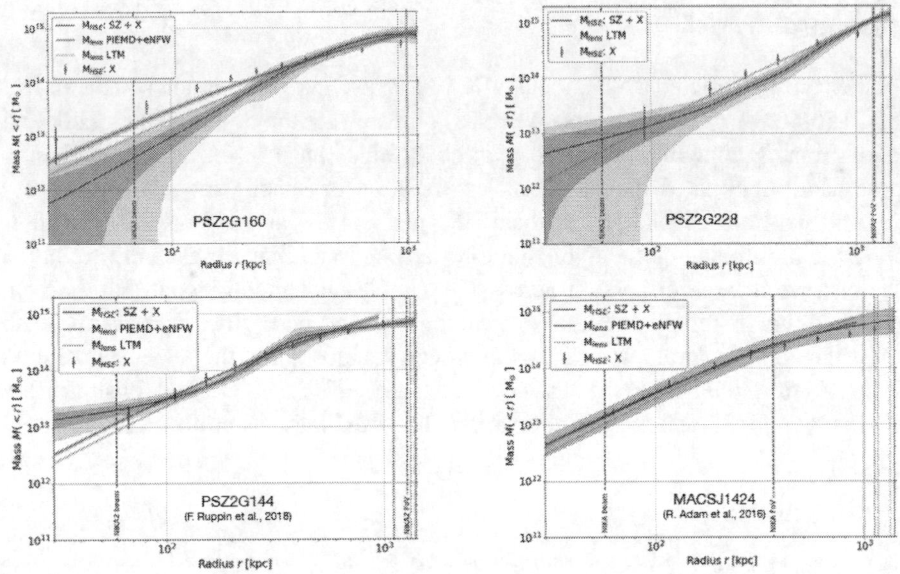

Figure 2. Reconstructed HSE mass profiles: in blue for SZ+X-ray combination and in white dots for X-ray only. Reconstructed lensing mass profiles for the two κ-map models in green and yellow. Vertical colored dashed lines represent the R_{500} for each mass profile and black dashed lines NIKA(2) instrumental beam and field of view.

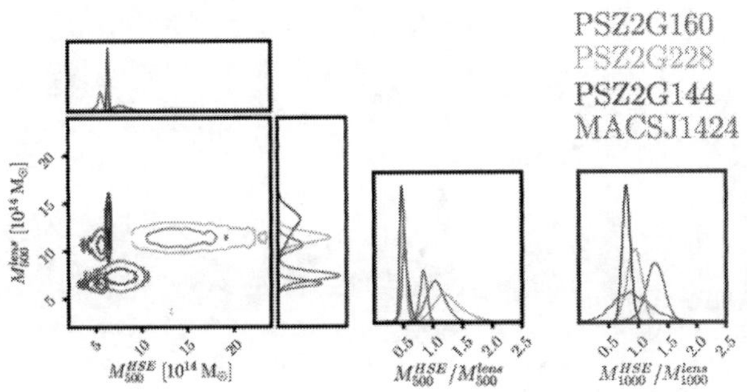

Figure 3. *Left:* marginalized distributions of the density probability for the lensing and HSE mass at R_{500}. *Center:* hydrostatic-to-lensing mass ratio for each cluster at R_{500}. *Right:* hydrostatic-to-lensing mass ratio for each cluster at a fixed R_{1000}^{HSE}.

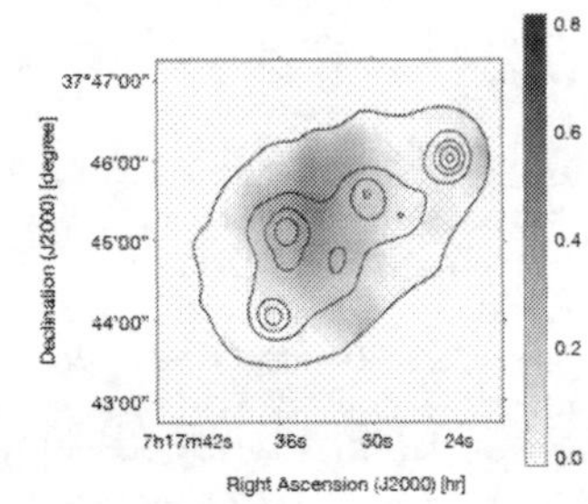

Figure 4. tSZ/κ map for MACS J0717.5+3745. Map pixels for which $\kappa < 0.5$ are set to zero. Contours correspond to [0.5, 1, 1.5, 2, 2.5] levels in κ-map. The map has been smoothed with a 18" Gaussian kernel.

map comparison for complex clusters. We take as a test case the MACS J0717.5+3745 galaxy cluster. Previous studies [18, 26–28] have revealed its complex morphology and dynamical state: at least 4 sub-clusters interacting in a complex merger scenario, as well as sub-clusters moving along the line-of-sight direction. We combine here the tSZ map from [18] and the κ-map from CLASH [11]. In Figure 4 we present the tSZ/κ ratio map (in an arbitrary color scale), with contours showing the levels in the κ-map. The tSZ map has been normalized to Y_{500} and the κ-map to M_{500}. The darkest green regions indicate the areas where the relative contribution of the thermal pressure from the hot gas to the total mass budget is greater than average. The presence of such hot region is consistent with outcomes from X-rays and from the combination of X-rays and tSZ data [27].

6 Conclusions

In this work we have constructed a NIKA and NIKA2 LPSZ-CLASH common sample of five high redshift (0.55 - 0.89) galaxy clusters. We have compared for the first time their HSE and lensing mass estimates, showing the impact that observables have in the mass calculation. Moreover, we have highlighted the important contribution of lensing models to the mass uncertainties and we have accounted for them. From the HSE and lensing mass profiles reconstruction, we have obtained the hydrostatic-to-lensing mass bias for each cluster in the sample. We observe a large dispersion across the sample and hints of contribution from systematics. It will be, therefore, essential to treat these effects carefully when dealing with larger samples. From this sample, we can not draw conclusions about the redshift evolution of the bias. In addition, the analysis performed on MACS J0717.5+3745 has illustrated the potential of map to map tSZ and lensing comparison for revealing cluster physics.

Acknowledgements

We would like to thank the IRAM staff for their support during the campaigns. The NIKA2 dilution cryostat has been designed and built at the Institut Néel. In particular, we acknowledge the crucial contribution of the Cryogenics Group, and in particular Gregory Garde, Henri Rodenas, Jean Paul Leggeri, Philippe Camus. This work has been partially funded by the Foundation Nanoscience Grenoble and the LabEx FOCUS ANR-11-LABX-0013. This work is supported by the French National Research Agency under the contracts "MKIDS", "NIKA" and ANR-15-CE31-0017 and in the framework of the "Investissements d'avenir" program (ANR-15-IDEX-02). This work has benefited from the support of the European Research Council Advanced Grant ORISTARS under the European Union's Seventh Framework Programme (Grant Agreement no. 291294). F.R. acknowledges financial supports provided by NASA through SAO Award Number SV2-82023 issued by the Chandra X-Ray Observatory Center, which is operated by the Smithsonian Astrophysical Observatory for and on behalf of NASA under contract NAS8-03060.

References

[1] S. Bocquet *et al.*, The Astrophysical Journal **878**, 55 (2019)
[2] Planck Collaboration *et al.*, Astronomy & Astrophysics **594**, A24 (2016)
[3] M. Lubini *et al.*, MNRAS **437**, 2461 (2014)
[4] G. W. Pratt *et al.*, Space Sci. Rev. **215**, 25 (2019)
[5] F. Mayet *et al.*, EPJ Web of Conferences **228**, 00017 (2020)
[6] R. Adam *et al.*, Astronomy & Astrophysics **609**, A115 (2018)
[7] M. Calvo *et al.*, Journal of Low Temperature Physics **184**, 816 (2016)
[8] O. Bourrion *et al.*, Journal of Instrumentation **11**, 11001 (2016)
[9] L. Perotto *et al.*, Astronomy & Astrophysics **637**, A71 (2020)
[10] M. Postman *et al.*, The Astrophysical Journal Supplement Series **199**, 25 (2012)
[11] A. Zitrin *et al.*, The Astrophysical Journal **806**, 4 (2015)
[12] A. Zitrin *et al.*, MNRAS **396**, 1985 (2009)
[13] E. Jullo *et al.*, New Journal of Physics **9**, 447 (2007)
[14] I. Bartalucci *et al.*, Astronomy & Astrophysics **617**, A64 (2018)
[15] F. Ruppin *et al.*, Astronomy & Astrophysics **615**, A112 (2018)
[16] A. Catalano *et al.*, Astronomy & Astrophysics **569**, A9 (2014)
[17] R. Adam *et al.*, Astronomy & Astrophysics **586**, A122 (2016)
[18] R. Adam *et al.*, Astronomy & Astrophysics **598**, A115 (2017)
[19] F. Keruzore *et al.*, Astronomy & Astrophysics **644**, A93 (2020)
[20] A. Ferragamo *et al.*, Astronomy & Astrophysics, submitted
[21] J. F. Navarro, C. S. Frenk, S. D. M. White, The Astrophysical Journal **462**, 563 (1996)
[22] M. Bartelmann, Astronomy & Astrophysics **313**, 697 (1996)
[23] G. Gianfagna *et al.*, MNRAS **502**, 5115–5133 (2021)
[24] Planck Collaboration *et al.*, Astronomy & Astrophysics **571**, A20 (2014)
[25] Planck Collaboration *et al.*, Astronomy & Astrophysics **594**, A27 (2016)
[26] R. Adam *et al.*, Astronomy & Astrophysics **614**, A118 (2018)
[27] R. Adam *et al.*, Astronomy & Astrophysics **606**, A64 (2017)
[28] A. Bonafede *et al.*, MNRAS **478**, 2927–2938 (2018)
[29] A. Bonafede *et al.*, MNRAS **426**, 40-56 (2012)
[30] K. Umetsu *et al.*, The Astrophysical Journal **795**, 163 (2014)
[31] K. Umetsu *et al.*, The Astrophysical Journal **860**, 104 (2018)
[32] M. Penna-Lima *et al.*, Astronomy & Astrophysics **604**, A89 (2017)

Modelling the cold dust in nearby spiral galaxies with radiative transfer

Angelos Nersesian[1,*], *Maarten* Baes[1], and *Suzanne C.* Madden[2]

[1]Sterrenkundig Observatorium Universiteit Gent, Krijgslaan 281 S9, B-9000 Gent, Belgium
[2]Laboratoire AIM, CEA/DSM - CNRS - Université Paris Diderot, IRFU/Service d'Astrophysique, CEA Saclay, 91191, Gif-sur- Yvette, France

Abstract. Cosmic dust grains are one of the fundamental ingredients of the interstellar medium (ISM). In spite of their small contribution to the total mass budget, dust grains play a significant role in the physical and chemical evolution of galaxies. Over the past decades, a plethora of multi-wavelength data, from UV to far-infrared, has increased substantially our knowledge on the dust properties of nearby galaxies. Nevertheless, one regime of the spectrum, the mm range, remains relatively unexplored. Thanks to the new, high-resolution data in the mm range observed with the NIKA2 instrument and our radiative transfer framework, we aim to firmly characterise the physical properties of the very cold dust (<15K), and to quantify the importance of different emission mechanisms in the mm. So far, we have developed a methodology to use dust radiative transfer modelling and applied it to a small group of face-on spiral galaxies. The combination of the new NIKA2 data with our radiative transfer techniques would provide the right conditions to generate an accurate model of the interplay between starlight and dust in a sizeable sample of spatially-resolved nearby galaxies.

1 Introduction

Galaxies are complex structures of multiple components: dark matter, stars, interstellar gas and dust. In particular, dust grains are carbonaceous or silicate based particles in the interstellar medium (ISM) of galaxies, and are responsible for the attenuation and reddening effects at ultraviolet (UV) and optical wavelengths. The role of dust does not stop there, since it is involved in several physical and chemical processes, by catalysing the formation of molecular hydrogen, and regulating the heating or cooling of the interstellar gas. One of the more pronounced characteristics of dust is that it absorbs roughly 30% of the light emitted by stars and redistributes this light back in the mid- and far-infrared wavelengths [1–4]. The emission in that wavelength regime is often associated and used as a tracer of the star-formation activity in galaxies [5–7]. However, the contribution of stars of age greater than 8 Gyr to the dust heating can be non-negligible [e.g. 8, 9] and needs to be considered while estimating the rate which galaxies form their stars.

One of the more advanced methods at our disposal to retrieve the distribution and physical properties of dust in galaxies, is by modelling the interplay between starlight and dust

*e-mail: angelos.nersesian@ugent.be

through radiative transfer (RT). RT modelling allows the generation of a realistic model of the radiation field and its interaction with dust, by quantifying the relative contribution of each stellar population to the dust heating. The advantage of using RT in relation to other methods, such as pixel-by-pixel SED (spectral energy distribution) fitting, is that it takes into account the effect of non-local heating in a 3D geometrical setting. In a recent series of radiative transfer studies, [10] developed a method to quantify the dust heating fraction related to the star-formation activity in face-on spiral galaxies, by constructing the RT model of M 51a (Paper I). The authors showcased the importance of the older stellar population (~10 Gyr) to the heating of the diffuse dust in the ISM, donating up to 35% of its total energy budget to that cause. Within the DustPedia[1] framework [11] the methodology of constructing an RT model was optimised and applied to the early-spiral galaxy M 81 [Paper II, 12], to a sample of face-on barred galaxies [Paper III, 13], to NGC 1068 which hosts an active galactic nucleus (AGN) [Paper IV, 14], and to the M 51 interacting system [Paper V, 15]. Two additional studies followed the same strategy of building models for two galaxy members of the Local Group, M 31 [16], and M 33 [17].

In this proceedings paper, we would like to present few of the highlight-results of this series of radiative transfer papers, to inform about the limitations of our methodology, and how we can treat them using the newly available observational data in the mm regime from the NIKA2 instrument.

2 Modelling approach

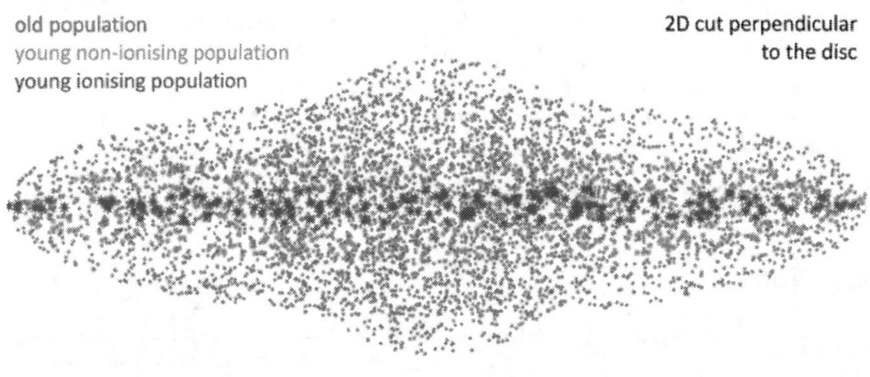

Figure 1. A schematic of the standard geometric model of spiral galaxies adopted in our studies. The model consists of an old stellar bulge and disk component, and two young stellar populations: a non-ionising (100 Myr), and an ionising (10 Myr). The dust distribution is assumed to have the same scale-height as the young non-ionising stars. Image Credit: [12].

In order to generate a self-consistent 3D model for a specific galaxy, we use the state-of-the-art RT code SKIRT [18–20]. SKIRT embraces the Monte-Carlo method, taking into account all relevant physical processes such as scattering, absorption and thermal re-emission by dust. Figure 1 shows a schematic representation of the standard geometric model of spiral galaxies adopted in our work. The model consists of four stellar components and a dust component. We consider a central stellar bulge, and a disk populated by an old and young (non-ionising and ionising) stellar populations. We model the old and young non-ionising

[1] http://dustpedia.astro.noa.gr

stellar populations using the Bruzual & Charlot [21] SSP of ages 8 Gyr and 100 Myr respectively, while for the young ionising stellar population we adopt the SED templates from MAPPINGS III [22] assuming an age of 10 Myr.

Broadband images at different wavelengths and at their native resolution were used to retrieve the geometric distributions of the various stellar and dust components. A brief description of the modelling steps follows. First, a bulge-to-disk decomposition is performed using an IRAC 3.6 μm image, tracing the emission of old stars. The bulge is approximated with a flattened Sérsic profile [23]. Then the old stellar population disk is obtained by subtracting the bulge from the total observed 3.6 μm emission. Similarly, we retrieve the distributions of the young non-ionising and ionising stars in the disk from the GALEX far-ultraviolet (FUV) image, and the combination of the MIPS 24 μm image with an Hα map, respectively. The dust component is constrained through a FUV attenuation map, while for the dust composition we used the dust model `THEMIS` [24].

The 3D distribution of each of the disk components is introduced by assigning an exponential profile of different scale-heights based on previous estimates of the vertical extent of edge-on galaxies [25]. Based on the dust map a binary tree dust grid is generated [26], through which photons propagate in our simulations. Each galaxy's dust grid has approximately 3 million dust cells. In general, we have three free parameters in our model determined via a χ^2 optimisation procedure. These parameters are: the FUV luminosities of the young and ionising stellar populations and the dust mass. In special occasions, e.g. in the case of AGN emission, more free parameters are introduced in the model. Initial guesses for these parameters are based on global SED fitting results [27].

3 Results

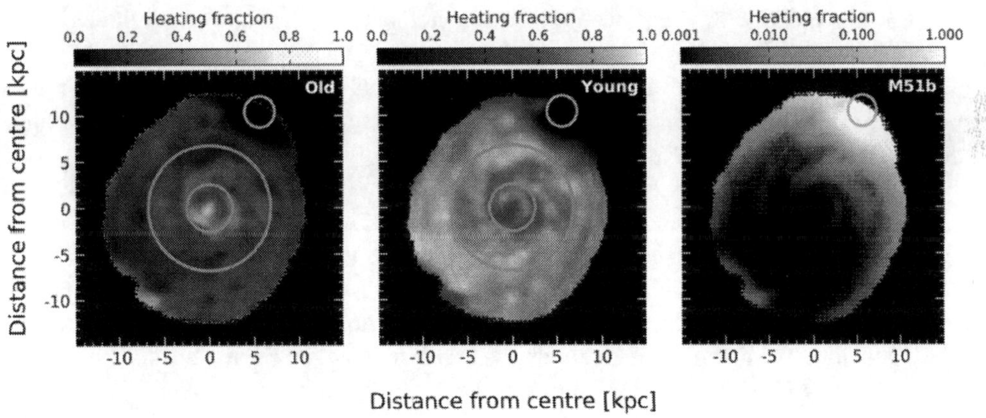

Figure 2. Dust heating maps of a face-on view of M 51. Left panel: dust heating fraction by the old stellar population; middle panel: dust heating fraction by the young stellar populations; and right panel: dust heating fraction by M 51b. The gold circle indicates the extent of the dusty disk of M 51b. Image Credit: [15].

Throughout this series of radiative transfer papers new insights on the dust heating processes were presented. [12] showed that the contribution of the old stellar population to the dust heating of M 81 is approximately 50%, highlighting the importance of old stars to the FIR emission. Furthermore, [13] studied a small sample of barred galaxies (M 83, M 95, M 100, and NGC 1365) and presented evidence of how the galactic bars may affect the processes of star formation and dust heating. The estimated contribution of the old stellar population in those galaxies remains considerably high, yet not as not as significant as in M 81.

Viaene et al. [14] demonstrated that the AGN in NGC 1068 (M 77) is only important to dust heating within the inner few hundred parsecs from the centre of the galaxy, while most of the FIR emission is powered by star formation. The most recent RT study by [15] investigated how the companion galaxy (M 51b) of the M 51 interacting system affects the SED, and consequently the dust heating of the main galaxy (M 51a). For that reason, a 3D RT model of the system was constructed. Figure 2 depicts the various dust-heating fractions from a face-on view of M 51. The left and middle panels show the dust-heating maps for the old and young stellar populations, respectively. The right panel depicts the dust-heating fraction provided by the older stellar population of the companion galaxy. The authors found that for the combined M 51 system, on average, 5.8% of the dust emission is attributed to the dust heating by the old stellar population of M 51b, of which 4.8% occurs in the M 51a subsystem; 23% of the dust heating is supplied by the old stellar population of M 51a, while the remaining 71.2% arises from the heating by the young stellar populations.

From the 3D model of each galaxy we can immediately quantify the contribution of different sources (young stars, evolved stellar populations) to the dust. Our RT models suggest a strong correlation between the dust heating fraction and the specific star-formation rate (sSFR). In Fig. 3 we plot the heating fraction for six galaxies, as a function of the sSFR. It is immediately evident that there is an increasing trend between the heating fraction and the sSFR. This, in principle, enables us to quantify the heating fraction based on sSFR measurements in other galaxies. As we discuss later, our plan is to uniformly model a significant sample of galaxies which will allow us to confirm these trends and to firmly quantify this relation.

In spite of its clear advancement compared to the previous generation of galaxy dust radiative transfer models, this approach still has a number of persistent limitations. One of the most important limitations is the insensitivity of the DustPedia data set to cold dust. Although the *Herschel* Space Observatory delivered new data that expanded our view on the dust properties, they are still limited to 500 μm which may not be sufficient to trace any cold dust ($<$ 15 K). Of course, *Planck* data cover longer wavelengths, but unfortunately they lack the necessary spatial resolution. The lack of high-resolution observations at submm/mm wavelengths hinders our ability to firmly determine the dust emissivity index, and to trace the cold dust component assumed to carry most of the dust mass in galaxies [28, 29]. Apart from cold dust emission, synchrotron emission (dominated by emission from supernova remnants) and free-free emission (from ionised gas in star-forming regions) are the two other dominant processes, which we need to quantitatively assess and disentangle from the dust emission.

Fortunately, this issue can now be addressed thanks to a new development. New observational imaging data on nearby galaxies are currently being taken in the critical submm/mm wavelength range. The most important data are coming from the NIKA2 camera (1.2 and 2 mm) on the IRAM 30 m telescope, and the SCUBA2 camera (850 μm) on the JCMT telescope. In particular, the IMEGIN project (PI: Suzanne Madden) is an ongoing 200 h program dedicated to deep NIKA2 continuum imaging of a set of 22 nearby spiral galaxies. All IMEGIN galaxies are taken from the DustPedia sample and hence have excellent ancillary UV-FIR data available, as well as complementary radio, molecular and atomic gas observations. Moreover, many of the IMEGIN galaxies are also targeted with SCUBA2 in the frame of the DOWSING survey, and plans are currently made to also observe the remaining galaxies with SCUBA2. Combining all these ingredients will allow us to move to a next level of 3D dust radiative transfer modelling of nearby galaxies, and to use these models to address a number of the fundamental questions on interstellar dust in galaxies.

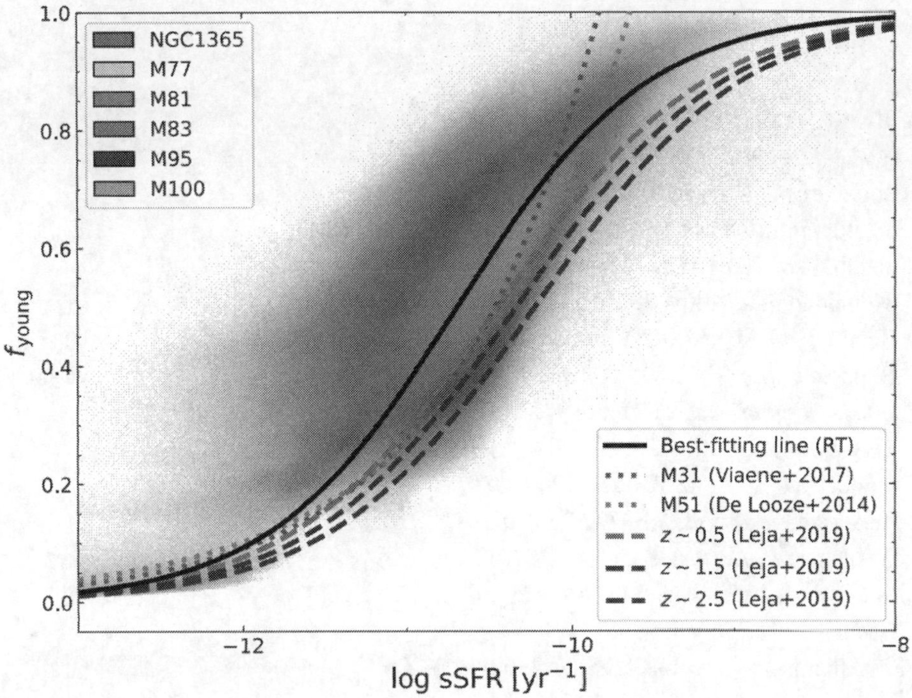

Figure 3. The relation between the fraction of dust heating by young stars and the specific star formation rate of galaxies. Image Credit: [13].

4 Conclusions

We have constructed highly detailed 3D RT models for a small group of face-on galaxies to investigate the dust heating processes. We find that the contribution from the older stellar population to the dust emission at the infrared bands is more significant than previously thought. This result questions the use of infrared luminosity as a proxy for the star formation activity in star-forming galaxies. Furthermore, we confirm a tight correlation between the dust heating fraction and the sSFR. The full description of our methodology and the results of the modelling of M 81 are presented in [12], while the detailed results for the barred galaxies are discussed in [13]. Finally, the modelling of a galaxy with an AGN component, NGC 1068 is presented in [14], and the results of the M 51 interacting system in [15].

Acknowledgements

AN, gratefully acknowledges the support of the Research Foundation - Flanders (FWO Vlaanderen). DustPedia is a collaborative focused research project supported by the European Union under the Seventh Framework Programme (2007-2013) call (proposal no. 606847). The participating institutions are: Cardiff University, UK; National Observatory of Athens, Greece; Ghent University, Belgium; Université Paris Sud, France; National Institute for Astrophysics, Italy and CEA, France. This work was supported by the Programme National "Physique et Chimie du Milieu Interstellaire" (PCMI) of CNRS/INSU with INC/INP co-funded by CEA and CNES.

References

[1] C.C. Popescu, *et al.*, ApJ **567**, 221 (2002)
[2] R.A. Skibba, *et al.*, ApJ **738**, 89 (2011)
[3] S. Viaene, *et al.*, A&A **586**, A13 (2016)
[4] S. Bianchi, *et al.*, A&A **620**, A112 (2018)
[5] D. Calzetti, *et al.*, ApJ **666**, 870 (2007)
[6] D. Calzetti, *et al.*, ApJ **714**, 1256 (2010)
[7] R.C. Kennicutt, *et al.*, ApJ **703**, 1672 (2009)
[8] G.J. Bendo, F. Galliano, S.C. Madden, MNRAS **423**, 197 (2012)
[9] G.J. Bendo, *et al.*, MNRAS **448**, 135 (2015)
[10] I. De Looze, *et al.*, A&A **571**, A69 (2014)
[11] J.J. Davies, *et al.*, PASP **129**, 044102 (2017)
[12] S. Verstocken, *et al.*, A&A **637**, A24 (2020)
[13] A. Nersesian, *et al.*, A&A **637**, A25 (2020)
[14] S. Viaene, *et al.*, A&A **638**, A150 (2020)
[15] A. Nersesian, *et al.*, A&A **643**, A90 (2020)
[16] S. Viaene, *et al.*, A&A **599**, A64 (2017)
[17] T.G. Williams, *et al.*, MNRAS **487**, 2753 (2019)
[18] M. Baes, *et al.*, ApJS **196**, 22 (2011)
[19] P. Camps, M. Baes, Astronomy and Computing **9**, 20 (2015)
[20] P. Camps, M. Baes, Astronomy and Computing **31**, 100381 (2020)
[21] G. Bruzual, S. Charlot, MNRAS **344**, 1000 (2003)
[22] B. Groves, *et al.*, ApJS **176**, 438 (2008)
[23] J.L. Sérsic, Boletin de la Asociacion Argentina de Astronomia La Plata Argentina **6**, 41 (1963)
[24] A.P. Jones, *et al.*, A&A **602**, A46 (2017)
[25] G. De Geyter, *et al.*, MNRAS **451**, 1728 (2015)
[26] W. Saftly, M. Baes, P. Camps, A&A **561**, A77 (2014)
[27] A. Nersesian, *et al.*, A&A **624**
[28] R. Siebenmorgen, E. Krügel, R. Chini, A&A **351**, 495 (1999)
[29] G.J. Bendo, *et al.*, ApJ **652**, 283 (2006)

EPJ Web of Conferences **257**, 00035 (2022) https://doi.org/10.1051/epjconf/202225700035
mm Universe @ NIKA2

Pressure profiles of distant Galaxy clusters with *Planck*-SPT data

Filippo Oppizzi[1,*], *Federico* De Luca[1], *Hervé* Bourdin[1], *Pasquale* Mazzotta[1], and the CHEX-MATE collaboration

[1]Università degli studi di Roma 'Tor Vergata', Via della ricerca scientifica, 1; 00133 Roma, Italy

Abstract. We present a full set of numerical tools to extract Galaxy Cluster pressure profiles from the joint analysis of *Planck* and South Pole Telescope (SPT) observations. Pressure profiles are powerful tracers of the thermodynamic properties and the internal structure of the clusters. Tracing the pressure over the cosmic times allows one to constraints the evolution of the cluster structure and the contribution of astrophysical phenomena. SPT and Planck are complementary to constrain the cluster structure at various spatial scales. The SPT cluster catalogue counts 677 cluster candidates up to redshift 1.7, it is a nearly mass-limited sample, an ideal benchmark to test cluster evolution. We developed a pipeline to first separate the cluster signal from the background and foreground components and then jointly fit a parametric profile model on a combination of Planck and SPT data. We validate our algorithm on a subsample of six clusters, common to the SPT and the CHEX-MATE catalogues, comparing the results with the profiles obtained from X-ray observations with XMM-*Newton*.

1 Introduction

Galaxy clusters form from the largest gravitational overdensities in the density field and track the evolution of the Universe at the largest scales [1]. A large amount of ionised hot gas lies in hydrostatic equilibrium in the potential well of the Dark Matter (DM) halos of the clusters. As a consequence of the tight correlation between the gas and the DM density field, the scale invariance of the latter reflects in the self-similarity of the pressure profiles. The intra-cluster gas pressure is thus a sensitive tracer of the internal structure of galaxy clusters.

The Sunyaev-Zeldovich effect (SZ) arises from the interaction of the free electrons in the ionised gas with the Cosmic Microwave Background (CMB) photons. It is a direct tracer of the gas pressure and is independent of the redshift. [2]. The gas also emits bremsstrahlung radiation in X-ray, providing a complementary probe of the cluster structure that is sensitive to different density regimes.

In this work, we present a pipeline to extract cluster pressure profiles from a combination of *Planck* and South Pole Telescope (SPT) data. The high sensitivity of Planck and the SPT high resolution make the two instruments complementary in constraining the cluster structure over a wide range of scales. We compare the results from this SZ analysis with the profiles derived from X-ray measurements. We analyse six clusters extracted from the sample of the Cluster

*e-mail: filippo.oppizzi@gmail.com

HEritage project with XMM-Newton-Mass Assembly and Thermodynamics at the Endpoint of structure formation (CHEX-MATE) [3].
In section 2 we present the data-sets and the cluster sample, in section 3 we describe the analysis techniques and in section 4 we present our results. Finally, section 5 is dedicated to the conclusions.

2 Data-sets and Cluster Sample

SPT and *Planck* represent the state of the art of cosmological surveys at millimetre and sub-millimetre wavelengths. Combining the information from both data-sets it is possible to solve the inner core of all clusters in the SPT catalogue and at the same time to map the faint peripheries with high precision.
We use the public data from the *Planck* High-Frequency Instrument (HFI) [4]. The six sky maps with nominal frequencies 100, 143, 217, 353, 545 and 857GHz and resolution 9.66, 7.22, 4.90, 4.92, 4.67, 4.22 arcmin respectively.
SPT observes the microwave sky in three frequency bands centred at 95, 150 and 220 GHz with 1.7, 1.2, and 1.0 arcmin resolution, respectively over an area of 2540 $\deg^2$ [5]. Thanks to its resolution it can solve the core of distant galaxy clusters, inaccessible by *Planck* [6]. In this work, we use the public SPT maps[1] that are convolved with a common Gaussian beam with 1.75 arcmin (FWHM).
CHEX-MATE is a program of X-ray observations with the XMM-*Newton* satellite of a selection of 118 galaxy clusters of the *Planck* PSZ2 catalogue designed to be unbiased and signal-to-noise-limited [3]. To test our technique we use a set of clusters representing the overlap between the SPT and the CHEX-MATE catalogue. We use the X-ray counterparts as a benchmark for the pressure profile derived from our analysis of millimetric data. The sample covers a redshift range from $z = 1.6$ to $z = 0.6$, with angular dimension from $\theta_{500} = 2.8$ to $\theta_{500} = 7.9$ arcmin.

3 Method

We first perform a component separation to remove the contaminant emissions on the data from the two instruments separately and then we do a join-fit of the combined data-set, hence fitting a single pressure profile on both data-sets. We refer to [7] for detailed description of the method.

3.1 Processing of *Planck* data

We project the full sky *Planck* maps on tiles of 512×512 pixels, with the resolution of 1 arcmin/pixel centred on the cluster. Each image is then processed with the technique developed in [8] to isolate the cluster signal from the background and foreground components. The technique consists of a four-components fit taking into account the cluster SZ signal, the primary CMB anisotropies, the Galactic Thermal Dust and a correction for the intra-cluster Dust emission. We use a wavelet thresholding algorithm to obtain the spatial templates of the CMB and the Galactic dust directly from the 857 GHz and the 217 GHz *Planck* channels respectively. To model the thermal dust spectral energy distribution we use the two-temperature grey body

[1] https://pole.uchicago.edu/public/data/chown18/index.html

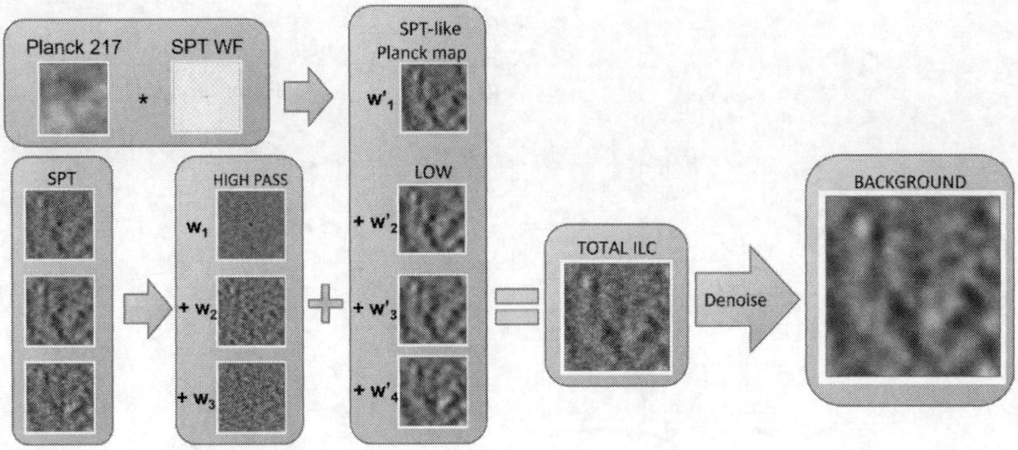

Figure 1. Flow chart that describe the SPT component separation pipeline described in section 3.2

proposed in [9]. We compute the correction for intra-cluster dust assuming a Navarro-Frenk-White profile (NFW) and a grey body spectral energy distribution (SED). All the templates are then convolved with the proper beams to match the resolution of each channel, fitted and subtracted to the raw data to obtain the cleaned maps where we will perform the cluster fit.

3.2 Processiong of SPT data

As for *Planck* we select patches of 512×512 pixels around the XMM cluster centre, but with resolution of 0.5 arcmin/pixel instead of 1 arcmin/pixel. We do not use the same technique applied in the *Planck* data reduction, since SPT lacks the high-frequency channels necessary to model the thermal dust. On the other hand, SPT is insensitive to the large spatial scales where the dust emission is relevant, our clusters are far from the dusty galactic plane, and SPT observe at low frequencies where the dust SED is low. Thus, we do not expect to observe significant dust contamination, and we do not find any evidences of it in our analysis. Therefore, we focus on the modelling of the CMB signal using a linear combination (LC) of the three SPT channels and the 217 GHz channel of *Planck* that we re-project on a map with the same characteristic than the SPT ones. We compute the weights of the LC by minimising the variance with respect to a signal with a flat SED (*i.e.* the CMB) and simultaneously to null the non-relativistic SZ component. We split then each map in a low-pass filtered map matching the *Planck* 217 resolution, and in a high pass filtered map containing the residual small scale features. We then perform the LC separately on the low-passed maps and high-passed maps, including the *Planck map* in the combination of the large scales. We obtain two maps representing the large and small scales CMB fluctuations, that we re-combine into a single map. Finally, we denoise the CMB template with a wavelet thresholding algorithm. Fig. 1 shows the flow chart of the process just described.

3.3 Joint fit

We combine the SPT and *Planck* data in a joint fit of a cluster thermal SZ template derived from the projection along the line of sight of a generalised NFW (gNFW) pressure profile:

$$P(r) = P_0 \times \frac{P_{500}}{x^{\gamma}(1 + x^{\alpha})^{(\beta-\gamma)/\alpha}}, \tag{1}$$

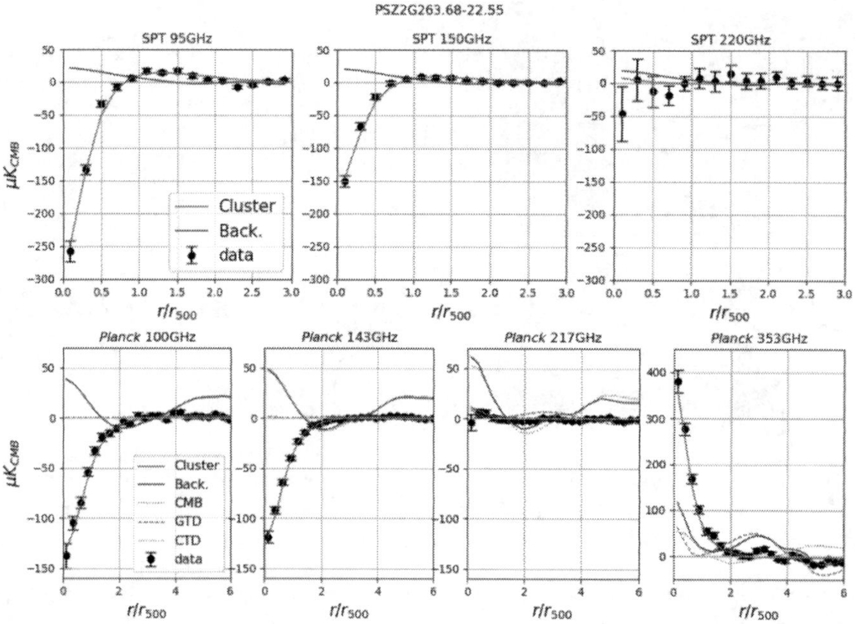

Figure 2. Intensity profiles around PSZ2G259.98-63.43 for the SPT channels (upper panels) and *Planck* channels (lower panels). Black dots represent the radial average on the cleaned maps, the red line is the cluster profile, the blue line is the background. For *Planck* the dotted cyan line is the CMB, the dashed red line is the GTD and the dash-dotted orange line is the cluster dust.

where $x = c_{500}r/r_{500}$, r_{500} is the radius where the density is 500 times the critical density of the Universe at the cluster redshift, c_{500} is the concentration parameter and α, β and γ are the slopes at $r \sim r_{500}/c_{500}$, $r \gg r_{500}/c_{500}$ and $r \ll r_{500}/c_{500}$ respectively. The template is then converted in SZ emissivity and convolved with the beam and the SPT window function to build a multi channel model of the cluster signal. We compute the likelihood on the radially averaged profiles of the cleaned data and the model. We use circular bins of width $0.2r_{500}$ for SPT, and of 2 arcmin for *Planck*, up to $3r_{500}$ for SPT and $6r_{500}$ for *Planck*. We assume a Gaussian Likelihood and we perform a MCMC fit with the Cobaya MCMC sampler [10]. For *Planck* channels, we obtain the covariance matrices from Monte-Carlo simulations of the noise in the surroundings of the cluster. For SPT channels instead, we use the sample covariance of the noise maps provided by the SPT collaboration [11]. We try different combinations of parameters to optimise the convergence of our chain. We conclude to leave free to vary the amplitude P_0 and the inner and outer slopes γ and β and to keep the concentration parameter and the intermediate slope fixed to their universal values identified by [12] $c_{500} = 1.177$ and $\alpha = 1.0501$. We fix the characteristic gas pressure P_{500} and the radius r_{500} to the values derived from the M_{500} and redshifts reported in the PSZ2 *Planck* catalogue by using the self-similar model [13, 14] and assuming a ΛCDM cosmology with $H_0 = 70$ Km s^{-1}Mpc^{-1}, $\Omega_M = 0.3$ and $\Omega_\Lambda = 0.7$.

3.4 X-ray analysis

We based our X-ray analysis on the observations of the European Photon Imaging Camera (EPIC) public data of the XMM-*Newton* telescope. We assume a background model taking

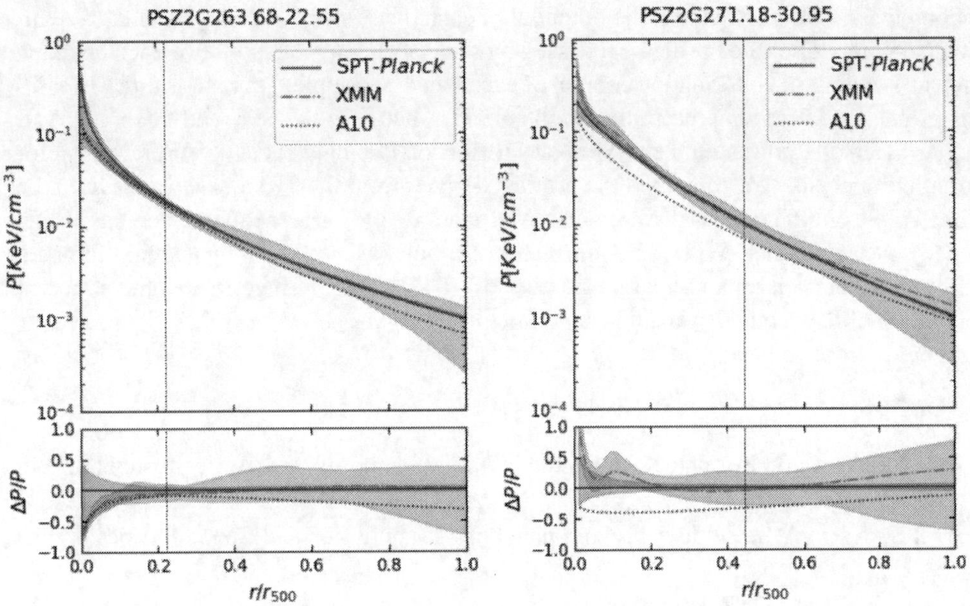

Figure 3. Profiles from XMM-*Newton* data (dash dotted red lines) and from SPT-Planck joint fit (blue lines). The shaded regions correspond to the 68% credible intervals. The dotted black lines show the universal profile from [12]. The dashed lines indicate the FWHM of the SPT beam.

into account the quiescent particle background, the cosmic X-ray background, and the thermal emission associated with our Galaxy. We threat flares and high-background periods as in [15]. We estimate the X-ray thermodynamical profiles as done in [8]. We use the expressions introduced in [16] to parametrise the emission measure $[n_p n_e](r)$ and temperature $T(r)$. We then integrate the templates along the line of sight and we fit them to the background-subtracted X-ray observable profiles. We refer to [17] for a complete description of our X-ray data reduction technique.

4 Results

We show in Fig. 2 the intensity profile of the estimated cluster signal and the contaminant components at each frequency for one of the clusters in our sample, compared with the cleaned data. We compare the pressure profile derived from the SZ maps with the expectations from X-ray observations. Since X-ray and SZ data probe the cluster structure at different regimes, we limit our comparison to the overlapping region between the two, *i.e.* inside r_{500}. We show a preliminary comparison between the two fits on two clusters of our sample in Fig. 3. The agreement is remarkable: the credibility intervals overlap along the whole radius range tested. Some degree of divergence can be observed in the very inner regions that are not solved by SPT. The agreement between the two analysis confirms the consistency of our techniques.

5 Conclusions

We presented the preliminary results of the measurement of the pressure profile of galaxy clusters from the joint analysis of SPT and *Planck* data. We developed two independent data reduction pipelines to fully exploit the potential of each instrument. We fit a spherically symmetric gNFW model of the profiles, processed by the instrumental response of each channel, to efficiently exploit the information content of each band. We applied our algorithm to a sample of six galaxy clusters representing the intersection between the SPT and CHEX-MATE catalogues. Here we presented the preliminary results on two clusters and we refer to [7] for the complete analysis. We compared the profiles derived from the SPT-*Planck* joint fit to the ones obtained from the XMM-*Newton* data. We found a good agreement between the results of the two measurements. With the confirmation of the consistency of our method in hand, we will expand our analysis to a larger number of SPT clusters to investigates the structure and evolution of the profiles through cosmic time.

References

[1] A.V. Kravtsov, S. Borgani, Annual Review of Astronomy and Astrophysics **50**, 353 (2012), 1205.5556

[2] R.A. Sunyaev, Y.B. Zeldovich, Comments on Astrophysics and Space Physics **4**, 173 (1972)

[3] The CHEX-MATE Collaboration, Arnaud, M., Ettori, S., Pratt, G. W., Rossetti, M., Eckert, D., Gastaldello, F., Gavazzi, R., Kay, S.T., Lovisari, L. et al., A&A **650**, A104 (2021)

[4] Planck Collaboration, N. Aghanim, Y. Akrami, M. Ashdown, J. Aumont, C. Baccigalupi, M. Ballardini, A.J. Banday, R.B. Barreiro, N. Bartolo et al., A&A **641**, A3 (2020), 1807.06207

[5] J.E. Carlstrom, P.A.R. Ade, K.A. Aird, B.A. Benson, L.E. Bleem, S. Busetti, C.L. Chang, E. Chauvin, H.M. Cho, T.M. Crawford et al., PASP **123**, 568 (2011), 0907.4445

[6] L.E. Bleem, B. Stalder, T. de Haan, K.A. Aird, S.W. Allen, D.E. Applegate, M.L.N. Ashby, M. Bautz, M. Bayliss, B.A. Benson et al., The Astrophysical Journal Supplement Series **216**, 27 (2015), 1409.0850

[7] F. Oppizzi, F. De Luca, H. Bourdin, P. Mazzotta, The CHEX-MATE Collaboration, in prep (2022)

[8] H. Bourdin, P. Mazzotta, A. Kozmanyan, C. Jones, A. Vikhlinin, The Astrophysical Journal **843**, 72 (2017), 1707.02248

[9] A.M. Meisner, D.P. Finkbeiner, The Astrophysical Journal **798**, 88 (2015), 1410.7523

[10] J. Torrado, A. Lewis, JCAP **2021**, 057 (2021), 2005.05290

[11] R. Chown, Y. Omori, K. Aylor, B.A. Benson, L.E. Bleem, J.E. Carlstrom, C.L. Chang, H.M. Cho, T.M. Crawford, A.T. Crites et al., The Astrophysical Journal Supplement Series **239**, 10 (2018), 1803.10682

[12] M. Arnaud, G.W. Pratt, R. Piffaretti, H. Böhringer, J.H. Croston, E. Pointecouteau, A&A **517**, A92 (2010), 0910.1234

[13] N. Kaiser, MNRAS **222**, 323 (1986)

[14] D. Nagai, A.V. Kravtsov, A. Vikhlinin, The Astrophysical Journal **668**, 1 (2007), astro-ph/0703661

[15] H. Bourdin, P. Mazzotta, M. Markevitch, S. Giacintucci, G. Brunetti, The Astrophysical Journal **764**, 82 (2013)

[16] A. Vikhlinin, A. Kravtsov, W. Forman, C. Jones, M. Markevitch, S.S. Murray, L. Van Speybroeck, The Astrophysical Journal **640**, 691 (2006), astro-ph/0507092

[17] F. De Luca, H. Bourdin, P. Mazzotta a, F. Oppizzi, The CHEX-MATE Collaboration, in prep (2022)

The Three Hundred–NIKA2 Sunyaev–Zeldovich Large Program twin samples: Synthetic clusters to support real observations

A. Paliwal[1,*], *E.* Artis[2], *W.* Cui[3], *M.* De Petris[1], *F.-X.* Désert[4], *A.* Ferragamo[1], *G.* Gianfagna[1], *F.* Kéruzoré[2], *J.-F.* Macías-Pérez[2], *F.* Mayet[2], *M.* Muñoz-Echeverría[2], *L.* Perotto[2], *E.* Rasia[5,6], *F.* Ruppin[7], and *G.* Yepes[8]

[1]Dipartimento di Fisica, Sapienza Università di Roma, Piazzale Aldo Moro 5, 00185 Roma, Italy
[2]Univ. Grenoble Alpes, CNRS, Grenoble INP, LPSC-IN2P3, 53, avenue des Martyrs, 38000 Grenoble, France
[3]Institute for Astronomy, University of Edinburgh, Royal Observatory, Edinburgh EH9 3HJ, United Kingdom
[4]Univ. Grenoble Alpes, CNRS, IPAG, 38000 Grenoble, France
[5]National Institute for Astrophysics, Astronomical Observatory of Trieste (INAF-OATs), via Tiepolo 11, 34131 Trieste, Italy
[6]Institute for Fundamental Physics of the Universe (IFPU), via Beirut 2, 34014 Trieste, Italy
[7]Kavli Institute for Astrophysics and Space Research, Massachusetts Institute of Technology, Cambridge, MA 02139, USA
[8]Departamento de Física Teórica and CIAFF, Módulo 8, Facultad de Ciencias, Universidad Autónoma de Madrid, 28049 Madrid, Spain

Abstract. The simulation database of The Three Hundred Project has been used to pick synthetic clusters of galaxies with properties close to the observational targets of the NIKA2 camera Sunyaev–Zeldovich (SZ) Large Program. Cross–matching of cluster parameters such as mass and redshift of the cluster in the two databases has been implemented to generate the so–called twin samples for the Large Program. This SZ Large Program is observing a selection of galaxy clusters at intermediate and high redshift ($0.5 < z < 0.9$), covering one order of magnitude in mass. These are SZ–selected clusters from the Planck and Atacama Cosmology Telescope catalogs, wherein the selection is based on their integrated Compton parameter values, Y_{500}: the value of the parameter within the characteristics radius R_{500}.

The Three Hundred hydrodynamical simulations provide us with hundreds of clusters satisfying these redshift, mass, and Y_{500} requirements. In addition to the standard post-processing analysis of the simulation, mock observational maps are available mimicking X–ray, optical, gravitational lensing, radio, and SZ observations of galaxy clusters. The primary goal of employing the twin samples is to compare different cluster mass proxies from synthetic X–ray, SZ effect and optical maps (via the velocity dispersion of member galaxies and lensing κ-maps) of the clusters. Eventually, scaling laws between different mass proxies and the cluster mass will be cross–correlated to reduce the scatter on the inferred mass and the mass bias will be related to various physical parameters.

*e-mail: aishwarya.paliwal@uniroma1.it

1 Introduction

Galaxy clusters are the most massive gravitationally bound objects in the Universe, forming at the nodes of the cosmic web [1, 2]. They are excellent tools for both astrophysical and cosmological studies (see *e.g.* [3–5]). The imprint of structure formation is carried by the galaxy cluster number density and spatial distribution of galaxy clusters. Hence, the abundance of galaxy clusters with mass and redshift is a well–known cosmological probe. The mass function of galaxy clusters describes the number density of clusters above a given mass M. The sensitivity of the cluster distribution, to the underlying cosmology, is well reflected by the mass function of galaxy clusters, which is exponentially sensitive to cosmological parameters [6].

Understandably, the cluster mass is a key parameter for studies that aim to constrain cosmological parameters using galaxy cluster distribution with mass and redshift, *i.e.* galaxy cluster number density. The cluster mass, however, is not directly observable and there are several ways to infer it via observations under simple theoretical assumptions. Each of these ways to infer the mass is prone to a bias and intrinsic scatter associated with the observations being used. The cluster mass can be estimated using the hydrostatic equilibrium (HE) equation under the assumption of the cluster having a spherical Dark Matter halo; the Intra Cluster Medium (ICM) profiles, namely, the density profile from X–ray and the pressure profile from the Sunyaev–Zeldovich (SZ) effect [7] observations, can be combined to infer the cluster mass (see *e.g.* [8]). This inference is prone to mass bias associated to the assumptions made for the HE equation. The kinematic studies of the member galaxies can also be used to estimate the mass of galaxy clusters (see *e.g.* [9]). These calculations require spectral information of a lot of member galaxies to get an accurate mass estimate with good precision. Estimating the cluster mass through measuring the gravitational lensing effect of clusters on the background sources is one of the least biased methods but, comes with high dispersion on the mass estimate [10]. Finally, another way to infer the cluster mass is through scaling laws, which, when correctly calibrated, relate the mass to observational quantities, such as X–ray luminosity, SZ brightness, etc. [4, 11–14]. An alternative and orthogonal approach to refine the cluster mass estimates is by using state–of–the–art N–body hydrodynamical simulations as useful and efficient test–beds to increase our theoretical understanding [15]. Evidently, it is critical to understand astrophysical details that may affect the mass estimation via the observational methods mentioned above and to properly calibrate scaling relations between observables and the mass of the cluster.

Establishing the relations between observables and the cluster mass from direct observations is, however, difficult and not a straightforward process. The combination of high sensitivity and high angular resolution observations with data from detailed hydrodynamical simulations of clusters could alleviate these problems. This combination will enable one to accurately estimate the cluster mass and minimise the scatter on the inferred value. This work aims to achieve this goal by the use of observations from the NIKA2 SZ Large Program [16] along with simulated data from The Three Hundred Project [17].

This paper is structured as follows: Section 2 gives a brief overview of the observational catalog (SZ Large Program catalog) along with the simulation database (The Three Hundred). Section 3 describes how the simulation database has been used to build twin samples of the SZ Large Program and mentions the associated data products. Section 4 summarises the twin samples and discusses how their data products may be utilised to perform analyses that support observational studies.

2 Overview of the observational and simulated database

2.1 Observations: The NIKA2 Sunyaev–Zeldovich Large Program

The NIKA2 Sunyaev–Zeldovich Large Program (LPSZ) observes galaxy clusters with the NIKA2 camera [18, 19], which is a dual–band millimeter continuum instrument installed at the IRAM 30–meter telescope located at Pico Veleta, Spain. It is capable of mapping the sky simultaneously at 150 GHz and 260 GHz using kinetic inductance detectors or KIDs [20]. The camera has a field of view of 6.5′ and a high angular resolution of 17.6″ and 11.1″ at 150 GHz and 260 GHz, respectively. The sensitivity at the two respective wavelengths is also high, standing at 9 and 30 mJy $s^{\frac{1}{2}}$, respectively. Details of the performance of the camera can be found in [21]. The instrument has been proven to be well–designed for mapping the ICM of galaxy clusters and their thermodynamic properties at intermediate to high redshift (see *e.g.* [22, 23]).

The LPSZ is one of the four Large Programs that are a part of the NIKA2 guaranteed time. It is a high–resolution follow–up of approximately 50 SZ–selected galaxy clusters from the Planck [24] and Atacama Cosmology Telescope (ACT) [25] catalogs. The sample is representative of these clusters in mass and redshift. It spans an order of magnitude in mass with the cluster mass in the range $3 \leqslant M_{500}/10^{14}\ \mathrm{M}_{\odot} \leqslant 11$ and a relatively high redshift range of $0.5 < z < 0.9$, which is distributed over 4 equally spaced redshift bins.

The prior knowledge of the integrated SZ Compton parameters of these clusters from Planck and ACT catalogs, in combination with high–resolution data from NIKA2, allows the precise inference of the dynamical state of these clusters through their morphology and pressure profiles. The simulation data, explained in the next section, further strengthens this knowledge, creating room for improvement on current standards of ICM profile evaluation of clusters and consequently their mass estimation, in this redshift range.

2.2 Simulations: The Three Hundred

To completely exploit the high–quality observational data from NIKA2, this project picks complimentary data from The Three Hundred Project [17]. Few examples of detailed analyses of various aspects of the simulated clusters in The Three Hundred catalog can be found in [26–28].

These simulations are performed for a set of 324 Lagrangian regions of radius 15 h^{-1}Mpc, each centred on a galaxy cluster with $M_{500} > 4.6 \times 10^{14}\ h^{-1}\mathrm{M}_{\odot}$. These clusters are the ones originally recognised as the most massive galaxy clusters (virial mass $M > 8 \times 10^{14}\ h^{-1}\mathrm{M}_{\odot}$) identified at $z = 0$ in the *MultiDarkPlanck2* box of the simulation *MultiDark*[1] [29]. This is a dark–matter (DM)–only simulation and the *MultiDarkPlanck2* box is a periodically defined cube with size ~ 1.5 Gpc comoving length. This cube contains 3840^3 DM particles, each of mass $1.5 \times 10^9\ h^{-1}\mathrm{M}_{\odot}$. As suggested by its name, this simulation assumes a cosmology based on the best–fit model and cosmological parameter estimates from the Planck 2015 cosmological results [30].

All the regions are re-simulated at high resolution in 128 snapshots between redshifts $z = 0$ and $z = 17$ using three hydrodynamical codes: `GADGETMUSIC` [15], `GADGET-X` [31], and `GIZMO-SIMBA` [32]. The first of the three is a standard smooth–particle–hydrodynamic (SPH) code while the other two are modern SPH codes. Apart from hydrodynamic simulations with astrophysical details, all the regions are simulated with a semi–analytical model or SAM using the codes `Galacticus`, `SAG`, and `SAGE`. The technical details such as initial conditions and box sizes related to these codes can be found in [17]. The Three Hundred Project also

[1] Publically available at: https://www.cosmosim.org/

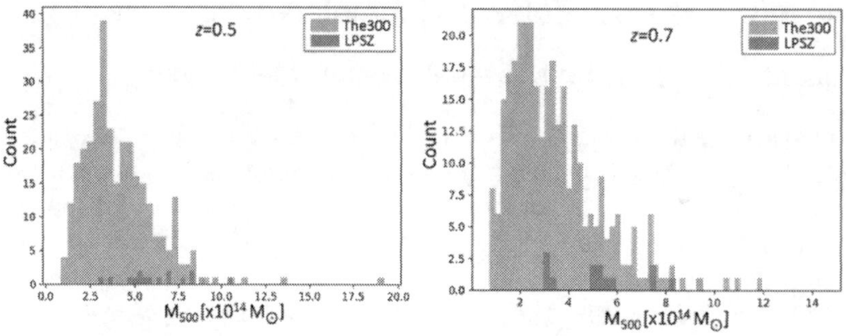

Figure 1. Redshift distribution of target clusters in two of the four LPSZ redshift bins, compared to the redshift distribution of clusters in the closest identified redshift snapshot of The Three Hundred Project.

provides multi–wavelength information of the clusters at the centre of each of the 324 regions, and their environment. These include mock X–ray, optical, gravitational lensing, radio, and SZ Compton parameter maps of each cluster. Clearly, this simulation provides a very detailed, all–rounded picture of clusters and as such, is a mass–complete, volume–limited sample for objects with $M_{500} > 4.6 \times 10^{14}\ h^{-1}\mathrm{M}_{\odot}$, over a wide range of redshift.

3 The twin samples

Some of the simulated clusters in The Three Hundred Project have been hand–picked to generate the so–called twin samples of the LPSZ. This has been done by cross–matching different parameters of the clusters in LPSZ and The Three Hundred Project. At first, redshift snapshots were identified in the simulation such that the median redshift of each LPSZ redshift bin is the closest to the redshift of the selected snapshot. Samples of the redshift distribution of clusters in both catalogs can be seen in Fig. 1. Evidently, the simulation offers a wide range of clusters to create twin samples at each of these redshifts.

After identifying simulated clusters at optimal redshifts, their following properties were cross–matched with the LPSZ clusters to, so far, generate 3 twin samples (TS): total mass ($TS_{\rm TM}$), hydrodynamic mass ($TS_{\rm HEM}$), and integrated Compton parameter at R_{500} (TS_Y). Each of these has generated a unique sample of clusters that is, in the given criteria, equivalent to the LPSZ. This uniqueness can be visualised through the redshift–mass distribution of the twin samples compared to the LPSZ target clusters, shown in Fig. 2.

The average ratio of masses for $TS_{\rm TM}$ is $\langle M_{\rm LPSZ}/M_{TS_{\rm TM}}\rangle = 1.01 \pm 0.02$, indicating the availability of twins for each cluster in the LPSZ. On the other hand, the mass bias for the other two samples is estimated to be $\langle M_{\rm LPSZ}/M_{TS_{\rm HEM}}\rangle = 0.93 \pm 0.02$ and $\langle M_{\rm LPSZ}/M_{TS_Y}\rangle = 0.9 \pm 0.1$, respectively. These values are consistent with mass bias values estimated in [33]. The matching criterion for the integrated Compton parameter is also satisfied for TS_Y with $\left\langle Y_{500_{\rm LPSZ}}/Y_{500_{TS_Y}}\right\rangle = 1.01 \pm 0.03$. As mentioned in the previous section, The Three Hundred Project provides a multitude of multi–wavelength data products for each cluster, some of which are also provided along different lines–of–sight (see Section 4). An example of the Y map and X–ray photon count map of two clusters can be seen in Fig. 3.

4 Summary and outlook

Twin samples have successfully been generated from The Three Hundred project by identifying equivalent clusters that accurately represent the NIKA2 LPSZ clusters, based on different

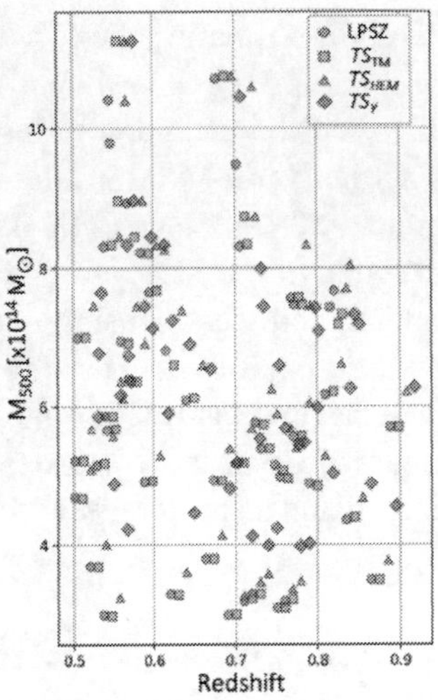

Figure 2. Redshift–mass distribution of the target LPSZ clusters compared to the twin samples generated using The Three Hundred Project. Each twin sample uniquely represents the LPSZ catalog.

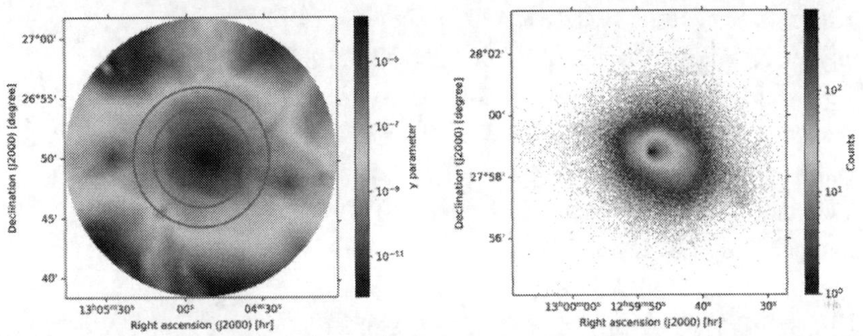

Figure 3. Multi–wavelength maps of two clusters in The Three Hundred simulation database. *Left:* Compton parameter map for a cluster with $\log[M_{500}/h^{-1}\,M_{\odot}] = 14.661$, $R_{500} = 1229.5\ h^{-1}$kpc, at $z = 0.817$. Map size: $\sim 23.2' \times 23.3'$. Pixel size: $\sim 5''$. The two concentric circles represent R_{500} and R_{200}, respectively. *Right:* Mock X-Ray photon count image of a cluster with $\log[M_{200}/h^{-1}M_{\odot}] = 14.824$, $R_{200} = 1693.4\ h^{-1}$kpc, at $z = 0.82$. Map size: $\sim 9.2' \times 9.2'$. Pixel size: $\sim 0.7''$.

selection criteria discussed in Section 3. These twin samples serve well for a wide range of analyses that support and test observational inferences. Below, we briefly discuss some of these undertaken analyses.

The Compton parameter maps from the simulations will be utilised to calculate Y_{500} by directly using the projected data, to be compared to the same quantity derived from de–projected pressure profiles. Calculations will be done along different lines–of–sight in a given mass and redshift range to assess the dynamical state of clusters and its impact on the estimation of the ICM profiles of the cluster. These estimated values will also be used to calibrate the $Y_{500} - M_{500}$ and $Y_{500} - M_{\mathrm{HE}}$ scaling relations in the LPSZ redshift range, extending the laws to higher redshifts than existing standards. Additionally, to monopolise on the high angular resolution of the NIKA2 camera, the maps will be injected with artificial point sources of varying positions and fluxes to estimate a threshold S/N and position of a point source that needs to be considered in estimating the ICM profile of the cluster, without bias.

The use of the above ICM profiles will be extended to infer the impact of redshift and morphology on mass estimation, using the X–ray maps available in the simulations. Further, the X–ray maps will be used to calibrate the $L_X - M_{500}$ and $L_X - M_{HE}$ scaling laws. Eventually, these scaling laws will be cross–correlated to the SZ scaling laws to minimise the scatter on inferred mass. Parallel use of the optical and weak lensing maps will be done for inferring observational estimates of the hydrostatic mass bias. These twin samples will also be used for cosmological analyses such as forecasting gas fraction measurements and the inference of the impact of systematics on the Hubble Constant (H_0) measurements. Clearly, these twin samples are viable, potent tools to extend and test inferences from high–quality observations.

References

[1] J.R. III Gott, A.L. Melott, & M. Dickinson, ApJ **306**, 341 (1986)
[2] L. Guzzo *et al.*, A&A **566**, A108 (2014)
[3] Planck Collaboration *et al.*, A&A **594**, A24 (2016)
[4] G. Schellenberger & T.H. Reiprich, MNRAS **469**, 3738 (2017)
[5] T.H. Reiprich *et al.*, A&A **647**, A2 (2021)
[6] W.H. Press & P. Schechter, ApJ **187**, 425 (1974)
[7] R.A. Sunyaev & Y.B. Zeldovich, ARA&A **18**, 537 (1980)
[8] A. Lindley *et al.*, American Astronomical Society Meeting Abstracts **220**, 507.03 (2012)
[9] E.L. Łokas & G.A. Mamon, MNRAS **343**, 401 (2003)
[10] B.R. Matthew & A.V. Kravtsov, ApJ **740**, 25 (2011)
[11] Planck Collaboration *et al.*, A&A **571**, A20 (2014)
[12] E. Bulbul *et al.*, ApJ **871**, 50 (2019)
[13] A. Nagarajan *et al.*, MNRAS **488**, 1728 (2019)
[14] I.–N. Chiu *et al.*, arXiv e-prints, arXiv:2107.05652 (2021)
[15] F. Sembolini *et al.*, MNRAS **429**, 323 (2013)
[16] F. Mayet *et al.*, EPJ Web Conf. **228**, 00017 (2020)
[17] W. Cui *et al.*, MNRAS **480**, 2898 (2018)
[18] R. Adam *et al.*, A&A **609**, A115 (2018)
[19] O. Bourrion *et al.*, JINST **11**, P11001 (2016)
[20] S. Shu *et al.*, IEEE Transactions on Terahertz Science and Technology **8**, 605 (2018)
[21] L. Perotto *et al.*, A&A **637**, A71 (2020)
[22] R. Adam *et al.*, A&A **614**, A118 (2018)
[23] F. Ruppin *et al.*, A&A **615**, A112 (2018)
[24] Planck Collaboration *et al.*, A&A **594**, A27 (2016)
[25] M. Hasselfield *et al.*, J. Cosmology Astropart. Phys. **2013**, 008 (2013)
[26] Y. Wang *et al.*, ApJ **868**, 130 (2018)
[27] R. Haggar *et al.*, MNRAS **492**, 6074 (2020)
[28] F. De Luca *et al.*, MNRAS **504**, 5383 (2021)
[29] A. Klypin *et al.*, MNRAS **457**, 4340 (2016)
[30] Planck Collaboration *et al.*, A&A **594**, A13 (2016)
[31] E. Rasia *et al.*, ApJ **813**, L17 (2015)
[32] R. Davé *et al.*, MNRAS **486**, 2827 (2019)
[33] G. Gianfagna *et al.*, MNRAS **502**, 5115 (2021)

Galactic star formation with NIKA2 (GASTON): Filament convergence and its link to star formation

N. Peretto[1,*], *R.* Adam[2], *P.* Ade[1], *H.* Ajeddig[3], *P.* André[3], *E.* Artis[4], *H.* Aussel[3], *A.* Bacmann[10], *A.* Beelen[5], *A.* Benoît[6], *S.* Berta[7], *L.* Bing[8], *O.* Bourrion[4], *M.* Calvo[6], *A.* Catalano[4], *M.* De Petris[9], *F.-X.* Désert[10], *S.* Doyle[1], *E. F. C.* Driessen[7], *A.* Gomez[11], *J.* Goupy[6], *F.* Kéruzoré[4], *C.* Kramer[12], *B.* Ladjelate[12], *G.* Lagache[8], *S.* Leclercq[7], *J.-F.* Lestrade[13], *J.-F.* Macías-Pérez[4], *A.* Maury[3], *P.* Mauskopf[1,14], *F.* Mayet[4], *A.* Monfardini[6], *M.* Muñoz-Echeverría[4], *L.* Perotto[4], *G.* Pisano[1], *N.* Ponthieu[10], *V.* Revéret[3], *A.* Rigby[1], *I.* Ristorcelli[15], *A.* Ritacco[5,16], *C.* Romero[17], *H.* Roussel[18], *F.* Ruppin[19], *K.* Schuster[7], *S.* Shu[20], *A.* Sievers[12], *C.* Tucker[1], and *R.* Zylka[7]

[1]School of Physics and Astronomy, Cardiff University, Queen's Buildings, The Parade, Cardiff, CF24 3AA, UK
[2]LLR (Laboratoire Leprince-Ringuet), CNRS, École Polytechnique, Institut Polytechnique de Paris, Palaiseau, France
[3]AIM, CEA, CNRS, Université Paris-Saclay, Université Paris Diderot, Sorbonne Paris Cité, 91191 Gif-sur-Yvette, France
[4]Univ. Grenoble Alpes, CNRS, Grenoble INP, LPSC-IN2P3, 53, avenue des Martyrs, 38000 Grenoble, France
[5]Institut d'Astrophysique Spatiale (IAS), CNRS, Université Paris Sud, Orsay, France
[6]Institut Néel, CNRS, Université Grenoble Alpes, France
[7]Institut de RadioAstronomie Millimétrique (IRAM), Grenoble, France
[8]Aix Marseille Univ, CNRS, CNES, LAM (Laboratoire d'Astrophysique de Marseille), Marseille, France
[9]Dipartimento di Fisica, Sapienza Università di Roma, Piazzale Aldo Moro 5, I-00185 Roma, Italy
[10]Univ. Grenoble Alpes, CNRS, IPAG, 38000 Grenoble, France
[11]Centro de Astrobiología (CSIC-INTA), Torrejón de Ardoz, 28850 Madrid, Spain
[12]Instituto de Radioastronomía Milimétrica (IRAM), Granada, Spain
[13]LERMA, Observatoire de Paris, PSL Research University, CNRS, Sorbonne Université, UPMC, 75014 Paris, France
[14]School of Earth and Space Exploration and Department of Physics, Arizona State University, Tempe, AZ 85287, USA
[15]Univ. Toulouse, CNRS, IRAP, 9 Av. du colonel Roche, BP 44346, 31028, Toulouse, France
[16]Laboratoire de Physique de l'École Normale Supérieure, ENS, PSL Research University, CNRS, Sorbonne Université, Université de Paris, 75005 Paris, France
[17]Department of Physics and Astronomy, University of Pennsylvania, 209 South 33rd Street, Philadelphia, PA, 19104, USA
[18]Institut d'Astrophysique de Paris, Sorbonne Université, CNRS (UMR7095), 75014 Paris, France
[19]Kavli Institute for Astrophysics and Space Research, Massachusetts Institute of Technology, Cambridge, MA 02139, USA
[20]Caltech, Pasadena, CA 91125, USA

Abstract. In the past decade filaments have been recognised as a major structural element of the interstellar medium, the densest of these filaments hosting

*e-mail: Nicolas.Peretto@astro.cf.ac.uk

the formation of most stars. In some star-forming molecular clouds converging networks of filaments, also known as hub filament systems, can be found. These hubs are believed to be preferentially associated to massive star formation. As of today, there are no metrics that allow the systematic quantification of a filament network convergence. Here, we used the IRAM 30m NIKA2 observations of the Galactic plane from the GASTON large programme to systematically identify filaments and produce a filament convergence parameter map. We use such a map to show that: i. hub filaments represent a small fraction of the global filament population; ii. hubs host, in proportion, more massive and more luminous compact sources that non-hubs; iii. hub-hosting clumps are more evolved that non-hubs; iv. no discontinuities are observed in the properties of compact sources as a function of convergence parameter. We propose that the rapid global collapse of clumps is responsible for (re)organising filament networks into hubs and, in parallel, enhancing the mass growth of compact sources.

1 Introduction

The existence of interstellar filaments has been known for more than four decades [1]. Their ubiquity within the interstellar medium, however, has only been recognised with the launch in 2009 of the *Herschel* mission, and its mapping of the sky in the far-infrared [2, 3]. Since then filaments are front and centre of a large fraction of star formation research, and numerous breakthroughs have been made. It has been shown, for instance, that the majority of prestellar cores, and, by extension, the majority of stars form as the result of the fragmentation of individual 0.1pc-wide self-gravitating filaments [e.g. 4, 5]. Massive star-forming cores, on the other hand, seem to preferentially find themselves at the centre of converging networks of filaments, or hubs [e.g. 6, 7]. This raises a number of questions on the physical relationship between filament convergence and the evolution of star-forming clouds. It has been proposed that hub formation is a two step process, with the formation of individual filaments first and then the collision of filament pairs in a second step [8]. One key aspect of such scenario is the existence of two distinct formation path for individual and hub filaments. In the study presented here, we use the very sensitive 1.15mm continuum data of the GASTON large programme on the IRAM 30m to determine the time evolution of compact millimetre sources as a function of the local filament network convergence.

2 GASTON observations and catalogue

GASTON is a 200h guaranteed-time large programme on the IRAM 30m telescope (PI: N. Peretto). About a third of the allocated time has been used to map a $\sim 2\ \mathrm{deg}^2$ region of the Galactic plane around $\ell = 24°$ at 1.15mm and 2mm using the NIKA2 instrument [9, 10]. In the present study, we only make use of the 1.15mm image which has an angular resolution of 12″, a pixel size of 3″, and a rms sensitivity of 3.45 mJy/beam (see Fig. 1), achieved with only half of the allocated time. In addition, we also make use of the GASTON catalogue of 1.15mm compact source presented in [11]. In that catalogue, the distances and infrared brightness of all $\sim$ 1400 sources are given. The latter provides an indication on how bright the source parent clump is at 8μm, and is used here as a time evolution tracer [11, 12]. Finally, all GASTON sources have been cross-correlated with the Hi-GAL source catalogue [13].

3 Quantifying filament convergence and hub definition

Even though the study of hub filament systems has become, in the past 10 years, relatively widespread, there is no clear definition of what a hub is. For the first time, we propose to

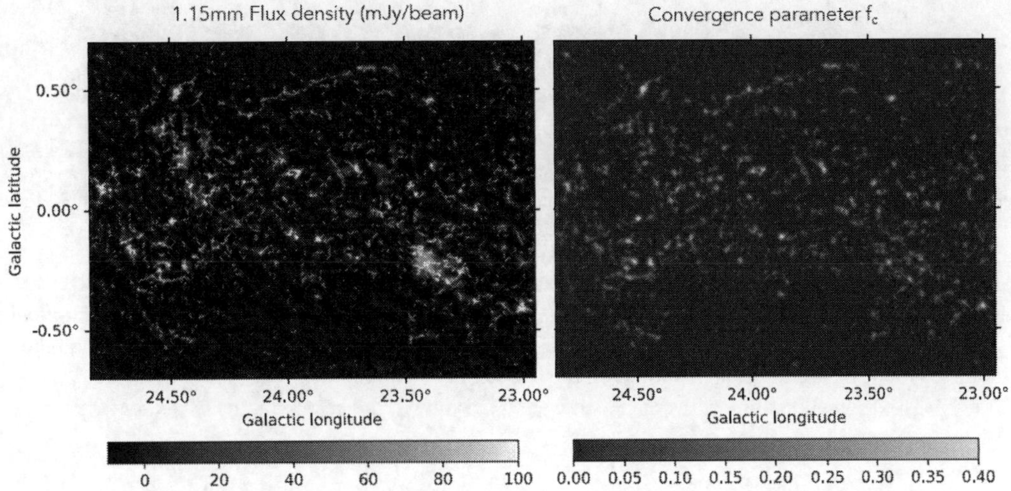

Figure 1. (left): GASTON 1.15mm image of the $\ell = 24°$ region (colour scale). The identified filament skeleton is overlaid on top as grey solid lines. (right): Map of the convergence parameter f_c. Same as in the left hand side panel, the skeleton is overlaid on top of the image.

quantify the convergence of filaments in star-forming regions through the construction of a new convergence parameter, f_c. In order to do this, we first need to build the skeleton of the filament network within the GASTON 1.15mm image of the $\ell = 24°$ region. For that purpose, we use the same second derivative method as presented in [e.g. 14]. Skeletons are obtained by first using a local thresholding on the eigenvalue images, selecting only regions that have negative eigenvalues lower than -3 times the local standard deviation. The python *skeletonize* function is then used to thin each region to a 1-pixel width skeleton. We set a lower limit on the length of individual filaments to twice the beam size ($\sim 24''$ - 0.6pc at 5.2 kpc, the median cloud distance in the $\ell = 24°$ field). Figure 1 shows the resulting skeleton overlaid on top of the GASTON 1.15mm image. A total of $\sim$ 2600 filaments are identified in the $\ell = 24°$ field. Compared to published catalogues of Galactic plane filaments, the GASTON filament density is larger, however those studies were typically geared towards the detection of larger-scale filaments [e.g. 15, 16]. Since hubs have been in the past mostly visually identified in mid-infrared extinction, a visual check between our GASTON skeleton and the *Spitzer* 8μm image of the field is performed (see Fig. 2). The agreement is excellent and shows the power of our GASTON 1.15mm image in identifying small scale filamentary structures.

Next we want to construct a new metric allowing to quantify how hub-like a network of filaments really is. This is the first ever attempt to provide an objective definition to what a hub is. For that purpose we define the following convergence parameter f_c:

$$f_c(x, y) = N_{\rm fil} \frac{\sum_{i=1}^{N_{\rm pix}} \cos(\Delta\theta)}{C_n} \tag{1}$$

where (x, y) is the pixel coordinate in the image, $N_{\rm fil}$ is the number of unique filaments that enter the search radius r_s, $N_{\rm pix}$ is the total number of skeleton pixels that are within the search radius, $\Delta\theta$ is the angle difference between the direction defined by the pixel (x, y) and the skeleton pixel i, and the tangential direction of the skeleton at pixel i. Finally, C_n is a normalisation constant such that a hub of six perfectly radial parsec-long connected filaments has a f_c value of 1. The presence of the $N_{\rm fil}$ is needed to ensure that, for a fixed value of $N_{\rm pix}$, networks of converging filaments have larger f_c values than single filaments. This definition

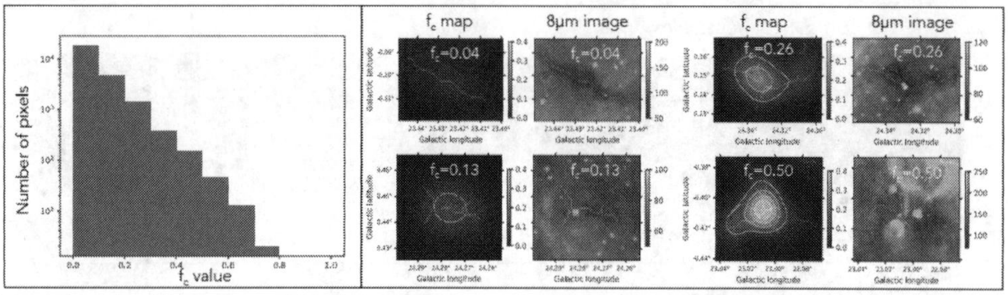

Figure 2. (left): Histogram of the f_c values presented in Fig. 1. (right): zoom-ins of 4 different sources exhibiting the range of f_c values with their peak value indicated on the image. For each of the four pairs of images, we display the convergence map on the left and the corresponding *Spitzer* 8μm image on the right hand side. The skeletons are overlaid on top of each panel, and f_c contours from 0.1 to 0.5 by step of 0.1 are also displayed on the convergence images. For context, the search radius r_s is about 0.01°.

of the convergence parameter implies that high f_c value positions exhibit a high level of filament convergence within a radius r_s, while low f_c value implies the opposite. In this study we chose $r_s = 39''$, which corresponds to ~ 1 pc at the cloud median distance of 5.2 kpc. Figure 1(right) shows the resulting convergence map obtained from the underlying filament skeleton where high convergence spots can clearly be seen.

Figure 2(left) shows the histogram of f_c values presented in Fig. 1, ranging from 0 to 0.75. One can see that the distribution can be described by a single power law, with no obvious breaks. Therefore the segmentation of clumps between hubs and non-hubs is somehow arbitrary. However one can still investigate what f_c values typical hub filament systems have in order to separate hubs from non-hubs. Figure 2 shows four sources with increasing peak f_c values and increasing filament complexity, the last two of which have peak f_c values of 0.26 and 0.50 and exhibit a typical hub morphology. We therefore decide to define hubs as connected groups of pixels with $f_c \geq 0.2$. By doing so, we identify 63 hubs and we find that only 5% of skeleton pixels are located within hubs. Hub filament systems are rare objects.

4 Source properties as a function of convergence parameter

Now that we have established a working definition of a hub, we can investigate the properties of compact GASTON sources as a function of the type of clump they find themselves into, and more generally, as a function of their convergence parameter. For that purpose we use the catalogue of compact sources identified by [11] and split the ~ 1400 compact 1.15mm GASTON sources in four f_c bins: $f_c < 0.1$, $0.1 < f_c < 0.2$, $0.2 < f_c < 0.3$, and $f_c > 0.3$, in which there are 1186, 170, 56 and 19 compact sources, respectively. Based on this splitting, we can already see that only 8% of GASTON compact sources are embedded within a hub.

First, we want to make sure that there is no distance bias when splitting out source sample into f_c bins. Figure 3(1st row) shows the distance histograms for each f_c bin. The vertical dash line shows the median value. One can see that there is nearly no change whether in the shape of the distribution or in the median distance values, all at about 5.2 kpc. Then we look at the distribution of the source masses. Figure 3 (2nd row) clearly shows that there is an increase in source mass as the convergence parameter increases. In fact the median mass for the $f_c < 0.1$ bin is at 135 $M_\odot$ while that of the $f_c > 0.3$ bin is at 1350 $M_\odot$, an order of magnitude larger. This shows that a larger fraction of compact sources associated with hubs are massive, although massive compact sources are also found in low f_c regions.

Next, we investigate how the infrared-bright fraction ($f_{\rm IRB}$) of the GASTON sources evolves as a function of the convergence parameter. The brightness parameter is a mea-

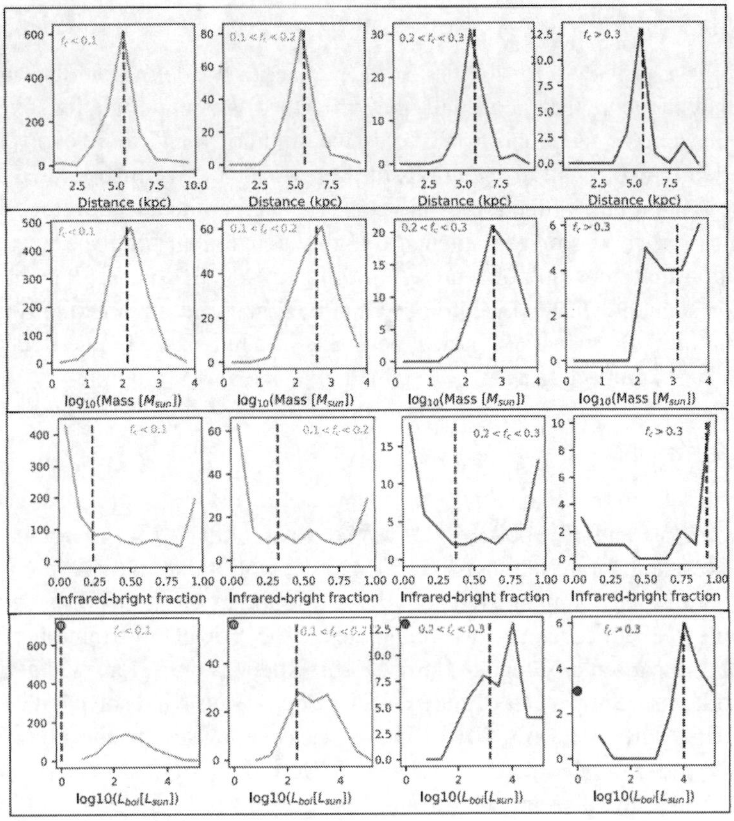

Figure 3. (1st row): Histogram of distances for the GASTON 1.15mm compact sources identified by [11]. Each panel corresponds to a range of f_c values indicated in each panel. The vertical dashed lines show the median values. (2nd row): same as top row but for the masses of the compact sources. (3rd row): same as top row but for the infrared darkness parameter. (4th row): Same as top row but for the 70μm-based bolometric luminosities. The circular symbols indicate the number of sources with no Hi-GAL 70μm sources, i.e. sources for which the bolometric luminosity could not be computed.

surement of how bright a cloud is as 8μm. For instance, infrared dark clouds that do not host any star formation activity yet will have an infrared brightness parameter of 0. The first of the four clumps shown in Fig. 2 is a good example of such pristine structure. On the other hand a cloud which is bright at 8μm, such as the last one of the four clumps presented in Fig. 2, will have an infrared brightness close to 1. Infrared brightness is believed to be a good tracer of clump evolution [11, 12]. In Fig.3(3rd row) we show the histograms of infrared brightness for the same four convergence parameter bins. There, we see a continuous shift from clumps that are mostly infrared dark at low f_c values, to clumps that are mostly infrared bright at large f_c values. One can interpret this in two ways: hubs are the late stages of clump evolution in which filament convergence increases with time; infrared dark hubs have a very short lifetime and become very quickly infrared bright via a strong star formation activity. Note that these two scenarios are not mutually incompatible. A similar picture can be drawn from Fig. 3(4th row) where the distribution of bolometric luminosities are presented. Here, the luminosities are being computed from the 70μm flux of internal protostellar sources. On that figure we can clearly see that hubs host the formation of more luminous sources, including all sources with $L_{\rm bol} \leq 10^5 L_{\odot}$, a result that was already found by [8].

5 Conclusion

In this study, we have provided, for the first time, a quantitative definition of what a hub is, and shown that they in fact represent a small fraction of the filament population. We have also shown that hubs have a higher proportion of massive, luminous, and more evolved compact sources within them. Also, in all distributions of properties we have investigated, we do not observe any discontinuity as a function of the convergence parameter but rather continuous and smooth transitions from single filaments to highly convergent filament networks. This is a clear indication that hubs do not represent a different population of filament with their specific formation scenario. Instead we propose that it is the rapid global collapse of clumps that is responsible for (re)organising filament networks into hubs and, in parallel, for leading to the mass growth of compact sources.

Acknowledgements

NP and AJR would like to thank the STFC for financial support under the consolidated grant numbers ST/N000706/1 and ST/S00033X/1, and the Royal Society for providing computing resources under Research Grant number RG150741. We would like to thank the IRAM staff for their support during the campaigns. The NIKA2 dilution cryostat has been designed and built at the Institut Néel. In particular, we acknowledge the crucial contribution of the Cryogenics Group, and in particular Gregory Garde, Henri Rodenas, Jean Paul Leggeri, Philippe Camus. This work has been partially funded by the Foundation Nanoscience Grenoble and the LabEx FOCUS ANR-11-LABX-0013. This work is supported by the French National Research Agency under the contracts "MKIDS", "NIKA" and ANR-15-CE31-0017 and in the framework of the "Investissements d'avenir" program (ANR-15-IDEX-02). This work has benefited from the support of the European Research Council Advanced Grant ORISTARS under the European Union's Seventh Framework Programme (Grant Agreement no. 291294). F.R. acknowledges financial supports provided by NASA through SAO Award Number SV2-82023 issued by the Chandra X-Ray Observatory Center, which is operated by the Smithsonian Astrophysical Observatory for and on behalf of NASA under contract NAS8-03060.

References

[1] Schneider & Elmegreen, ApJS, **41**, 87 (1979)
[2] André et al., A&A, **518**, 102 (2010)
[3] Molinari et al., A&A, **518**, 100 (2010)
[4] Arzoumanian et al., A&A, **529**, 6 (2011)
[5] Polychroni et al., ApJ, **777**, 33 (2013)
[6] Myers, ApJ, **700**, 1609 (2009)
[7] Peretto et al., A&A, **555**, 112 (2013)
[8] Kumar et al., A&A, **642**, 87 (2020)
[9] Adam et al., A&A, **609**, 115 (2018)
[10] Perotto et al., **637**, 71 (2020)
[11] Rigby et al., MNRAS, **502**, 4576 (2021)
[12] Watkins et al, A&A, in prep.
[13] Molinari et al., A&A, **591**, 149 (2016)
[14] Orkisz et al., A&A, **624**, 113 (2019)
[15] Schisano et al., MNRAS, **492**, 5420 (2020)
[16] Li et al., A&A, **591**, 5 (2016)

The NIKA2 Sunyaev-Zeldovich Large Program

Precise galaxy cluster physics for an accurate cluster-based cosmology

L. Perotto[1,*], *R.* Adam[2], *P.* Ade[3], *H.* Ajeddig[4], *P.* André[4], *M.* Arnaud[4], *E.* Artis[1], *H.* Aussel[4], *I.* Bartalucci[5], *A.* Beelen[6], *A.* Benoît[7], *S.* Berta[8], *L.* Bing[9], *O.* Bourrion[1], *M.* Calvo[7], *A.* Catalano[1], *M.* De Petris[10], *F.-X.* Désert[11], *S.* Doyle[3], *E. F. C.* Driessen[8], *A.* Ferragamo[10], *A.* Gomez[12], *J.* Goupy[7], *F.* Kéruzoré[1], *C.* Kramer[13], *B.* Ladjelate[13], *G.* Lagache[9], *S.* Leclercq[8], *J.-F.* Lestrade[14], *J.-F.* Macías-Pérez[1], *A.* Maury[4], *P.* Mauskopf[3,20], *F.* Mayet[1], *A.* Monfardini[7], *M.* Muñoz-Echeverría[1], *A.* Paliwal[10], *G.* Pisano[3], *E.* Pointecouteau[15], *N.* Ponthieu[11], *G. W.* Pratt[4], *V.* Revéret[4], *A. J.* Rigby[3], *A.* Ritacco[6,16], *C.* Romero[17], *H.* Roussel[18], *F.* Ruppin[19], *K.* Schuster[8], *S.* Shu[21], *A.* Sievers[13], *C.* Tucker[3], and *G.* Yepes[22]

[1]Univ. Grenoble Alpes, CNRS, Grenoble INP, LPSC-IN2P3, 38000 Grenoble, France
[2]LLR, CNRS, École Polytechnique, Institut Polytechnique de Paris, Palaiseau, France
[3]School of Physics and Astronomy, Cardiff University, Queen's Buildings, The Parade, Cardiff, CF24 3AA, UK
[4]AIM, CEA, CNRS, Université Paris-Saclay, Université Paris Diderot, Sorbonne Paris Cité, 91191 Gif-sur-Yvette, France
[5]INAF, IASF-Milano, Via A. Corti 12, 20133 Milano, Italy
[6]Institut d'Astrophysique Spatiale (IAS), CNRS, Université Paris Sud, Orsay, France
[7]Institut Néel, CNRS, Université Grenoble Alpes, France
[8]Institut de RadioAstronomie Millimétrique (IRAM), Grenoble, France
[9]Aix Marseille Univ, CNRS, CNES, LAM, Marseille, France
[10]Dipartimento di Fisica, Sapienza Università di Roma, Piazzale Aldo Moro 5, I-00185 Roma, Italy
[11]Univ. Grenoble Alpes, CNRS, IPAG, 38000 Grenoble, France
[12]Centro de Astrobiología (CSIC-INTA), Torrejón de Ardoz, 28850 Madrid, Spain
[13]Instituto de Radioastronomía Milimétrica (IRAM), Granada, Spain
[14]LERMA, Observatoire de Paris, PSL Research University, CNRS, Sorbonne Université, UPMC, 75014 Paris, France
[15]Univ. de Toulouse, UPS-OMP, CNRS, IRAP, 31028 Toulouse, France
[16]Laboratoire de Physique de l'École Normale Supérieure, ENS, PSL Research University, CNRS, Sorbonne Université, Université de Paris, 75005 Paris, France
[17]Department of Physics and Astronomy, University of Pennsylvania, 209 South 33rd Street, Philadelphia, PA, 19104, USA
[18]Institut d'Astrophysique de Paris, CNRS (UMR7095), 98 bis boulevard Arago, 75014 Paris, France
[19]Kavli Institute for Astrophysics and Space Research, Massachusetts Institute of Technology, Cambridge, MA 02139, USA
[20]School of Earth and Space Exploration and Department of Physics, Arizona State University, Tempe, AZ 85287, USA
[21]Caltech, Pasadena, CA 91125, USA
[22]Departamento de Física Teórica and CIAFF, Facultad de Ciencias, Modulo 8, Universidad Autánoma de Madrid, 28049 Madrid, Spain

*e-mail: laurence.perotto@lpsc.in2p3.fr

Abstract. The NIKA2 Guaranteed-Time SZ Large Program (LPSZ) is dedicated to the high-angular resolution SZ mapping of a representative sample of 45 SZ-selected galaxy clusters drawn from the catalogues of the Planck satellite, or of the Atacama Cosmology Telescope. The LPSZ sample spans a mass range from 3 to $11 \times 10^{14} M_{\odot}$ and a redshift range from 0.5 to 0.9, extending to higher redshift and lower mass the previous samples dedicated to the cluster mass calibration and universal properties estimation. The main goals of the LPSZ are the measurement of the average radial profile of the ICM pressure up to R_{500} by combining NIKA2 with Planck or ACT data, and the estimation of the scaling law between the SZ observable and the mass using NIKA2, XMM-Newton and Planck/ACT data. Furthermore, combining LPSZ data with existing or forthcoming public data in lensing, optical/NIR or radio domains, we will build a consistent picture of the cluster physics and further gain knowledge on the mass estimate as a function of the cluster morphology and dynamical state.
We give an overview of the LPSZ, present recent results and discuss the future implication for cosmology with galaxy clusters.

1 Introduction

Clusters of galaxies are powerful cosmological probes of the large-scale structure growth and of the Universe expansion over cosmic time (see e. g. [1]). However, galaxy cluster-based cosmology is currently limited by the accuracy with which the mass of these objects can be inferred [2]. Galaxy clusters being scale-invariant, scaling laws that relate their mass to an observable can be built from representative sub-samples of clusters. In particular, from X-ray photometric and spectroscopic observation towards clusters, the spherical profiles of both the density and the temperature of the Intra-Cluster Medium (ICM) can be reconstructed, enabling the estimation of the mass under the hydrostatic equilibrium assumption. The main drawback of this method lies in the high-exposure time needed to achieve the data quality that permits an accurate estimate of the temperature profile, which become prohibitive for faint or distant clusters. As a consequence, widely-used reference samples, from which the mass-observable relation and universal properties of clusters are measured, are limited to massive and low-redshift clusters [3, 4].

With the advent of high-angular-resolution millimeter-domain experiments, such as NIKA2, resolved observations of the Sunyaev-Zel'dovich effect (SZ [5]) towards clusters offer a way of improving the accuracy and extending the coverage in mass and redshift of the relation between the hydrostatic mass and the integrated Compton parameter. Such observations also enable measuring the mean pressure profile of clusters. The SZ effect being independent on the redshift and proportional to the line-of-sight integral of the pressure of the ICM, accurate pressure profiles can be inferred in clusters up to high-redshift. Pressure profiles drawn from resolved SZ observations can then be combined with density profiles extracted from photometric X-ray observations to derive a hydrostatic mass estimate. Moreover, the mean pressure profile of clusters can be estimated in cluster samples from a direct observable of the pressure.

NIKA2 is a millimetre-domain camera installed at the IRAM 30-m telescope and opened to the community since 2017 [6–8]. It operates in two frequency bands centered at 150 and 260 GHz, with an angular resolution of 17.6 and 11.1″, respectively, and offers high mapping speeds [9]. With an angular resolution comparable to that of the XMM-Newton space observatory, NIKA2 will therefore enable the full deployment of the combined SZ and X-ray methods to infer the hydrostatic mass, and unveil the pressure distribution of the ICM.

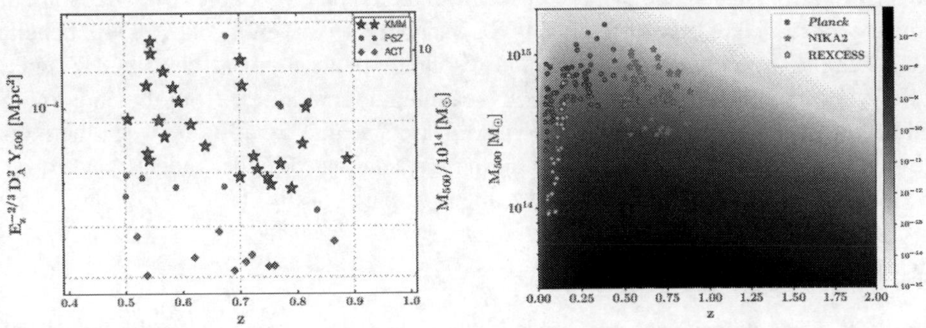

Figure 1. *Left:* Distribution of the LPSZ cluster sample in the plane of the redshift (abscissa) and the integrated Compton parameter (ordinate). The corresponding mass, as derived from the Planck Y-M scaling relation is given in the right ordinate-scale. The sample comprizes clusters drawn from the Planck (red) and ACT (blue) catalogs [10, 11], and already observed by XMM-Newton (stars). *Right:* Comparison of the LPSZ sample (red) with previous samples dedicated to the cluster mass calibration [3, 4] in the mass-redshift plane color-coded as a function of the predicted cluster abundance. Figure extracted from [18].

The NIKA2 SZ Large Program (LPSZ) consists of 300 hours of Guaranteed-Time dedicated to this goal.

2 Overview of the LPSZ

The LPSZ program relies on the state-of-the-art performance of the NIKA2 camera for the high-angular resolution mapping of the thermal SZ effect in a representative sample of galaxy clusters in view of improving the accuracy of cluster-based cosmology (P.I.: F. Mayet; co-P.I.: L. Perotto).

2.1 Sample

The LPSZ sample consists of 45 galaxy clusters selected in intervals of mass and redshift in the ACT [11] and Planck [10] cluster catalogs. Being SZ-selected, the sample is representative of the whole cluster population. It covers intermediate to high redshifts ($0.5 < z < 0.9$) and spans a large range of masses from 3 to $11 \times 10^{14} M_{\odot}$. The distribution of the sample as a function of the integrated Compton parameter and the redshift, as well as the mass and redshift bins defined for the sample selection, are shown in the left panel of Fig. 1. Most of the targeted Planck clusters were observed in the X-ray domain with XMM-Newton or Chandra. X-ray observations of the Planck and ACT LPSZ clusters will be completed by the end of the program.

2.2 Main Goals

The main goal of the LPSZ is providing the community with tools to improve the accuracy of the Cosmology with galaxy clusters, with a focus on the estimation of the mass. Combining SZ and X-ray observations at a similar angular resolution, we will perform in-depth characterizations of the thermodynamical properties of the ICM. For each cluster, we will provide spherical profiles of the ICM pressure, density, hydrostatic mass, entropy and temperature, independently of any X-ray spectroscopic data. From the complete sample, we will estimate the mean pressure profile of the clusters using SZ data only. In this endavour, we will benefit from NIKA2 unique combination of an angular resolution better than 18" at 150 GHZ and a 6.5' field-of-view to extend previous studies in spanning a radial range from the core to the typical radius of the clusters. Moreover we will release a scaling relation between the integrated Compton parameter and the hydrostatic mass extending to higher redshift and lower mass previous estimates [3, 4], as illustrated in the right panel of Fig. 1.

2.3 Complementary projects

Several multi-wavelength analyses will be conducted. First of all, combining the integrated line-of-sight mass density, which can be reconstructed from lensing data, with the LPSZ estimates of the hydrostatic mass, we will study the lensing-to-hydrostatic-mass bias as a function of the cluster morphology and dynamical state. The intersection of the LPSZ and CLASH [12] samples offers us a well-suited test bench of these studies [13]. Moreover about 50% of the LPSZ sample have deep-integration optical data (SDSS, PanStarrs) [14], allowing us to explore the connexions between the hydrostatic mass and the richness or the dynamical mass estimates. Finally, we will conduct dedicated studies of the non-thermal properties of the ICM to gain knowledge on turbulences, AGN feedback or shocks, using e. g. cross-correlation with radio-domain data.

We will use state-of-the-art hydrodynamical simulations, such as the Three-Hundred project [15], to select a series of synthetic samples of the same statistical or morphological properties as the LPSZ sample, and coming along with mock observations in SZ, X-ray and optical. These twin samples will be key to test novel methods, to assess the impact of dynamical state on cluster properties, to forecast the LPSZ sample sensitivity to redshift or mass evolution, to obtain accurate uncertainties on scaling laws and biases and to explore the LPSZ sample capability to measure the gas fraction or the Hubble parameter.

3 Status of the observation and the analysis

In order to achieve our main goals, the representativity of the sample must be preserved at the level of both the observation and the analysis. The observation time allocated to a cluster is defined so as to ensure a 3σ measurement of the Compton parameter profile at the angle R_{500}/D_A. Quantitatively, about half of the LPSZ Guaranteed Time has been spent to observe about three quaters of the sample. Qualitatively, preliminary analysis permit us to single out several interesting science case to study bimodality, merger events and shocks. A standard analysis method has been devised: at the map level, the calibration and map making use NIKA2 *Baseline* calibration method [9], while the angular scale filtering induced by the data reduction is accounted for following [17]. A novel method, dubbed PANCO2, has been developped for the reconstruction of the thermodynamical profiles from the maps, and further developments are being conducted in view of the statistical analysis of the sample [16].

4 First results

For the first LPSZ analysis, we chose PSZ2-G0144.83+25.11, a massive cluster of the first redshift-bin of the sample, making it an ideal target for observation as part of the Science Verification in April 2017. High-quality data were obtained enabling a 13.5 σ measurement of the SZ peak in the 150-GHz map [17]. An over-pression region was evidenced, as shown with the shaded area in Fig. 2 (left), which significatively impacts the pressure profile estimates inferred from the map. In turn, whether the over-pressure region is included or masked

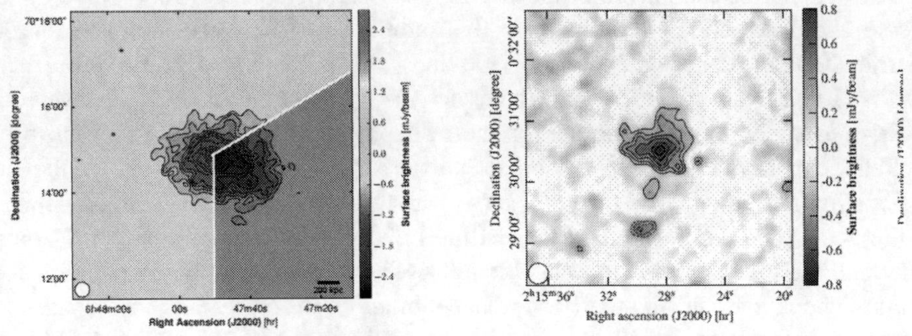

Figure 2. First LPSZ results. Left: NIKA2 map at 150-GHz towards the massive ($M_{500} = 7.8\times10^{14} M_{\odot}$) intermediate-redshift ($z \sim 0.6$) cluster PSZ2-G0144.83+25.11 after point source removal using NIKA2 260-GHz observation and ancillary data. The shaded sector shows the location of the over-pressure region that was evidenced as discussed in [17]; Right: NIKA2 150-GHz map of the low-mass ($M_{500} = 3.5 \times 10^{14} M_{\odot}$) high-redshift ($z \sim 0.9$) cluster ACT-CL J0215.4+0030 before point source treatment, as described in [19]. In both maps, contours show signal-to-noise levels starting from 3σ with a 1σ spacing.

out affects the hydrostatic mass profile estimate, and finally the measured integrated Compton parameter and mass of the cluster. As discussed in [17], this illustrates the capabilities of the LPSZ to assess the impact of the morphology on the mass-observable scaling relation.

By contrast, one of the most observationaly challenging cluster of the LPSZ sample was chosen as the object of the second LPSZ analysis: ACT-CL J0215.4+0030, a low-mass cluster in the highest redshift bin of the sample. While the SZ peak measurement reached 9σ, the contour of 3σ signal-to-noise level in the map at 150 GHz encloses only one square arcmin, as shown in Fig. 2 (right). The compactness of this cluster required special treatment of the point sources using NIKA2 observation at 260 GHz and ancillary data, as described in [19]. Combining NIKA2, ACT and XMM-Newton data, the thermodynamical profiles of the cluster ICM were estimated, and the integrated Compton parameter and hydrostatic mass were inferred, improving the uncertainties on the hydrostatic mass measurements with respect to both ACT and XMM-Newton measurements [19]. This shows that the powerful synergetic SZ/X-ray method can be fully deployed even for the faintest and more compact cluster of the LPSZ.

5 Conclusion

Improving the reliability of the cosmological model inferred from galaxy clusters requires a better knowledge of cluster physics and in particular, an accurate estimate of their mass and their universal properties. However the universal pressure profile and the hydrostatic mass of clusters are estimated from X-ray observation of massive low-redshift cluster samples. Using high-angular resolution observation of clusters in the millimeter-domain, we fully benefit from the redshift-independence property of the thermal SZ effect and from an observable directly proportional to the pressure of the ICM. Therefore combining X-ray and resolved thermal SZ observation of a representative sample of clusters is the key for extending to high redshift and low mass the estimation of the observable-to-mass scaling law and the universal pressure profile of clusters.

NIKA2 is perfectly suited for this endavour as it combines high-angular resolution at 150 and 260 GHz with a large field-of-view, enabling esquisite resolved SZ mapping of clusters.

This will be done in the framework of the LPSZ, a NIKA2 Guaranteed-Time Large Program dedicated to the observations of a representative sample of 45 SZ-selected clusters at $0.5 < z < 0.9$ from the Planck and ACT catalogs (300 hours, PI: F. Mayet, co-PI: L. Perotto).

The first LPSZ results demonstrate the synergetic capabilities of NIKA2 resolved SZ and XMM-*Newton* X-ray observations for the in-depth study of the ICM thermal properties, including for the faintest more compact cluster of the sample, and hint at the future implications for Cosmology of the whole sample analysis. As main LPSZ products, we will deliver the mean pressure profile and the scaling relation between the integrated Compton parameter and the mass, along with the surface brightness maps, the thermodynamical profiles and the codes that permits their inferrence. In addition to the X-ray/SZ analysis, we will perform multi-wavelength analysis combining LPSZ data with lensing, optical photometry/spectrometry, radio data to build a consistent picture of the cluster physics and further the LPSZ implication on cluster-based Cosmology.

Acknowledgements

We would like to thank the IRAM staff for their support during the campaigns. The NIKA2 dilution cryostat has been designed and built at the Institut Néel. In particular, we acknowledge the crucial contribution of the Cryogenics Group, and in particular Gregory Garde, Henri Rodenas, Jean Paul Leggeri, Philippe Camus. This work has been partially funded by the Foundation Nanoscience Grenoble and the LabEx FOCUS ANR-11-LABX-0013. This work is supported by the French National Research Agency under the contracts "MKIDS", "NIKA" and ANR-15-CE31-0017 and in the framework of the "Investissements d'avenir" program (ANR-15-IDEX-02). This work has benefited from the support of the European Research Council Advanced Grant ORISTARS under the European Union's Seventh Framework Programme (Grant Agreement no. 291294). F.R. acknowledges financial supports provided by NASA through SAO Award Number SV2-82023 issued by the Chandra X-Ray Observatory Center, which is operated by the Smithsonian Astrophysical Observatory for and on behalf of NASA under contract NAS8-03060.

References

[1] J. E. Carlstrom *et al.*, A. Rev. Astron. & Astrophys., **40**, 643 (2002)
[2] G. W. Pratt, M. Arnaud & A. Biviano, Space. Sci. Rev., **215**, 25 (2019)
[3] G. W. Pratt *et al.*, Astron. Astrophys., **498**, A361 (2009)
[4] Planck Collaboration, Astron. Astrophys., **550**, A131 (2013)
[5] R. A. Sunyaev &Y. B. Zeldovich, Comments on Astrophys. & Space Phys., **4**, 173 (1972)
[6] R. Adam *et al.*, Astron. Astrophys. **609**, A115 (2018)
[7] M. Calvo *et al.*, Journal of Low Temperature Physics, **184**, 816 (2016)
[8] O. Bourrion *et al.*, Journal of Instrumentation, **11**, 11001 (2016)
[9] L. Perotto, N. Ponthieu, J.-F. Macías-Pérez *et al.*, Astron. Astrophys. **637**, A71 (2020)
[10] Planck Collaboration, Astron. Astrophys., **594**, A27 (2016)
[11] M. Hasselfield *et al.*, J. Cosmol. Astropart. Phys. 2013.07 p. 008 (2013)
[12] M. Postman *et al.*, The Astrophysical Journal Supplement Series **199**, 25 (2012)
[13] M. Muñoz-Echeverría *et al.*, to appear in EPJ Web of Conferences (2021)
[14] R. F. J. van der Burg *et al.*, Astron. & Astrophys. **618** A140 (2018)
[15] W. Cui *et al.*, MNRAS **480** Issue 3 p.2898 (2018)
[16] F. Kéruzoré *et al.*, to appear in EPJ Web of Conferences (2021)
[17] F. Ruppin, F. Mayet, G. W. Pratt *et al.*, Astron. Astrophys. **615**, A112 (2018)
[18] F. Ruppin *et al.*, MNRAS p.2293 (2019)
[19] F. Kéruzoré, F. Mayet, G. W. Pratt *et al.*, Astron. Astrophys., **644**, A93 (2020)

EPJ Web of Conferences **257**, 00039 (2022) https://doi.org/10.1051/epjconf/202225700039
mm Universe @ NIKA2

Sunyaev Zel'dovich high resolution view of filamentary structures between galaxy clusters pairs

Federico Radiconi[1,2,*] for the ACT collaboration

[1]Sapienza University of Rome, Physics Department, Piazzale Aldo Moro 5, 00185 Rome, Italy
[2]INFN - Sezione di Roma, Piazzale Aldo Moro 5 - I-00185, Rome, Italy

Abstract. Strong observational and theoretical evidences suggest that about half of the baryons in the Universe should lie outside galaxy clusters in a low-density and hot phase in filaments connecting galaxy clusters. Due to the low density, most of this filamentary plasma can not be detected by X-ray observatories. In particular cases of low redshift cluster pairs in the pre-merging phase, the Sunyaev Zel'dovich (SZ) effect can be used to observe the intercluster regions and detect the imprint of missing baryons.
The Abell 399-401 (A399-401) system is the perfect laboratory to test our ability to detect filamentary structures via the SZ effect with $<\sim 1'$ angular resolution. This pair has been well studied at several frequencies: it exhibits double radio-halos, an excess of X-ray emission in the intercluster region and a synchrotron radio 'ridge' connecting the two clusters. Moreover the Planck satellite provided the first SZ detection of the gas between A399-401 despite the poor angular resolution ($\sim 10'$) of its SZ map.
We have used an Atacama Cosmology Telescope (ACT) and Planck satellite Compton-y map ($1.65'$ angular resolution) that combines ACT data from 2008 to 2019 with Planck maps to study the A399-401 system in detail. We present the data analysis and results.

1 Introduction

The observed baryon matter fraction in the local Universe is one of the open problems in Cosmology today. Observations of the baryonic matter based on HI absorption, X-ray emission, gas, stars, and galaxies counts provide a smaller baryonic fraction with respect to nucleosynthesis predictions, high-redshift measures of the Lyman-alpha forest absorption, and CMB power spectrum fits [10, 21]. Hydrodynamical simulations suggest that the fraction of the baryonic matter not directly observable, about half of the total budget, should lie outside galaxy clusters in the form of filamentary structures connecting the clusters forming the so-called cosmic web [7, 23]. These baryons must be characterized by a low-density phase and temperature in the range: $10^5 < T < 10^7\,K$.
The direct observation of missing baryons with X-rays and radio instruments is very challenging because:

- X-ray luminosity scales as the square of the electron density;

*e-mail: federico.radiconi@roma1.infn.it

- Outside of galaxy clusters, simulations predict few relativistic electrons and weak magnetic fields, thus making synchrotron emission very faint [24].

The Sunyaev-Zel'dovich (SZ) effect is another method for detecting these baryons; it is due to cosmic microwave background (CMB) radiation inverse Compton scattering off of electrons in a warm or hot gas. The SZ effect is mainly due to two separate components, the thermal and the kinematic one. The thermal component depends only on the gas pressure and it causes a distortion of the CMB spectrum. The kinematic effect is due to the overall peculiar motion of the gas with respect to the CMB. See Mroczkowski et al. (2019) [15] for a recent review about the SZ effect and its applications in astrophysics.
The amplitude of thermal SZ effect at any given frequency is proportional to the Compton-y parameter:

$$y = \frac{\sigma_{\mathrm{T}}}{m_{\mathrm{e}}c^2}\int P_{\mathrm{e}}(r)dr = \frac{\sigma_{\mathrm{T}}}{m_{\mathrm{e}}c^2}\int n_{\mathrm{e}}(r)k_{\mathrm{B}}T_{\mathrm{e}}(r)dr \tag{1}$$

where σ_{T} is the Thomson cross section, c is the speed of light, m_{e} is the electron mass and $P_{\mathrm{e}}(r)$ is the electron pressure along the line of sight. The thermal SZ effect is proportional to the electron density so SZ measurements of the missing baryons should be easier compared to X-rays and radio observations.

The galaxy cluster pair Abell 399 (z=0.071806, [17]) and Abell 401 (z=0.073664, [17]) is the perfect laboratory to investigate the missing baryon problem. The angular separation between the clusters is 36.9′. Assuming a flat ΛCDM Universe with H_0 = 67.6 km/s, Ω_m = 0.31 and Ω_Λ = 0.69 [1] and the mean redshift z = 0.072735, the separation on the plane of the sky between the two clusters is 3.2 Mpc. Due to the low redshift, it is possible to disentangle the two clusters at several wavelengths using wide images with ~arcmin angular resolution. Moreover, in the last decade several studies, at several wavelengths, have reported evidence of the presence of baryons between the two clusters (e.g. [9, 12, 19]). Recently Akamatsu et al. (2017) [3], using *Suzaku* data, confirmed the existence of hot gas (T ~ 6 – 7 keV) between A399 and A401. Govoni et al. (2019) [12], using Low Frequency Array (LOFAR) 140 MHz data, observed an excess of radio emission, a 'ridge', between the two clusters. Moreover, they were able to set an upper limit for the magnetic field in the intercluster region, $B < 1\mu G$. The Planck collaboration (2013) [19], using the MILCA Compton-y map with an angular resolution of 7.18′, detected an excess of Compton-y signal in the intercluster region despite the poor angular resolution. Bonjean et al. (2018) [4], using new *Planck's* y-map, confirmed previous results and estimated a density between the two clusters of n_e = $(4.3 \pm 0.7) \times 10^{-4}$ cm^{-3}.

In these proceedings, we summarise results that appeared in Hincks et al. (2021) [13] focusing on the model fitting and best fits comparison described in that work.

2 Data

In this work, we used a Compton-y map from the Atacama Cosmology Telescope combined with existing *Planck* data. ACT is a 6-metre off-axis Gregorian telescope that started observations in 2008. It is located at 5190 metres in altitude on the Chajnator plateau in the Atacama desert to minimize atmospheric contributions. ACT currently can observe the sky at several frequencies: 30, 40, 98, 150, and 220 GHz and it can reach an angular resolution between 1 and 2 arcmins at 98, 150 and 220 GHz. Together with the South Pole Telescope, ACT is the only high-resolution CMB survey telescope.
In this analysis we used the 98, 150, and 220 GHz data, from 2008 up to 2019, and *Planck* 100-545 GHz data to construct an ACT+*Planck* Compton-y map, as explained in Madhavacheril et al. (2020) [14]. The map is showed in Fig. 1 and its main difference with

Figure 1. Left: ACT Compton-y map of A399-401. The seven external black boxes enclose the regions we used to estimate the covariance matrix. The region surrounded by the red box has been excluded due to dust contamination. **Right**: Central region of the ACT map. It is the region we fitted with the MCMC in Sec. 3. Contour levels are 3, 5, and 7 σ. The effective beam of the map is showed on the bottom left side as a black circle with a radius equal to half beamwidth.

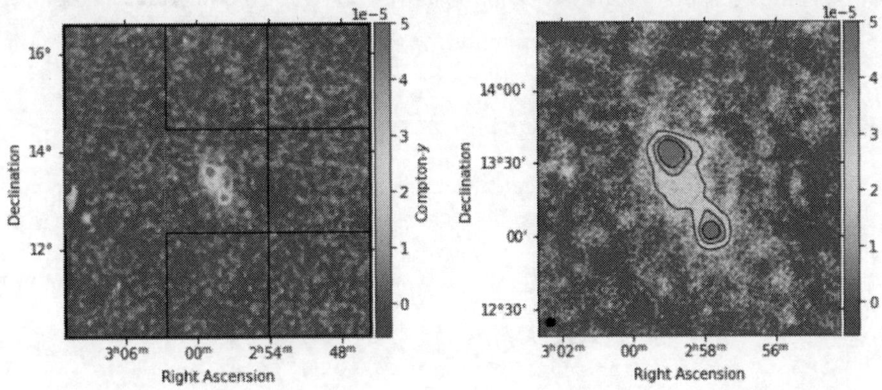

the ACT Data Release 5 (Naess et al. 2020, [16]) is the inclusion of the data acquired during 2019. The map is characterized by an angular resolution of 1.65′ and a noise of $\sigma = 5.9 \times 10^{-6} y$/pixel, where the pixel size is 0.5′. For further details about the procedure we followed to extract the Compton-y map, see Hincks et al. (2021) [13].

3 Fitted Models

In this section, we focus on the fitting procedure we used to model the two galaxy clusters and the intercluster region. We described the two clusters with the isothermal β-profile [6]:

$$P(r) = k_B T_e \frac{n_0}{[1+(\frac{r}{r_c})^2]^{\frac{3}{2}\beta}}, \tag{2}$$

where $k_B T_e$ is the temperature of the system, n_0 is the central electron density, r is the distance from the centre of the system, r_c is the core radius and β parameterizes the slope of the profile. The previous equation describes a spherical system; asphericity can be introduced replacing in Eq. 2:

$$\frac{r}{r_c} \longrightarrow \frac{\sqrt{x^2 + (y/R)^2}}{r_c}, \tag{3}$$

where x is the coordinate of the major axis, y that of the minor axis and R is the ratio between minor and major axis. To fit the intercluster region we adopted the following model to which we will refer as $mesa$-model:

$$g(l, w) = \frac{A_{fil}}{1 + \left(\frac{l}{l_0}\right)^8 + \left(\frac{w}{w_0}\right)^8}, \tag{4}$$

where l and w represent the axes of a reference frame centred in the center of the $mesa$, A_{fil} is the amplitude of the $mesa$, l_0 and w_0 are the typical sizes of the $mesa$ along the two axes. The $mesa$-model is an ad-hoc model without any physical meaning that has been used to simulate the apparent flatness of the Compton-y signal between the two clusters and sharp edges beyond the 3σ contour level, as it is possible to see in Fig. 1. We fitted four different models; each one includes also a plane to remove any residual offsets and gradients:

- 2 elliptical β-profiles for the two clusters;
- 3 elliptical β-profiles for the two clusters and the intercluster region. We fixed the β parameter for the intercluster region to 4/3, the value that describes an isothermal cylinder in hydrostatic equilibrium [18];
- 2 circular β-profiles for the two clusters and the *mesa*-model for the intercluster region;
- 2 elliptical β-profiles for the two clusters and the *mesa*-model for the intercluster region.

4 Results

We performed fits of the models described in Sec. 3 using the *Niagara* supercomputer at the SciNet[1] HPC Consortium (University of Toronto). We minimized the likelihood (Eq. 6) using *emcee* [8], a python implementation of the affine-invariant Monte Carlo Markov Chain algorithm [11] using the map region shown in the right panel of Fig. 1. The likelihood of a fitted model is:

$$\mathcal{L} = \frac{1}{2\pi|M|^{\frac{1}{2}}} \exp\left(-\frac{1}{2}\mathbf{m}^T M^{-1}\mathbf{m}\right), \tag{5}$$

where **m** is the map of residuals and M is the covariance matrix of the map.

We assumed that the fit converges when two conditions are satisfied [8] :

- the length of the chain is greater than 10τ, where τ is the maximum autocorrelation length for all the parameters;
- the autocorrelation length for each parameter does not increase more than 1% every 100 iterations.

Once each fit reached convergence, we burned the first 10τ iterations and we extracted our best fit results using only the remaining set of data. All the best-fit parameters are reported in Table 2 in Hincks et al., (2021) [13]. In Fig. 2 we show the posterior distribution for the '2 Ellip. β+*mesa*' in two different subsets to make the figure clearer.

We used two different statistical tests to compare the models: the likelihood ratio and the Akaike Information Criterion (AIC) [2]. The likelihood ratio W is defined as:

$$W = 2\log\frac{\max\mathcal{L}_2}{\max\mathcal{L}_1}, \tag{6}$$

where $\max\mathcal{L}$ is the likelihood estimate using best-fit parameters. The likelihood ratio compares two models only if model 2 adds additional parameters to model 1. In our case, the "base" model is the '2 Ellip. β' that is extended by the '3 Ellip. β' and the '2 Ellip. β+*mesa*'. The preferred model is the one that maximizes the W parameter. The AIC test is defined as:

$$AIC = 2K - 2\log(\max\mathcal{L}), \tag{7}$$

where K are the degree of freedom of the model. According to the Akaike Information Criterion, the preferred model is the one that minimizes the AIC value, which we will refer as AIC_0. If we define $\Delta_i = AIC_i - AIC_0$, Burnham & Anderson (2004) [5] suggest that, if $\Delta_i \lesssim 2$ the two models have "substantial support", if $\Delta_i \gtrsim 4$ the model i has "considerably less support", while if $\Delta_i \gtrsim 10$ the model i has "essentially no support". In Tab. 1 we report the results of the statistical tests we did. The likelihood ratio test shows that a bridge component is required to properly fit the data at a level greater than 5σ for the '2 Elliptical β-profiles+*mesa*' model. The same model is also preferred by the Akaike Information Criterion, having the

[1] https://www.scinethpc.ca/

Figure 2. Bottom Left: Posterior distributions for 13 of the 22 parameters that describe the '2 Elliptical β-profiles+$mesa$' model. **Top Right**: Posterior distributions for the remaining parameters. Contours levels describe the Gaussian confidence levels at 1, 2, and 3 σ.

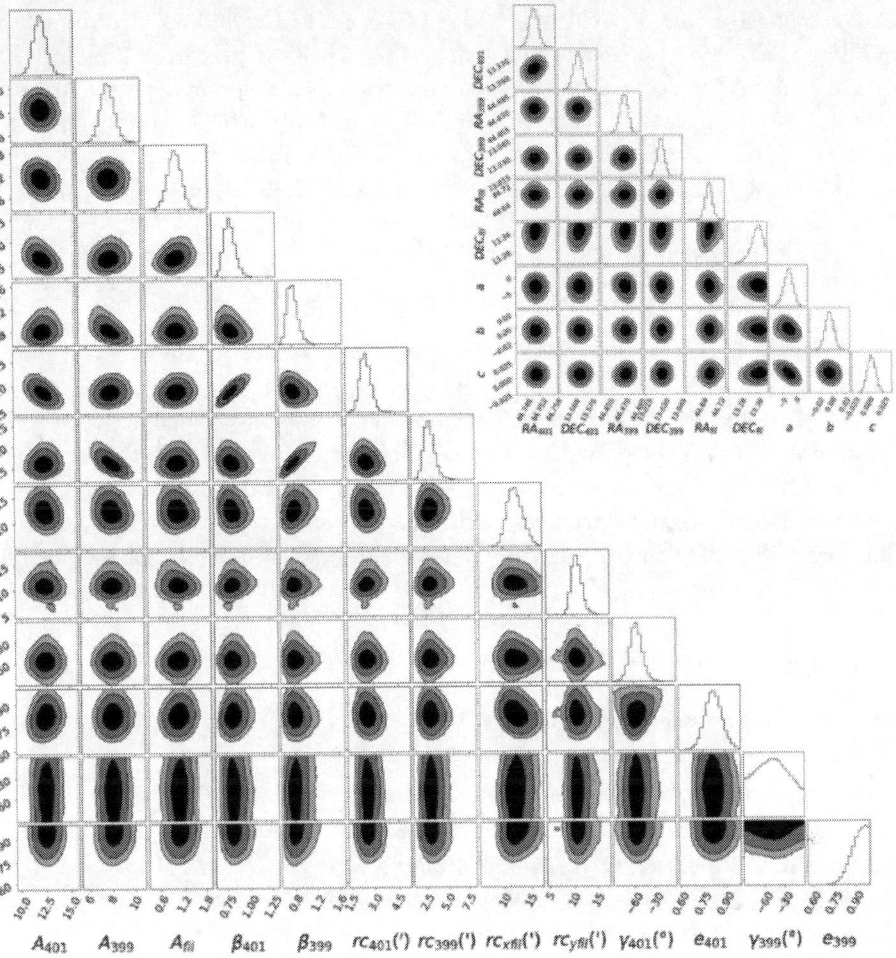

Table 1. The statistical information we used to compare the fitted models.

Data	Model	P	Likelihood Ratio		AIC	
			W	σ	AIC	Δ_i
ACT+*Planck*	2 Ellip. β	17	—	—	48523.2	33.4
	3 Ellip. β	22	33.2	4.6	48500.0	10.2
	2 Circ. β+$mesa$	18	—	—	48492.2	2.4
	2 Ellip. β+$mesa$	22	43.4	5.5	48489.8	0
Planck only	2 Ellip. β	17	—	—	2976.6	13.5
	2 Ellip. β+$mesa$	22	23.5	3.6	2961.1	0

lowest AIC (AIC_0). To estimate how higher angular resolution data can improve the detection of the intercluster signal, we fitted the *Planck* data [20] (PR2 MILCA, 10′ angular resolution) with the same '2 Elliptical β-profiles' and '2 Elliptical β-profiles+$mesa$' models.

We estimated the Compton-y signal at the centre of the $mesa$ from our fiducial 2 Ellip. β+$mesa$ model, $y_{tot} = (2.8 \pm 0.3) \times 10^{-5}$, which includes the $mesa$ itself and the clusters' halos. Assuming the temperature measured in the intercluster region by Akamatsu et al. 2017 [3] using *Suzaku* X-rays observations, $kT_{fil} = 6.5 \pm 0.5$ keV, they derive the relation that links the Compton-y signal with the distance integrated along the line of sight (r_{fil}), $y_{tot} = (8.0 \pm 1.0) \times \left(r_{fil}/Mpc\right)^{0.5} \times 10^{-6}$ (Note that we have corrected an algebraic error from their paper: see Appendix B of Hincks et al. [13]). With the previous y_{tot} we obtain a thickness along the line of sight of $r_{fil} = (12.1 \pm 3.9)$ Mpc and an electron density of $n_e = (0.88 \pm 0.24) \times 10^{-4}$ cm^{-3}. About half of this density comes from the cluster halos, while the remaining part is due to the filament, implying the density in the intercluster region to be $\sim$ 150 times the mean baryon density of the Universe at $z = 0.07$. Moreover, the typical sizes of the *mesa* on the plane of the sky are $l_0 \times w_0 = (2.2 \times 1.9)$ Mpc, much smaller than the thickness along the line of sight, indicating that the system does not lies entirely on the plane of the sky as it was previously assumed [3]. For a more detailed study about the geometry of the system see Hincks et al. (2021) [13].

5 Conclusions

We have shown that using ACT+*Planck* maps, with an effective angular resolution of 1.65′, we improved the significance level of the detection of a Compton-y signal between A399 and A401 with respect to *Planck* only data. Our best fit model consists of 2 Elliptical β-profiles for A399 and A401 and a *mesa* function (Eq. 4) between the two clusters. From our fiducial model combined with X-ray data we have evidence that the system lies mostly along the line of sight. We will further investigate this system with new ACT data, not included in this work.

References

[1] Aiola S., et al., *J. Cosmology Astropart. Phys.*, 047 (2020)
[2] Akaike H., *IEEE Trans. Automat. Contr.*, 19, 716 (1974)
[3] Akamatsu H., et al., *A&A*, 606, A1 (2017)
[4] Bonjean V., Aghanim N., Salomé P., Douspis M., Beelen A.,*A&A*, 609, A49 (2018)
[5] Burnham K. P., Anderson D. R., *Sociol. Methods Res.*, 33, 261 (2004)
[6] Cavaliere A., Fusco-Femiano R., *A&A*, 70, 677 (1978)
[7] Cen R., Ostriker J. P., *AJ*, 514, 1 (1999)
[8] Foreman-Mackey D., Hogg D. W., Lang D., Goodman J., *PASP*, 125, 306 (2013)
[9] Fujita Y., Koyama K., Tsuru T., Matsumoto H., *PASJ*, 48, 191 (1996)
[10] Fukugita M., Hogan C. J., Peebles P. J. E., *AJ*, 503, 518 (1998)
[11] Goodman J., Weare J., *CAMCoS*, 5, 65 (2010)
[12] Govoni F., et al., *Science*, 364, 981 (2019)
[13] Hincks A., et al., *arXiv:2107.04611* (2021)
[14] Madhavacheril M. S., et al., *Phys. Rev. D*, 102, 023534 (2020)
[15] Mroczkowski T., et al., *Space Sci. Rev.*, 215, 17 (2019)
[16] Naess S., et al., *J. Cosmology Astropart. Phys.*, 2020, 046 (2020)
[17] Oegerle W. R., Hill J. M., 2001, *AJ*, 122, 2858 (2001)
[18] Ostriker J., *ApJ*, 140, 1056 (1964)
[19] Planck Collaboration, *A&A*, 550, A134 (2013)
[20] Planck Collaboration, *A&A*, 594, A22 (2016)
[21] Planck collaboration, *A&A*, 641, A6 (2020)
[22] Sunyaev R. A., Zel'dovich Y. B., *CoASP*, 4, 173 (1972)
[23] Tuominen T., et al., *A&A*, 646, A156 (2021)
[24] Vazza F., Ferrari C., et al *A&A*, 580, A119 (2015)

EPJ Web of Conferences **257**, 00040 (2022) https://doi.org/10.1051/epjconf/202225700040
mm Universe @ NIKA2

Relativistic SZ maps and electron gas temperature spectroscopy

Mathieu Remazeilles[1,*]

[1]Jodrell Bank Centre for Astrophysics, Department of Physics and Astronomy, The University of Manchester, Manchester M13 9PL, UK

Abstract. While third-generation CMB experiments have allowed to release the first maps of Compton-y distortion due to thermal Sunyaev-Zeldovich (SZ) effect, next-generation CMB experiments should allow us to map also the electron gas temperature, T_e, across the sky through the detection of relativistic corrections to the thermal SZ effect. We discuss about experimental requirements to break the y-T_e degeneracy of the observed SZ intensity, and propose a new component separation approach based on moment expansion to disentangle the y and T_e observables of the relativistic SZ effect while mitigating foregrounds. We show how our approach offers a new spectroscopic view of the clusters not only across frequencies but now also across temperatures. We also show how the relativistic electron temperature power spectrum provides a new cosmological observable which may complement the Compton-y map power spectrum to break some of the parameter degeneracies in future cosmological SZ analyses.

1 Introduction

By crossing the hot gas of electrons in galaxy clusters and filaments, cosmic microwave background (CMB) photons get scattered while receiving a boost of energy as the result of inverse Compton scattering. This is known as the thermal Sunyaev-Zeldovich (SZ) effect [1], which causes y-type spectral distortions to the primary CMB blackbody spectrum prominently in the direction $\hat{n}$ of galaxy clusters:

$$I_\nu^{\rm SZ}(\hat{n}) \equiv \frac{\Delta I_\nu^{\rm CMB}}{I_\nu^{\rm CMB}}(\hat{n}) = g(\nu)\, y(\hat{n})\,. \tag{1}$$

The amplitude of the distortion given by the Compton parameter y depends on the integrated pressure profile P_e of the electron gas along the line-of-sight as

$$y = \frac{\sigma_T}{m_e c^2}\int P_e(l)dl\,, \tag{2}$$

where m_e is the electron mass, c is the speed of light, and σ_T is the Thomson scattering cross-section. The spectral shape $g(\nu)$ of the distortion across the frequencies ν is given in the *non-relativistic* limit, i.e. $kT_e/m_e c^2 \ll 1$, by [1]

$$g(\nu) \equiv f(\nu, T_e \simeq 0) = \frac{x^4 e^x}{(e^x-1)^2}\left[x\coth\left(\frac{x}{2}\right) - 4\right], \tag{3}$$

*e-mail: mathieu.remazeilles@manchester.ac.uk

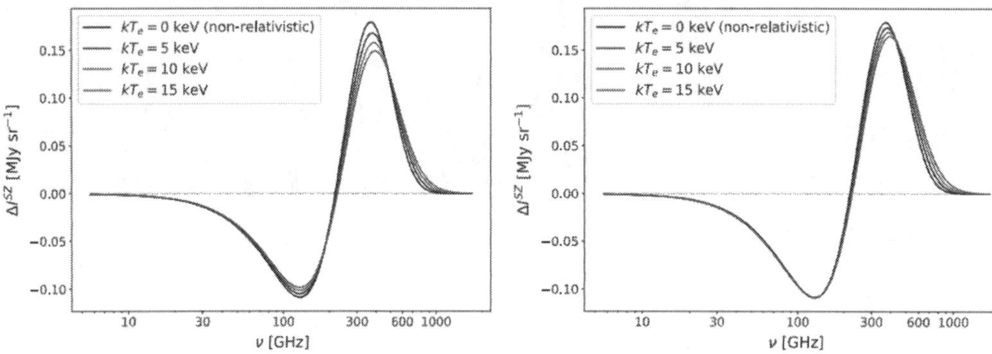

Figure 1. *Left*: Variations of the thermal SZ spectrum with the electron temperature T_e at fixed $y = 10^{-4}$, using SZPACK [2]. *Right*: A varying Compton-y parameter is used to rescale the amplitudes of the spectra and make them matching at low frequencies, which allows us to highlight that the distortion of the SZ spectral shape due to relativistic electron temperature corrections actually arises at high frequencies.

where T_e is the electron gas temperature, h is the Planck constant, k is the Boltzmann constant, $T_{\rm CMB} = 2.7255$ K is the CMB blackbody temperature, and $x \equiv h\nu/kT_{\rm CMB}$.

The peculiar spectral signature Eq. (3) of the thermal SZ effect, as illustrated in Fig. 1 by the black solid line, with a decrement at low frequencies, a null around 218 GHz and an increment at high frequencies, enables to detect galaxy clusters by spectroscopy across frequencies [3]. Hence, the thermal SZ effect has now become a routine tool of third-generation CMB experiments to detect thousands of galaxy clusters (e.g. [4]) and trace the hot gas in the Universe through the mapping of Compton-y parameter anisotropies across the sky [5].

However, galaxy clusters are massive, therefore they are hot objects with typical electron gas temperature of $kT_e \simeq 5$ keV for typical cluster mass of $M_{500} = 3 \times 10^{14}\, h^{-1} M_\odot$ [6]. Hence, the thermal velocity of electrons in the gas approach the speed of light, $v_e^{\rm th} = \sqrt{2kT_e/m_e} \gtrsim 0.1c$. Therefore, relativistic electron temperature corrections to the thermal SZ effect become relevant:

$$I_\nu^{\rm SZ}(\hat{n}) = f(\nu, T_e(\hat{n}))\, y(\hat{n})\,. \tag{4}$$

The spectral signature $f(\nu, T_e(\hat{n}))$ of the *relativistic* SZ (rSZ) effect thus varies depending on the local electron gas temperature in the sky (Fig. 1). While at *Planck* sensitivity relativistic temperature corrections are not perceptible on individual clusters, they become statistically significant for a large number of clusters [7–9]. With larger sensitivity and larger sets of frequencies, next-generation CMB experiments [10, 11] should allow us to not only map the thermal SZ Compton parameter $y(\hat{n})$ but also the relativistic electron temperature field $T_e(\hat{n})$ across the sky. In this proceeding paper we describe a new component separation approach allowing us to disentangle the y and T_e observables of the rSZ effect [12], and show how the extracted T_e field can complement the Compton-y observable to probe cosmology and astrophysics with galaxy clusters.

2 Disentangling the y and T_e observables of the relativistic SZ effect

As shown in the left panel of Fig. 1, relativistic corrections reduce the overall intensity of the thermal SZ signal. Therefore, for a measured intensity of the signal across the frequencies, the assumption of the non-relativistic limit (black line) of the spectrum to detect the SZ effect from hot galaxy clusters will result in underestimating the Compton-y parameter. As shown in [9], neglecting relativistic SZ corrections in the *Planck* data analysis may have led to an underestimation of the overall amplitude of the *Planck* y-map power spectrum, and thereby the inferred σ_8 value. Future cosmological analyses will thus have to account for rSZ effects.

2.1 Experimental requirements

In the right panel of Fig. 1, we rescaled the various rSZ spectra of different temperatures T_e by varying the parameter y so that they match at low frequency. This rescaling allows us to highlight that relativistic electron temperatures actually distort the spectral shape of the thermal SZ signal mostly in the high-frequency range $\gtrsim$ 300 GHz. At low frequencies < 200 GHz, the rSZ effect does not cause any spectral distortion to the signal (right panel of Fig. 1) but only a change in overall amplitude (left panel of Fig. 1). Therefore, there is a y-T_e degeneracy of the rSZ signal at low frequencies, where a lower amplitude of the measured intensity can either be attributed to lower y or larger T_e without any possible distinction. In order to break this spectral degeneracy and disentangle the y and T_e observables of the rSZ effect, it is thus recommended to observe the sky at high frequencies > 200 GHz. Fourth-generation CMB satellite concepts, e.g. *LiteBIRD* [10] or *PICO* [11], which are designed to observe the sky from space at frequencies above 300 GHz with high sensitivity, are thus best suited for our purpose.

2.2 Sky map simulations

We generated full-sky map simulations in 21 frequency bands ranging from 21 to 800 GHz using the instrumental specifications of *PICO* [11] in order to test the ability of our component separation method to disentangle the y and T_e observables of the rSZ effect. Our sky simulations include the relativistic SZ signal in each frequency channel, which we generated from all-sky template maps of y and T_e (see figure 3 in [12]) using the frequency scaling Eq. (4) computed with SZpack [2]. Our simulations also include several foreground emissions: kinetic SZ effect, CMB anisotropies, cosmic infrared background anisotropies, synchrotron, anomalous microwave emission, free-free and thermal dust (see [12] for details).

2.3 Component separation method

The observed sky data, $d_\nu(\hat{n})$, in the line-of-sight $\hat{n}$ and frequency channel ν can be written as

$$d_\nu(\hat{n}) = I_\nu^{\mathrm{SZ}}(\hat{n}) + n_\nu(\hat{n})\,, \tag{5}$$

where $I_\nu^{\mathrm{SZ}}(\hat{n})$ is the rSZ signal intensity (Eq. 4) and $n_\nu(\hat{n})$ is the overall nuisance term including foregrounds and instrumental noise. Following [2], we can perform the Taylor expansion of the rSZ spectral energy distribution (SED) around some pivot temperature $\overline{T}_e$ so that

$$d_\nu(\hat{n}) = f(\nu,\overline{T}_e)\,y(\hat{n}) + \frac{\partial f(\nu,\overline{T}_e)}{\partial \overline{T}_e}\left(T_e(\hat{n}) - \overline{T}_e\right)y(\hat{n}) + \tilde{n}_\nu(\hat{n})\,, \tag{6}$$

where the nuisance term $\tilde{n}_\nu$ now includes higher-order terms in $T_e(\hat{n})$ in addition to foregrounds and noise. The expansion Eq. (6) thus highlights two distinct components of emission, the Compton parameter component $y(\hat{n})$ and a *temperature-modulated* component $\left(T_e(\hat{n}) - \overline{T}_e\right)y(\hat{n})$, with different spectral signatures $f(\nu,\overline{T}_e)$ and $\partial f(\nu,\overline{T}_e)/\partial\overline{T}_e$, respectively. Therefore, it is possible in principle to disentangle the y and $y(T_e - \overline{T}_e)$ fields through multi-frequency observations and tailored component separation methods.

Given that both signals y and $y(T_e - \overline{T}_e)$ are correlated with each other, it is essential to fully deproject one signal from the estimated map of the other signal, i.e. nulling out y in the reconstructed map of the $y(T_e - \overline{T}_e)$ field and vice-versa, by using a *constrained* ILC approach [12–14]. Denoting the temperature-modulated Compton-y signal as $z(\hat{n}) \equiv y(\hat{n})\left(T_e(\hat{n}) - \overline{T}_e\right)$, we thus form our constrained-ILC estimate $\hat{z}(\hat{n})$ through the following weighted linear combination of the data across frequencies:

$$\hat{z}(\hat{n}) = \sum_\nu w_\nu\, d_\nu(\hat{n}) \quad \text{such that} \quad \begin{cases} \text{the variance } \langle\, \hat{z}^2 \rangle \text{ is minimum}\,, \\ \sum_\nu w_\nu \, \frac{\partial f(\nu, \overline{T}_e)}{\partial \overline{T}_e} = 1\,, \\ \sum_\nu w_\nu \, f(\nu, \overline{T}_e) = 0\,. \end{cases} \tag{7}$$

The unit constraint in Eq. (7) ensures the conservation of the signal of interest, $z \equiv y(T_e - \overline{T}_e)$, during variance minimization, while the null constraint guarantees full cancellation of y residuals in the recovered $y(T_e - \overline{T}_e)$-map. Finally, the minimum-variance condition in Eq. (7) ensures the mitigation of foregrounds and noise. Note that the y-map can be obtained in a similar way, simply by interchanging the unit and null constraints in Eq. (7). In addition to above constraints in Eq. (7) we add two extra nulling constraints to deproject the kinetic SZ effect (hence the CMB) and bulk of the thermal dust emission in the recovered rSZ maps (see [12] for details). The expression of the constrained-ILC weights $\boldsymbol{w} \equiv \{w_\nu\}$ for our specific component separation problem Eq. (7) can be derived using Lagrange multipliers (see [12]):

$$\boldsymbol{w}^{\mathrm{T}} = \boldsymbol{e}^{\mathrm{T}} \left(\mathrm{A}^{\mathrm{T}}\mathrm{C}^{-1}\mathrm{A}\right)^{-1} \mathrm{A}^{\mathrm{T}}\mathrm{C}^{-1}\,, \tag{8}$$

where the first two columns of matrix $\mathrm{A} = \left(\partial \boldsymbol{f}/\partial \overline{T}_e \;\; \boldsymbol{f} \;\; \boldsymbol{g}_{\mathrm{kSZ}} \;\; \boldsymbol{g}_{\mathrm{dust}}\right)$ collect the effective SED vectors of the rSZ signals (see Eq. 6) and the last two columns the SED vector $\boldsymbol{g}_{\mathrm{kSZ}}$ of the kinetic SZ effect and the average SED vector $\boldsymbol{g}_{\mathrm{dust}}$ of the thermal dust, while $\boldsymbol{e}^{\mathrm{T}} = (1\,0\,0\,0)$, and $C = \langle \boldsymbol{dd}^{\mathrm{T}} \rangle$ is the channel-to-channel covariance matrix of the data.

3 Overview of the results

3.1 Electron-temperature spectroscopy of galaxy clusters

The new rSZ observable $z(\hat{n}) \equiv y(\hat{n})(T_e(\hat{n}) - \overline{T}_e)$ is particularly interesting to find out the actual temperature $T_e(\hat{n})$ of a cluster. Indeed, using different pivot temperatures $\overline{T}_e$ for the Constrained-ILC filter Eq. (8) allows us to perform the temperature spectroscopy of the clusters: the recovered signal $z(\hat{n}) \equiv y(\hat{n})(T_e(\hat{n}) - \overline{T}_e)$ will show a *decrement* at the position of a cluster in the map if the actual temperature of that cluster is smaller than the pivot temperature, i.e. $T_e(\hat{n}) < \overline{T}_e$, while it will show an *increment* if $T_e(\hat{n}) > \overline{T}_e$ and a *null* if $T_e(\hat{n}) \simeq \overline{T}_e$. In the latter case the pivot temperature must be the actual temperature of the cluster.

Such temperature spectroscopy is illustrated in Fig. 2 where we show the reconstructed map of $y(\hat{n})\left(T_e(\hat{n}) - \overline{T}_e\right)$ around the Coma cluster as obtained for different pivots $k\overline{T}_e = 5$, 7, and 10 keV after component separation on the *PICO* map simulations. The recovered rSZ signal informs us that the temperature of Coma is hotter than 5 keV but colder than 10 keV, and closer to 7 keV (null), which is the fiducial temperature of Coma in the current simulation.

3.2 Electron gas temperature observables from the rSZ maps

From the reconstructed y and $y(T_e - \overline{T}_e)$ maps after component separation, we can easily create the yT_e-map simply by mutiplying the y-map by the pivot temperature $\overline{T}_e$ and adding

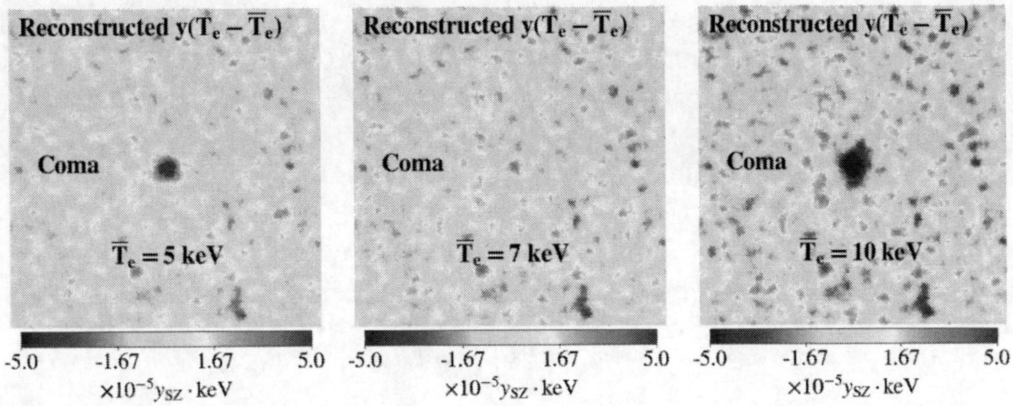

Figure 2. Temperature spectroscopy of galaxy clusters: Gnomonic view of the reconstructed $y(T_e - \overline{T}_e)$ all-sky map around Coma for different pivot temperatures $\overline{T}_e$.

it to the $y(T_e - \overline{T}_e)$-map: $yT_e(\hat{n}) = \overline{T}_e\, y(\hat{n}) + z(\hat{n}) = \overline{T}_e y(\hat{n}) + y(\hat{n})\left(T_e(\hat{n}) - \overline{T}_e\right)$. Using the y-map and the yT_e-map we can then derive several observables which we outline hereafter.

First, by computing the ratio of the flux in the yT_e-map and the flux in the y-map within R_{500} around a cluster, we obtain the y-weighted average temperature of that cluster over R_{500}:

$$T_e^y\,[R_{500}] = \frac{\langle yT_e(\hat{n})\rangle_{|\hat{n}-\hat{n}_c|\leq R_{500}}}{\langle y(\hat{n})\rangle_{|\hat{n}-\hat{n}_c|\leq R_{500}}}\,, \tag{9}$$

where $\hat{n}_c$ denotes the line-of-sight corresponding to the centre of the cluster. Figure 3 (left panel) shows our successful recovery of the electron temperatures of more than 800 galaxy clusters across the entire sky for the *PICO* sky simulation after foreground cleaning and component separation [12]. The recovered rSZ temperatures could be used in principle as a new proxy for determining cluster masses without relying on X-ray measurements.

Second, we can also derive the electron temperature profile of a cluster simply from the ratio of the cluster profile in the yT_e-map and the cluster profile in the y-map:

$$T_e^y(r) = \frac{yT_e(r)}{y(r)}\,, \tag{10}$$

where r is the radius from the centre of the cluster. Figure 3 (middle panel) shows as an example our recovery of the electron temperature profile of the Coma cluster up to large radii, with a 10σ measurement of the average temperature of Coma over R_{500} for *PICO* [12].

Finally, the cross-power spectrum between the reconstructed y- and yT_e-maps relative to the auto-power spectrum of the y-map measures the y^2-weighted average electron temperature over the full sky across multipoles:

$$T_e^{yy}(\ell) = \frac{\langle (yT_e)_{\ell m}\, y^*_{\ell m}\rangle}{\langle y_{\ell m}\, y^*_{\ell m}\rangle} = \frac{C_\ell^{y,yT_e}}{C_\ell^{yy}}\,, \tag{11}$$

thus providing an effective power spectrum of the electron temperature field [9]. Figure 3 (right panel) shows accurate recovery of the electron temperature power spectrum with *PICO* over a large range of multipoles after foreground cleaning and component separation [12].

Concluding remarks. The electron temperature power spectrum $T_e^{yy}(\ell)$ provides a new map-based observable, complementing the y-map power spectrum C_ℓ^{yy}, to constrain cosmological

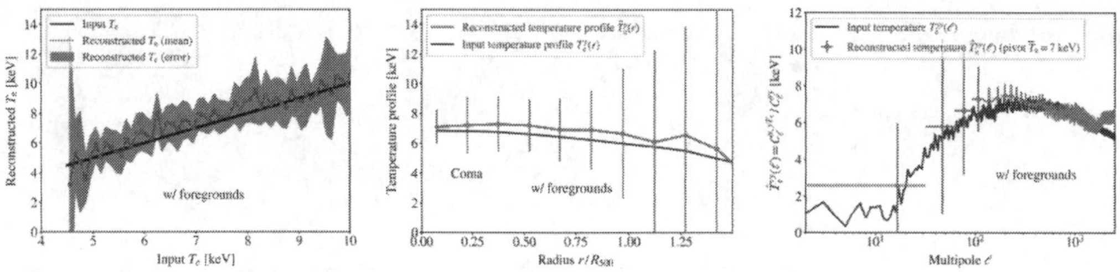

Figure 3. *Left*: Recovered electron temperatures $T_e[R_{500}]$ across the sky of a sample of 811 clusters after foreground cleaning. *Middle*: Reconstructed temperature profile $T_e(r)$ of Coma. *Right*: Reconstructed diffuse electron gas temperature power spectrum $T_e^{yy}(\ell)$ after foreground cleaning.

parameters, since the exact shapes of the power spectra T_e^{yy} and C_ℓ^{yy} have different scaling and dependence on cosmological parameters. To gain intuition on this, we may look into theoretical expressions:

$$C_\ell^{yy} = \int_0^{z_{\max}} dz \frac{dV}{dz} \int_{M_{\min}}^{M_{\max}} dM \, \frac{dn(M,z)}{dM} \, |y_\ell(M,z)|^2 \,, \tag{12}$$

$$C_\ell^{y,yT_e} = \int_0^{z_{\max}} dz \frac{dV}{dz} \int_{M_{\min}}^{M_{\max}} dM \, \frac{dn(M,z)}{dM} \, T_e(M,z) \, |y_\ell(M,z)|^2 \,. \tag{13}$$

While C_ℓ^{yy} and C_ℓ^{y,yT_e} have same scaling with the mass bias b through exact same dependence on the profile $|y_\ell(M,z)|^2$, they have different scaling with σ_8 due to the modulation of the halo mass function in C_ℓ^{y,yT_e} by the mass-dependent temperature T_e. Therefore, while the y-map power spectrum C_ℓ^{yy} depends on σ_8 and b in a degenerate form, the electron temperature power spectrum $T_e^{yy}(\ell)$, as the ratio of C_ℓ^{y,yT_e} and C_ℓ^{yy} (Eq. 11), must depend only on σ_8 and be quite insensitive to the mass bias [12]. Such new rSZ observable thus provides a possible avenue to break some of the parameter degeneracies with future cosmological SZ analyses.

Acknowledgements: This work is supported by the ERC Consolidator Grant CMBSPEC (No. 725456) as part of the European Union's Horizon 2020 research and innovation program.

References

[1] Zeldovich Ya. B., Sunyaev R. A., Astrophys. Space Sci., **4**, 301 (1969)
[2] Chluba J., Switzer E., Nelson K., Nagai D., MNRAS, **430**, 3054 (2013)
[3] Planck Collaboration, A&A, **536**, A8 (2011)
[4] Planck Collaboration, A&A, **594**, A27 (2016)
[5] Planck Collaboration, A&A, **594**, A22 (2016)
[6] Arnaud M., Pointecouteau E., Pratt G. W., A&A, **441**, 893 (2005)
[7] Hurier G., A&A, **596**, A61 (2016)
[8] Erler J., Basu K., Chluba J., Bertoldi F., MNRAS, **476**, 3360 (2018)
[9] Remazeilles M., Bolliet B., Rotti A., Chluba J., MNRAS, **483**, 3459 (2019)
[10] Hazumi M., et al., in SPIE Conference Series. p. 114432F (2020)
[11] Hanany S., et al., arXiv:1902.10541 (2019)
[12] Remazeilles M., Chluba J., MNRAS, **494**, 5734 (2020)
[13] Remazeilles M., Delabrouille J., Cardoso J.-F., MNRAS, **410**, 2481 (2011)
[14] Remazeilles M., Rotti A., Chluba J., MNRAS, **503**, 2478 (2021)

Galactic Star Formation with NIKA2 (GASTON):

Evidence of mass accretion onto dense clumps

A. J. Rigby[1,*], *R.* Adam[2], *P.* Ade[1], *H.* Ajeddig[3], *M.* Anderson[1], *P.* André[3], *E.* Artis[5], *H.* Aussel[3], *A.* Bacmann[4], *A.* Beelen[6], *A.* Benoît[7], *S.* Berta[8], *L.* Bing[9], *O.* Bourrion[5], *A.* Bracco[10], *M.* Calvo[7], *A.* Catalano[5], *M.* De Petris[11], *F.-X.* Désert[4], *S.* Doyle[1], *E. F. C.* Driessen[8], *P.* García[12,13], *A.* Gomez[14], *J.* Goupy[7], *F.* Kéruzoré[5], *C.* Kramer[15], *B.* Ladjelate[15], *G.* Lagache[9], *S.* Leclercq[8], *J.-F.* Lestrade[16], *J.-F.* Macías-Pérez[5], *A.* Maury[3], *P.* Mauskopf[1,17], *F.* Mayet[5], *A.* Monfardini[7], *M.* Muñoz-Echeverría[5], *N.* Peretto[1], *L.* Perotto[5], *G.* Pisano[1], *N.* Ponthieu[4], *V.* Revéret[3], *I.* Ristorcelli[18], *A.* Ritacco[6,19], *C.* Romero[20], *H.* Roussel[21], *F.* Ruppin[22], *K.* Schuster[8], *S.* Shu[23], *A.* Sievers[15], *C.* Tucker[1], *E. J.* Watkins[24], and *R.* Zylka[8]

[1]School of Physics and Astronomy, Cardiff University, Queen's Buildings, The Parade, Cardiff, CF24 3AA, UK
[2]LLR (Laboratoire Leprince-Ringuet), CNRS, École Polytechnique, Institut Polytechnique de Paris, Palaiseau, France
[3]AIM, CEA, CNRS, Université Paris-Saclay, Université Paris Diderot, Sorbonne Paris Cité, 91191 Gif-sur-Yvette, France
[4]Univ. Grenoble Alpes, CNRS, IPAG, 38000 Grenoble, France
[5]Univ. Grenoble Alpes, CNRS, Grenoble INP, LPSC-IN2P3, 53, avenue des Martyrs, 38000 Grenoble, France
[6]Institut d'Astrophysique Spatiale (IAS), CNRS, Université Paris Sud, Orsay, France
[7]Institut Néel, CNRS, Université Grenoble Alpes, France
[8]Institut de RadioAstronomie Millimétrique (IRAM), Grenoble, France
[9]Aix Marseille Univ, CNRS, CNES, LAM (Laboratoire d'Astrophysique de Marseille), Marseille, France
[10]Rudjer Bošković Institute, Bijenička cesta 54, 10000 Zagreb, Croatia
[11]Dipartimento di Fisica, Sapienza Università di Roma, Piazzale Aldo Moro 5, I-00185 Roma, Italy
[12]Chinese Academy of Sciences South America Center for Astronomy, National Astronomical Observatories, CAS, Beijing 100101, PR China
[13]Instituto de Astronomía, Universidad Católica del Norte, Av. Angamos 0610, Antofagasta 1270709, Chile
[14]Centro de Astrobiología (CSIC-INTA), Torrejón de Ardoz, 28850 Madrid, Spain
[15]Instituto de Radioastronomía Milimétrica (IRAM), Granada, Spain
[16]LERMA, Observatoire de Paris, PSL Research University, CNRS, Sorbonne Université, UPMC, 75014 Paris, France
[17]School of Earth and Space Exploration and Department of Physics, Arizona State University, Tempe, AZ 85287, USA
[18]Univ. Toulouse, CNRS, IRAP, 9 Av. du Colonel Roche, BP 44346, 31028, Toulouse, France
[19]Laboratoire de Physique de l'École Normale Supérieure, ENS, PSL Research University, CNRS, Sorbonne Université, Université de Paris, 75005 Paris, France
[20]Department of Physics and Astronomy, University of Pennsylvania, 209 South 33rd Street, Philadelphia, PA, 19104, USA

*e-mail: RigbyA@cardiff.ac.uk

[21]Institut d'Astrophysique de Paris, Sarbonne Université, CNRS (UMR7095), 75014 Paris, France
[22]Kavli Institute for Astrophysics and Space Research, Massachusetts Institute of Technology, Cambridge, MA 02139, USA
[23]Caltech, Pasadena, CA 91125, USA
[24]Astronomisches Rechen-Institut, Zentrum für Astronomie der Universit at Heidelberg, Monchhofsstraße 12-14, D-69120 Heidelberg, Germany

Abstract. High-mass stars ($m_* \gtrsim 8\,M_\odot$) play a crucial role in the evolution of galaxies, and so it is imperative that we understand how they are formed. We have used the New IRAM KIDs Array 2 (NIKA2) camera on the Institut de Radio Astronomie Millimétrique (IRAM) 30-m telescope to conduct high-sensitivity continuum mapping of ~ 2 deg^2 of the Galactic plane (GP) as part of the Galactic Star Formation with NIKA2 (GASTON) large program. We have identified a total of 1467 clumps within our deep 1.15 mm continuum maps and, by using overlapping continuum, molecular line, and maser parallax data, we have determined their distances and physical properties. By placing them upon an approximate evolutionary sequence based upon 8 μm Spitzer imaging, we find evidence that the most massive dense clumps accrete material from their surrounding environment during their early evolution, before dispersing as star formation advances, supporting clump-fed models of high-mass star formation.

1 Introduction

Obtaining an understanding of the origin of the stellar initial mass function (IMF) is one of the most important goals of contemporary astrophysics. Over the last decade, the advent of *Herschel* observations have highlighted the link between dense filamentary structures within the interstellar medium (ISM) and star formation [1, 2]. Thermally supercritical filaments are susceptible to gravitational fragmentation along their primary axis, resulting in strings of dense cores [e.g. 3] that may explain the shape of the IMF in the mass range $0.1 \lesssim m_* \lesssim 5\,\mathrm{M}_\odot$ [4, 5]. However, this *core-fed* scenario may be insufficient outside these mass limits; some observations [e.g. 6, 7] and simulations [e.g. 8, 9] suggest that high-mass star formation requires the large infall rates found at the centres of globally-collapsing molecular clouds in a *clump-fed* scenario, while brown dwarf-formation may require processes such as dynamical ejection from multiple star systems [10] or circumstellar disc fragmentation [11].

To tackle the origin of the IMF at the low- and high-mass extremes, the 200-hour GASTON large program (PI: N. Peretto) is conducting deep mapping of a region of the Galactic plane that is rich in high-mass star formation, in addition to surveying more nearby clouds for pre-brown dwarfs, whilst exploiting the dual waveband capabilities of NIKA2 on the IRAM 30-m telescope to study potential environmental variations in dust properties. In this proceeding, we summarise recent results from the GP survey [12], focusing on the transition between the core-fed and clump-fed modes of star formation.

2 Observations

We simultaneously observed a sector of the Galactic Plane in 1.15 and 2.00 mm continuum wavebands using NIKA2 [13–16], which have angular resolutions of 11.1 and 17.6 arcsec, respectively. The region is centred on $\ell = 23.9°, b = 0.05°$, and approximately covers an area of 2 deg^2 with fairly uniform sensitivity. The observations consist of pairs of scans centred on three different positions ($\ell = 23.3°, 23.9°, 24.5°$, with $b = 0.05°$) with position angles alternating between $\pm 30°$ from the normal to the Galactic plane. Flux calibration is

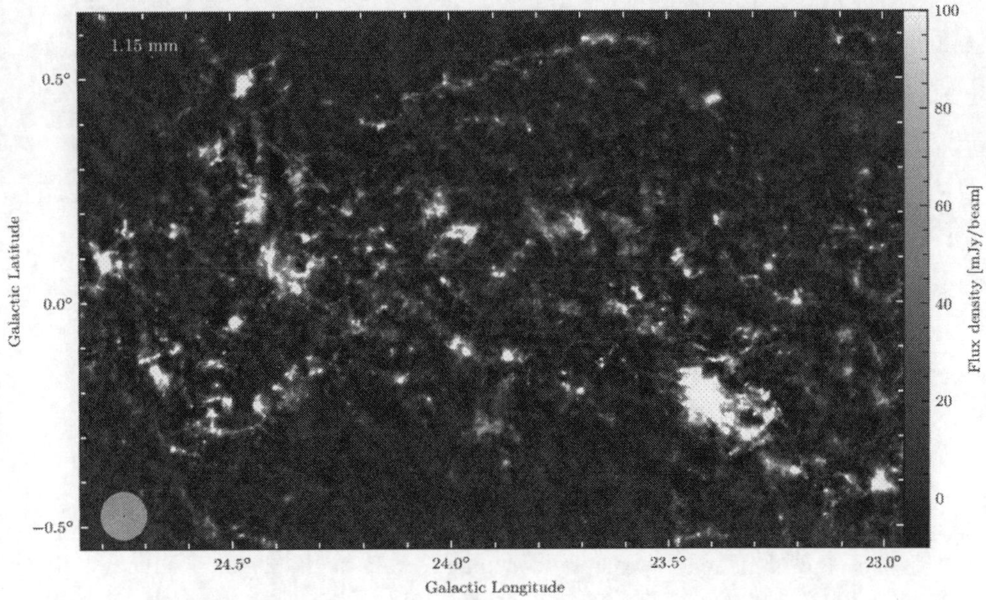

Figure 1. The GASTON GP field at 1.15 mm. The NIKA2 field of view and beam size are indicated by the grey and black circles, respectively, in the lower-left corner.

performed via observations of primary calibrators, mainly Uranus and Neptune, and pointing uncertainties are $<$ 3 arcsec. The data were reduced using the NIKA2 collaboration's IDL pipeline. The results presented here were obtained from the 1.15 mm image obtained after approximately 30 hours of observations between October 2017 and February 2019. As of August 2021, the GP observations are 92% complete, out of their total 73.6 hours of allocated time. We display the 1.15 mm map in Figure 2.

3 Source extraction and determination of physical properties

We extracted sources within the 1.15 mm map using the Python ASTRODENDRO implementation of dendrograms [17], after smoothing the image to 13 arcsec resolution to slightly increase the signal-to-noise ratio. The global rms of the 1.15 mm map, determined from the null maps produced by the IDL pipeline, is 4.18 mJy per 13-arcsec beam in 2.5-arcsec pixels. We catalogued 488 independent emission structures (dendrogram trunks), containing a total 1467 compact sources (leaves) that contain no further discernible substructures, that will be the focus of our analysis.

We determined systemic velocities for all sources by first creating an average spectrum within each source footprint in overlapping 21 arcsec-resolution ^{13}CO (1–0) data from the FUGIN survey [18], and assigned the velocity corresponding to the brightest emission. Secondly, we revised these velocities for sources which also have overlapping coverage in the RAMPS pilot study data [19]. These observations of the NH_3 (1,1) inversion transition are more appropriate for tracing dense gas, and the observations show better morphological agreement with the 1.15 mm structures than the FUGIN data, although the lower angular resolution (32 arcsec) will probably result in incorrect velocities in some cases. We will eventually revise these velocities with dedicated follow-up observations of dense gas tracers such as N_2H^+ (2–1). Velocities were then converted to kinematic distances by adopting a Galactic rotation curve model [20], and the assumption that the sources are following circular orbits around the Galactic centre. We used version 2.4.1 of the Bayesian distance calculator [21], to determine kinematic distance solutions for all of our sources, though we removed the spiral arm model's influence on the resulting distances since we do not know *a priori* that all our

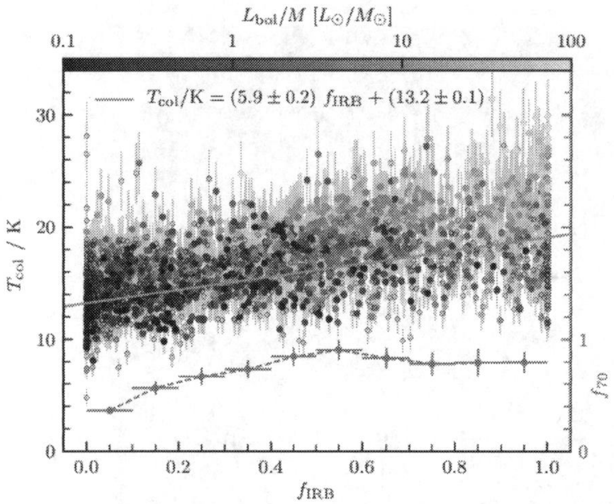

Figure 2. The infrared-bright fraction, f_{IRB}, of all compact sources as a function of 160/250 μm temperature, coloured by the bolometric luminosity to mass ratio. Sources with no compact 70 μm counterpart are shown as black circles. On the secondary axis, we show the fraction of sources within 0.1-wide f_{IRB} bins that have a counterpart compact 70 μm source.

sources belong to the spiral arms. We also directly adopted maser parallax distances where possible, in place of the less accurate kinematic distances. Finally, we use a combination of friends-of-friends algorithm, and information contained within the hierarchical structure of the dendrogram, to refine the distances for sources whose velocities were less well defined. In total, we place 92 percent of the compact sources at the near kinematic distance solution, with a median distance of 5.6 kpc. Around 90% of the sample lies between 3.5 and 7.0 kpc.

Masses were determined by assuming a modified blackbody for the continuum emission, with dust temperatures estimated using the *Herschel* 160/250 μm flux density ratio method [22], and a dust absorption coefficient of $0.1(\nu/\mathrm{THz})^{1.8}$ [23]. The integrated flux densities of each source were first background-subtracted to remove the contribution from overlapping structures in complex regions. We also define a new quantity for each source, the *infrared-bright fraction* (f_{IRB}), giving the fraction of pixels within the source that are brighter in 8 μm *Spitzer*/GLIMPSE [24] imaging than the median value within a 4.8-arcmin box (the size of which approximates the largest structures within the GLIMPSE data). We use f_{IRB} as a tracer of the relative evolutionary stage of each source, since very young sources tend to be infrared-dark (i.e. $f_{IRB} \approx 0$), while sources in an advanced stage of star formation tend to be infrared-bright ($f_{IRB} \approx 1$). In Figure 3, we demonstrate the utility of f_{IRB} as an evolutionary tracer, which correlates well with temperature, as well as the bolometric luminosity-to-mass ratio, and likelihood of a coincident compact 70 μm source [25].

4 Mass and temperature evolution

We refer to the compact sources within our catalogue as *clumps*, though we note that, given the range in deconvolved sizes (~ 0.05–2 pc) and background-subtracted masses (~ 1–5 × 10^3 $M_\odot$), our sample suggests that it contains a mixture of both cores (i.e. the precursors of individual star systems) and clumps (i.e. the precursors of stellar clusters).

To quantify the evolution of the properties of the sources, we create a distance-limited sample of the clumps, selecting only those with distances of $3.5 \leq d \leq 7.0$ kpc, and then break that sample into four equal-sized subsamples (labelled i–iv) depending on their f_{IRB} values. In Figure 4, we display the (logarithmic) mass–temperature distributions for four subsamples, and a number of important points are evident. Firstly, the mean temperature increases monotonically as we progress through the samples from infrared-dark to infrared-bright, reflecting the evolution of the underlying star formation. Secondly, we notice that the mean value of the mass is significantly enhanced (when considering the errors on the mean) in stages ii) and iii) compared to i) and iv). Finally, the scatter also increases monotonically.

We compare what we see in Figure 4 to models of core- and clump-fed evolution [26] in mass–temperature space. For core-fed evolution, we would expect to see sources maintaining

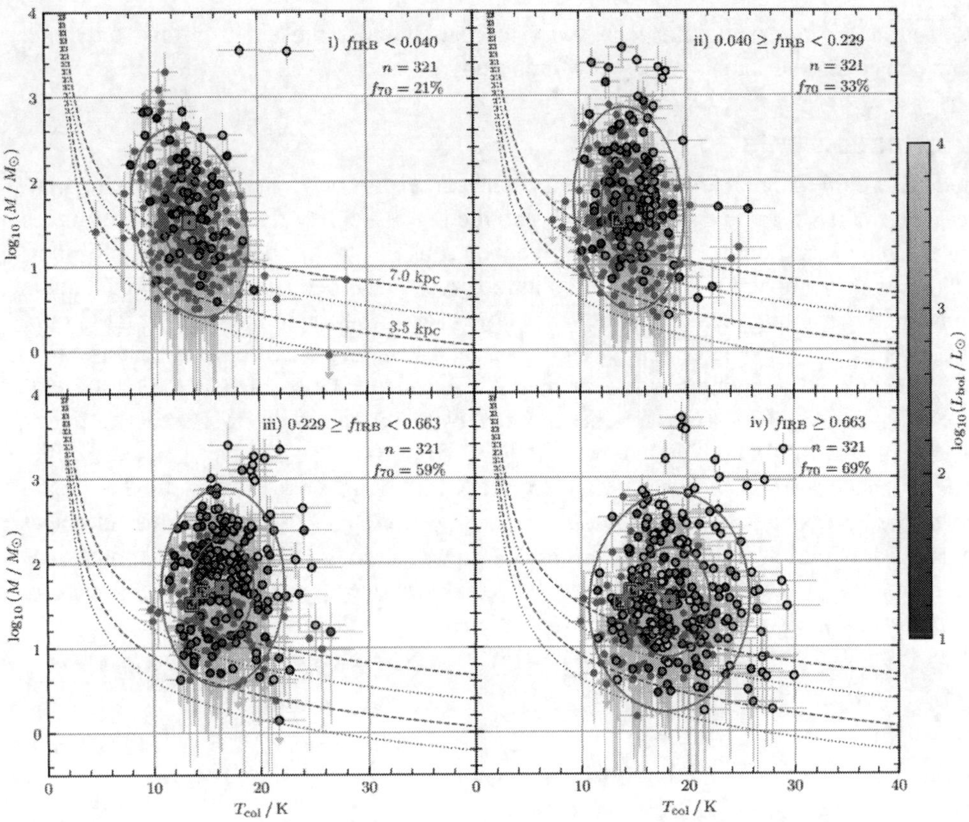

Figure 3. Clump mass as a function of dust temperature for the four distance-limited subsamples of clumps. For each subsample, the mean value in both axes is shown as an orange square, along with the associated error, and approximate 1- and 2-σ confidence ellipses are shown in red. Sources with a compact 70 μm counterpart are coloured by the corresponding bolometric luminosity, while those without are shown as grey circles. The sensitivity limits for point-like and extended sources are shown as purple dashed and dotted curves, respectively.

a constant mass at early stages in their evolution, before being dispersed by feedback as star formation progresses. By contrast, if the evolution is clump-fed, we expect to see sources gaining mass at early stages, before losing mass at the later stages. The increases of the mean mass in stages (ii) and (iii) compared to (i), which are significant at the 3.1- and 4.3-σ levels, respectively, suggests that the populations, as a whole, are gaining mass between these phases, and then losing mass, on average, between stage (iii) and (iv). This would be inconsistent with a population of sources undergoing purely core-fed star-formation, and so we interpret this as evidence that sources are accreting material from their wider environments in a clump-fed scenario.

5 Summary

We have presented an analysis of deep 1.15 mm continuum mapping of ~2 deg^2 of the Galactic plane centred on $\ell = 24°$ using NIKA2. After approximately 40% of the target integration time, we have identified almost 1500 compact sources after smoothing to 13 arcsec resolution, and identified systemic velocities and distances, and thereby calculated physical properties. By splitting a distance-limited sample into subsamples based on the $f_{\rm IRB}$ evolutionary indicator, we find evidence for the mass growth of the compact sources, supporting models of clump-fed star formation in which material is accreted from the wider environment. We expect observations for the GP survey to be concluded during the Winter of 2021–22, enabling the production of maps that are at least twice as deep as those presented in this study, and therefore enhancing the statistical power of this study.

Acknowledgements

AJR and NP would like to thank the STFC for financial support under the consolidated grant numbers ST/N000706/1 and ST/S00033X/1, and the Royal Society for computing resources under Research Grant number RG150741. We would like to thank the IRAM staff for their support during the campaigns. The NIKA2 dilution cryostat has been designed and built at the Institut Néel. In particular, we acknowledge the crucial contribution of the Cryogenics Group, and in particular Gregory Garde, Henri Rodenas, Jean Paul Leggeri, Philippe Camus. This work has been partially funded by the Foundation Nanoscience Grenoble and the LabEx FOCUS ANR-11-LABX-0013. This work is supported by the French National Research Agency under the contracts "MKIDS", "NIKA" and ANR-15-CE31-0017 and in the framework of the "Investissements d'avenir" program (ANR-15-IDEX-02). This work has benefited from the support of the European Research Council Advanced Grant ORISTARS under the European Union's Seventh Framework Programme (Grant Agreement no. 291294). F.R. acknowledges financial supports provided by NASA through SAO Award Number SV2-82023 issued by the Chandra X-Ray Observatory Center, which is operated by the Smithsonian Astrophysical Observatory for and on behalf of NASA under contract NAS8-03060.

References

[1] André P. *et al.*, A&A **518**, L102 (2010)
[2] Molinari S. *et al.*, A&A **518**, L100 (2010)
[3] Inutsuka S.-i., Miyama S. M., ApJ **480**, 681 (1997)
[4] Lee Y.-N., Hennebelle P., Chabrier G., ApJ **847**, 114 (2017)
[5] André P. *et al.*, A&A **629**, L4 (2019)
[6] Peretto N. *et al.*, A&A **555**, A112 (2013)
[7] Chen H.-R. V. *et al.*, ApJ **875**, 24 (2019)
[8] Wang P. *et al.*, ApJ **709**, 27 (2010)
[9] Vázquez-Semadeni E. *et al.*, MNRAS **490**, 3061 (2019)
[10] Bate M. *et al.*, MNRAS **332**, 65 (2002)
[11] Thies I. *et al.*, ApJ **717**, 577 (2010)
[12] Rigby A. J. *et al.*, MNRAS **502**, 4576 (2021)
[13] Bourrion O. *et al.*, J. Instrum. **11**, 11001 (2016)
[14] Calvo M. *et al.*, J. Low Temp. Phys **184**, 816 (2016)
[15] Adam R. *et al.*, A&A **609**, A115 (2018)
[16] Perotto L., Ponthieu N., Macías-Pérez *et al.*, A&A **637**, A71 (2020)
[17] Rosolowsky, E. W. *et al.*, ApJ **679**, 1338 (2018)
[18] Umemoto T. *et al.*, PASJ **69**, 78 (2017)
[19] Hogge T. *et al.*, ApJS **237**, 27 (2018)
[20] Reid M. J. *et al.*, ApJ **885**, 131 (2019)
[21] Reid M. J. *et al.*, ApJ **823**, 77 (2016)
[22] Rigby A. J. *et al.*, A&A **615**, A18 (2018)
[23] Beckwith S. V. W. *et al.* AJ **99**, 924 (1990)
[24] Churchwell E. *et al.*, PASP **121**, 213 (2009)
[25] Elia D. *et al.*, MNRAS **471**, 100 (2017)
[26] Peretto N. *et al.*, MNRAS **496**, 3482 (2020)

Crab nebula at 260 GHz with the NIKA2 polarimeter

Implications for the polarization angle calibration of future CMB experiments

A. Ritacco[1,2,*], *R.* Adam[3], *P.* Ade[4], *H.* Ajeddig[5], *P.* André[5], *E.* Artis[6], *J.* Aumont[7], *H.* Aussel[5], *A.* Beelen[1], *A.* Benoît[8], *S.* Berta[9], *L.* Bing[10], *O.* Bourrion[6], *M.* Calvo[8], *A.* Catalano[6], *M.* De Petris[11], *F.-X.* Désert[12], *S.* Doyle[4], *E. F. C.* Driessen[9], *A.* Gomez[13], *J.* Goupy[8], *F.* Kéruzoré[6], *C.* Kramer[14], *B.* Ladjelate[14], *G.* Lagache[10], *S.* Leclercq[9], *J.-F.* Lestrade[15], *J.-F.* Macías-Pérez[6], *A.* Maury[5], *P.* Mauskopf[4,16], *F.* Mayet[6], *A.* Monfardini[8], *M.* Muñoz-Echeverría[6], *L.* Perotto[6], *G.* Pisano[4], *N.* Ponthieu[12], *V.* Revéret[5], *A. J.* Rigby[4], *C.* Romero[17], *H.* Roussel[18], *F.* Ruppin[19], *K.* Schuster[9], *S.* Shu[20], *A.* Sievers[14], *C.* Tucker[4], and *R.* Zylka[9]

[1]Institut d'Astrophysique Spatiale (IAS), CNRS, Université Paris Sud, Orsay, France
[2]Laboratoire de Physique de l'École Normale Supérieure, ENS, PSL Research University, CNRS, Sorbonne Université, Université de Paris, 75005 Paris, France
[3]LLR (Laboratoire Leprince-Ringuet), CNRS, École Polytechnique, Institut Polytechnique de Paris, Palaiseau, France
[4]School of Physics and Astronomy, Cardiff University, Queen's Buildings, The Parade, Cardiff, CF24 3AA, UK
[5]AIM, CEA, CNRS, Université Paris-Saclay, Université Paris Diderot, Sorbonne Paris Cité, 91191 Gif-sur-Yvette, France
[6]Univ. Grenoble Alpes, CNRS, Grenoble INP, LPSC-IN2P3, 53, avenue des Martyrs, 38000 Grenoble, France
[7]CNRS, IRAP, 9 Av. colonel Roche, BP 44346, 31028 Toulouse cedex 4, France
[8]Institut Néel, CNRS, Université Grenoble Alpes, France
[9]Institut de RadioAstronomie Millimétrique (IRAM), Grenoble, France
[10]Aix Marseille Univ, CNRS, CNES, LAM (Laboratoire d'Astrophysique de Marseille), Marseille, France
[11]Dipartimento di Fisica, Sapienza Università di Roma, Piazzale Aldo Moro 5, I-00185 Roma, Italy
[12]Univ. Grenoble Alpes, CNRS, IPAG, 38000 Grenoble, France
[13]Centro de Astrobiología (CSIC-INTA), Torrejón de Ardoz, 28850 Madrid, Spain
[14]Instituto de Radioastronomía Milimétrica (IRAM), Granada, Spain
[15]LERMA, Observatoire de Paris, PSL Research University, CNRS, Sorbonne Université, UPMC, 75014 Paris, France
[16]School of Earth and Space Exploration and Department of Physics, Arizona State University, Tempe, AZ 85287, USA
[17]Department of Physics and Astronomy, University of Pennsylvania, 209 South 33rd Street, Philadelphia, PA, 19104, USA
[18]Institut d'Astrophysique de Paris, CNRS (UMR7095), 98 bis boulevard Arago, 75014 Paris, France
[19]Kavli Institute for Astrophysics and Space Research, Massachusetts Institute of Technology, Cambridge, MA 02139, USA
[20]Caltech, Pasadena, CA 91125, USA

Abstract. The quest for primordial gravitational waves enclosed in the Cosmic Microwave Background (CMB) polarization B-modes signal motivates the

*e-mail: alessia.ritacco@phys.ens.fr

development of a new generation of high sensitive experiments (e.g. CMB-S4,LiteBIRD), thus allowing to probe the inflationary epoch in the early Universe. However, this will be only possible by ensuring a high control of the instrumental systematic effects and an accurate absolute calibration of the polarization angle.
The Crab nebula is known to be a polarization calibrator on the sky for CMB experiments. Already used for the Planck satellite it exhibits a high polarized signal at microwave wavelengths. In this work we present Crab polarization observations obtained, in the 260 GHz frequency band, with the NIKA2 instrument. Furthermore, we discuss the accuracy needed on such a measurement to improve the constraints on the absolute angle calibration for CMB experiments.

1 Introduction

Cosmic Microwave Background (CMB) observations have allowed us to establish the current standard model for cosmology, a.k.a the ΛCDM model. The inflation theory [1, 2] has been introduced to explain an observational evidence that the Universe appears to be almost exactly spatially-flat and uniform in all directions. This theory postulates that the Universe exponentially expanded in a tiny fraction of a second in the early times and predicts the existence of a relic background of primordial gravitational waves, which produces a distinct, curl-like signature in the polarization of the CMB, a.k.a CMB B-modes. The search for primordial gravitational waves B-modes, which are a direct probe of inflation, is one of the main ambitious goals of the modern cosmology and it motivates the development of several facilities from ground (e.g. BICEP3+Keck, CLASS, SO, SPT3G, S4, AdvACTPOL, QUIJOTE), balloon-borne to space experiments (PIPER, PICO, LiteBIRD). These experiments share the goal of detecting the tensor-to-scalar ratio r, which gives the relative amplitude of the primordial tensor (B-modes) and scalar perturbations (E-modes) of the CMB. This parameter gives the energy scale of inflation and in many well-motivated models, still consistent with observations, is predicted to be between 10^{-2} and 10^{-4} [3, 4].

In order to reach the sensitivity required for an unambiguous detection of the primordial B-modes we need i) to increase the signal-to-noise ratio by scaling up the number of detectors; ii) to use optical systems as half-wave-plates, which modulates the sky polarization signal and filters out low frequency noise; iii) and to reduce residual systematic effects [5, 6]. Furthermore the determination of the polarization angle absolute calibration is crucial to ensure an unbiased detection of the polarization orientation of the CMB.

A recent study [7, 8] addressed this issue by using the brightest polarized astrophysical object in the microwave sky, the Crab nebula (or Tau A). They point out that the polarization angle is constant and stable in a frequency range of 23-353 GHz consisting with a mean angle of $-88.26° \pm 0.27°$ (Galactic coordinates), which would allow us to calibrate experiments with an accuracy enabling the measurement of $r \sim 0.01$, far from inflationary theoretical models predictions of at least $r \sim 10^{-3}$.

In the following we present the preliminary results from Crab nebula polarization observations obtained with NIKA2 during the commissioning campaign of November 2020. We will discuss how improving the precision on these measurements could allow us to reach the sensitivity requirements for the detection of the primordial CMB B-modes.

2 NIKA2 observations

The NIKA2 instrument [10] performes high angular resolution simultaneous observations at 1.15 and 2.05 mm, from the IRAM 30m telescope, being the 1.15 mm (260 GHz) channel also able to detect the linear polarization. In total intensity it is operating for open time observations since 2017 [9], when it also started the commissioning phase of its polarization facilities. The NIKA2 polarimeter consists of a continuously rotating Half-Wave-Plate (HWP) [5] placed at ambient temperature in front of the NIKA2 cryostat window. The detection of the linear polarization expressed through the Stokes parameters Q and U is possible thanks to two detector arrays placed at ~100 mK in the 260 GHz channel inside the cryostat [11]. For more details on the polarization system see [5].

2.1 Crab nebula observations

The Crab nebula has an extension of 5x7 arcminutes in the sky and it exhibits a high polarization, of the order of ~ 20% of the total intensity at the peak emission position of the nebula. [8] presented a compendium of all avalaible observations in a large range of frequencies (23-353 GHz) and demonstrated that the synchrotron emission is responsible for both the total intensity and polarization observed. Moreover they found a strong case for stability of the polarization angle, which places the Crab nebula as a good candidate for a polarization angle calibrator for CMB experiments. However at higher frequencies (above 200 GHz) the sample is smaller, contributing to increase the uncertainty on the angle stability.

Observations performed with the NIKA2 polarimeter would increase the precision on the polarization angle estimation. Observations on the the Crab nebula have been performed with the NIKA2 instrument during three commissioning campaigns, on December 2018, February and November 2020. The polarization Stokes Q and U maps obtained from 70 minutes of observation performed during the campaign of November 2020 are shown in figure 1.

The left panel of figure 2 shows the Crab polarization angles estimated during the three campaigns. The averaged polarization angle (solid black line) is computed as median of the polarization angles, estimated as $\psi_{Gal} = 0.5 \arctan U/Q$ (where the Stokes Q and U have been computed by integrating the emission flux over a 7′ diameter where SNR > 3 in Stokes Q and U maps) of the three data sets and corresponds to -87.195 ± 0.806°, expressed in galactic coordinates. Notice that the angles have been corrected for a -5.6 ± 1 ° offset in absolute angle calibration on the sky as measured in OMC1 and 3C286 targets, being 1° the uncertainty accounting for the stability of this offset over almost two years of observations. Taking only the November 2020 data, which are the most reliable across the two years we obtain the median value of ψ = -87.23 ± 0.29 deg. On the right panel of the figure we compare this result to other experiments [8]. Making the assumption that in the future we will be able to constrain the absolute offset angle and reduce the 1° uncertainty to negligible values, we use the November data to calculate the total weighted polarization angle (including the results of all experiments). In this case we obtain ψ = -87.55 ± 0.22 °, which would strongly reduce the uncertainty on the polarization angle reconstruction at higher frequencies.

Notice that we consider equations reported in [8] to compute unbiased estimates for polarization quantities given in this work.

3 Polarized Spectral Energy Distribution

The polarized spectral energy distribution of the Crab nebula has been determined for the first time in [8]. Figure 3 left panel shows the polarized intensity map obtained as $P = \sqrt{Q^2 + U^2}$ at 260 GHz with NIKA2 and on right the spectral energy distribution as estimated by [8] accounting also for the NIKA2 value (red), which is P = 13.22 ± 0.15 (dev. standard) ± 1.2 (10% calibr. error). We notice that the polarized flux measured is in very good agreement with expectations. In polarization there is no filtering of the large angular scales [9] and thus, the whole intensity is recovered.

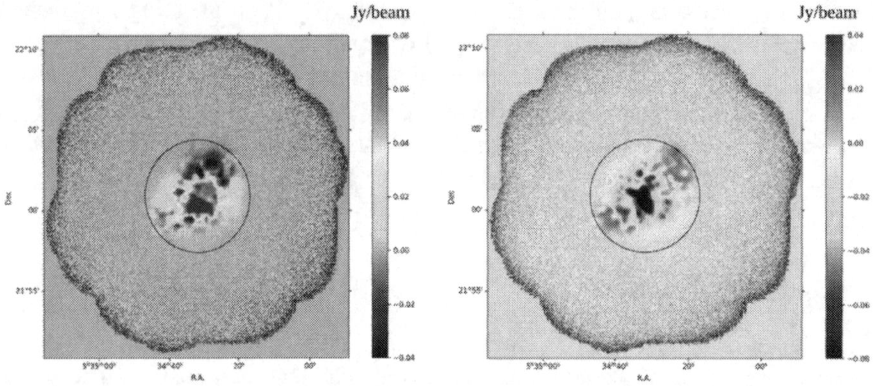

Figure 1. From left to right: Stokes Q and U maps of the Crab nebula observed at 260 GHz during the commissioning campaign of November 2020. The circle overplotted indicates the region, within a diameter of 7′ arcmin, where the polarization angle is computed.

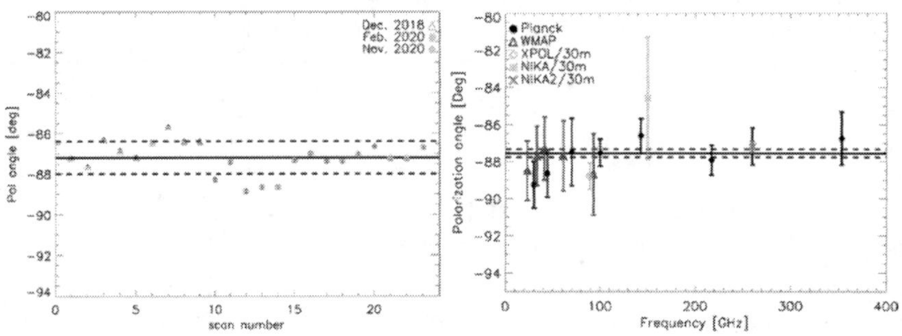

Figure 2. Left: Variation of polarization angles measured with NIKA2-POL for the mean angle over the Crab nebula during three commissioning campaigns. The thick black line shows the median value, the dashed lines delineate the 0.8° standard deviation over all scans, spanning ~ 2 years. Right: NIKA2-POL mean polarization angle over the Crab nebula, obtained in the last campaign, vs other experiments [8]. The red error bar represents the uncertainty on the NIKA2 value, computed as dispersion within the observing scans of Nov. 2020, and the black error bar shows the absolute 1° uncertainty due to the absolute angle calibration offset.

4 CMB experiment absolute calibration challenge

The stability of the polarization angle with frequency is the key to use the Crab nebula as a sky calibrator. The absolute calibration of the polarization angle is of a particular importance for a CMB experiment targeting the measurement of the primordial gravitational waves through the CMB polarization B-modes. Indeed, a miscalibration of an angle $\Delta\psi$ is equivalent to a rotation of the polarization plane by the same angle, mixing the true linear polarization Q and U and consequently the polarized angular power spectra that describe the amplitude of the signal. In the CMB and because the CMB polarization E-modes power spectrum C_l^{EE} is greater than B-modes one C_l^{BB}, this is often referred to as an "E to B leakage" and reads:

$$\tilde{C}_\ell^{BB} = C_\ell^{BB} \cos^2 2\Delta\psi + C_\ell^{EE} \sin^2 2\Delta\psi \Leftrightarrow \Delta C_\ell^{BB} \simeq (2\Delta\psi)^2 C_\ell^{EE} \tag{1}$$

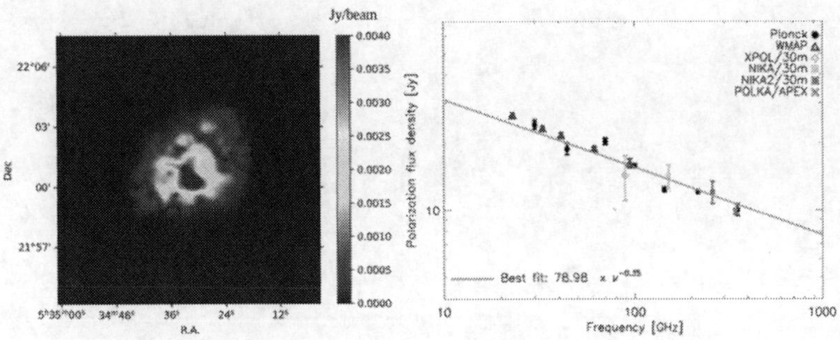

Figure 3. Left: NIKA2 polarization intensity P map. Right: spectral energy distribution obtained by previous measurements [8] accounting for the new value obtained with NIKA2 (red).

where ΔC_ℓ^{BB} represents the spurious bias component. In order to use the Crab nebula polarization angle ψ as an absolute angle calibrator for CMB experiments, we are interested in its uncertainty $\Delta\psi$. In the following we present preliminary results on the bias computed for the parameter r as depending by the uncertainty on ψ. We update a previous study [7] with the new value given by NIKA2 at 260 GHz. Notice that we make the strong assumption that we would be able to reduce the uncertainty on the polarization angle calibration offset of NIKA2 to a negligible value. In other words, we assume here that the NIKA2 polarization angle uncertainty is only due to instabilities of the instrument itself. Figure 4 shows on left panel the $\Delta D_\ell^{BB} = \ell(\ell+1)/(2\pi)\,\Delta C_\ell^{BB}$ power spectrum bias from $E-B$ mixing due to miscalibration of the absolute polarization angle. On the right panel we show the likelihood analysis on the r parameter. The different cases considered are taken from [7] updating the two plots with the value measured at 260 GHz by NIKA2 in its best configuration taking only the data from the November 2020 commissioning campaign.

Figure 4 shows that NIKA2 polarization angle stability improves the uncertainty Δr by 30% w.r.t previous studies [7] enabling to approach the requirements for the measurement of the r parameter with a total uncertainty due to miscalibration on the polarization angle of $\Delta r = 5\times10^{-4}$. This is promising for CMB experiments development but we need to carefully assess the NIKA2 absolute polarization angle offset.

5 Conclusion

In this paper we show the results obtained on Crab nebula polarization observations during the commissioning phase of the NIKA2-Pol instrument. Comparing the three commissioning campaigns across 2018 and 2020 we have found that the data sets on the Crab and others sky calibrators, e.g. Orion OMC-1 and 3c286, are consistent with an absolute polarization angle calibration offset of -5.6 ± 1 deg. This systematic uncertainty prevents us, for instance, to use NIKA2 Crab nebula polarization angle value as reference for CMB experiments. Though, the NIKA2 Crab nebula observations appears to be very consistent between one scan to another, especially in optimal weather conditions reducing dramatically the statistical error associated. Propagating this value to the likelihood analysis in order to estimate the total bias on the detection of the r parameter for the CMB B-modes detection, we see that we could reach the requirements for the next generation CMB experiments. This is a strong assumption on NIKA2 polarization performances and we are currently investigating possible ways to reduce the uncertainties on the NIKA2 polarization angle offset.

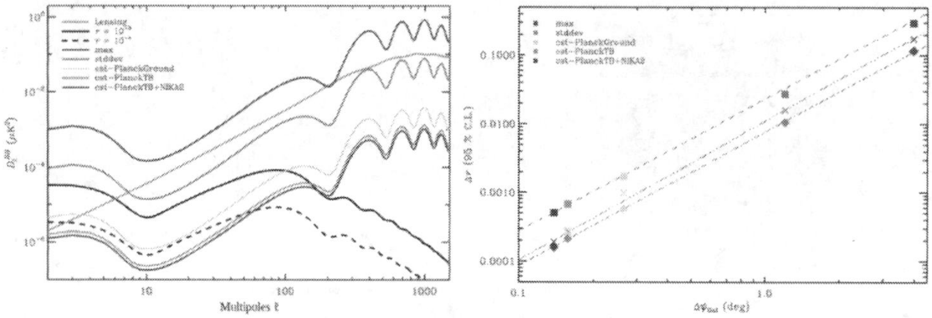

Figure 4. Left panel: power spectrum bias from $E - B$ mixing due to the miscalibration of the absolute polarization angle. This bias is plotted for the different absolute calibration errors $\Delta\psi$ presented in [7]. The ΛCDM best fit D_ℓ^{BB} primordial tensor model for $r = 10^{-3}$ and $r = 10^{-4}$ (solid and dashed black lines, respectively) and D_ℓ^{BB} lensing model (gray line) are also displayed. Right panel: likelihood posterior on r biases (with respect to an input signal of r = 0) for the different cases presented in [7], from 10 000 Monte-Carlo simulations, as a function of the combined error on the angle $\Delta\psi_{Gal}$ in degrees. They are computed independently for the recombination bump ($30 < \ell < 300$, squares), the reionization bump ($2 < \ell < 30$, diamonds) and the combination of both ($2 < \ell < 300$, crosses).

Acknowledgements

We would like to thank the IRAM staff for their support during the campaigns. The NIKA2 dilution cryostat has been designed and built at the Institut Néel. In particular, we acknowledge the crucial contribution of the Cryogenics Group, and in particular Gregory Garde, Henri Rodenas, Jean Paul Leggeri, Philippe Camus. This work has been partially funded by the Foundation Nanoscience Grenoble and the LabEx FOCUS ANR-11-LABX-0013. This work is supported by the French National Research Agency under the contracts "MKIDS", "NIKA" and ANR-15-CE31-0017 and in the framework of the "Investissements d'avenir" program (ANR-15-IDEX-02). This work has benefited from the support of the European Research Council Advanced Grant ORISTARS under the European Union's Seventh Framework Programme (Grant Agreement no. 291294). F.R. acknowledges financial supports provided by NASA through SAO Award Number SV2-82023 issued by the Chandra X-Ray Observatory Center, which is operated by the Smithsonian Astrophysical Observatory for and on behalf of NASA under contract NAS8-03060.

References

[1] Guth, Alan H., Phys. Rev. D **23**, 347–356 (1981)
[2] Linde, A.D., Physics Letters B **116**, 335-339 (1982)
[3] Hazumi M. and LiteBIRD collaboration, Space Telescopes and Instrumentation 2020: Optical, Infrared, and Millimeter Wave **11443**, 431 – 450 (2020)
[4] Zhilei Xu and Simons Observatory collaboration, AAS **5**, 100 (2021)
[5] Ritacco A., Ponthieu N., Catalano A., *et al.* A&A **599**, A34 (2017)
[6] O'Dea, Daniel and Challinor, Anthony and Johnson, Bradley MNRAS **376**, (2006)
[7] Aumont J., Macías-Pérez J.-F., Ritacco A., *et al.* A&A **634**, A100 (2020)
[8] Ritacco A., Macías-Pérez J.-F., Ponthieu N., *et al.* A&A, **616**, A35 (2018)
[9] Perotto L., Ponthieu N., Macías-Pérez J.-F., *et al.*, Astron. Astrophys. **637**, A71 (2020)
[10] Adam R. *et al.*, Astron. Astrophys. **609**, A115 (2018)
[11] Calvo M. *et al.*, Journal of Low Temperature Physics, **184**, 816 (2016)
[12] Bourrion O. *et al.*, Journal of Instrumentation, **11**, 11001 (2016)

Mapping the intracluster medium temperature in the era of NIKA2 and MUSTANG-2

F. Ruppin[1,*], *R.* Adam[2], *P.* Ade[3], *H.* Ajeddig[4], *P.* André[4], *E.* Artis[5], *H.* Aussel[4], *A.* Beelen[6], *A.* Benoît[7], *S.* Berta[8], *L.* Bing[9], *O.* Bourrion[5], *M.* Brodwin[10], *M.* Calvo[7], *A.* Catalano[5], *B.* Decker[10], *M.* De Petris[11], *F.-X.* Désert[12], *S.* Doyle[3], *E. F. C.* Driessen[8], *P. R. M.* Eisenhardt[13], *A.* Gomez[14], *A. H.* Gonzalez[15], *J.* Goupy[7], *F.* Kéruzoré[5], *C.* Kramer[16], *B.* Ladjelate[16], *G.* Lagache[9], *S.* Leclercq[8], *J.-F.* Lestrade[17], *J.-F.* Macías-Pérez[5], *A.* Maury[4], *P.* Mauskopf[3,18], *F.* Mayet[5], *M.* McDonald[1], *A.* Monfardini[7], *E.* Moravec[15], *M.* Muñoz-Echeverría[5], *L.* Perotto[5], *G.* Pisano[3], *N.* Ponthieu[12], *V.* Revéret[4], *A. J.* Rigby[3], *A.* Ritacco[6,19], *C.* Romero[20], *H.* Roussel[21], *K.* Schuster[8], *S.* Shu[22], *A.* Sievers[16], *S. A.* Stanford[23], *D.* Stern[13], *C.* Tucker[3], and *R.* Zylka[8]

[1]Kavli Institute for Astrophysics and Space Research, Massachusetts Institute of Technology, Cambridge, MA 02139, USA
[2]LLR (Laboratoire Leprince-Ringuet), CNRS, École Polytechnique, Institut Polytechnique de Paris, Palaiseau, France
[3]School of Physics and Astronomy, Cardiff University, Queen's Buildings, The Parade, Cardiff, CF24 3AA, UK
[4]AIM, CEA, CNRS, Université Paris-Saclay, Université Paris Diderot, Sorbonne Paris Cité, 91191 Gif-sur-Yvette, France
[5]Univ. Grenoble Alpes, CNRS, Grenoble INP, LPSC-IN2P3, 53, avenue des Martyrs, 38000 Grenoble, France
[6]Institut d'Astrophysique Spatiale (IAS), CNRS, Université Paris Sud, Orsay, France
[7]Institut Néel, CNRS, Université Grenoble Alpes, France
[8]Institut de RadioAstronomie Millimétrique (IRAM), Grenoble, France
[9]Aix Marseille Univ, CNRS, CNES, LAM (Laboratoire d'Astrophysique de Marseille), Marseille, France
[10]Department of Physics and Astronomy, University of Missouri, 5110 Rockhill Road, Kansas City, MO 64110, USA
[11]Dipartimento di Fisica, Sapienza Università di Roma, Piazzale Aldo Moro 5, I-00185 Roma, Italy
[12]Univ. Grenoble Alpes, CNRS, IPAG, 38000 Grenoble, France
[13]Jet Propulsion Laboratory, California Institute of Technology, Pasadena, CA 91109, USA
[14]Centro de Astrobiología (CSIC-INTA), Torrejón de Ardoz, 28850 Madrid, Spain
[15]Department of Astronomy, University of Florida, 211 Bryant Space Center, Gainesville, FL 32611, USA
[16]Instituto de Radioastronomía Milimétrica (IRAM), Granada, Spain
[17]LERMA, Observatoire de Paris, PSL Research University, CNRS, Sorbonne Université, UPMC, 75014 Paris, France
[18]School of Earth and Space Exploration and Department of Physics, Arizona State University, Tempe, AZ 85287, USA
[19]Laboratoire de Physique de l'École Normale Supérieure, ENS, PSL Research University, CNRS, Sorbonne Université, Université de Paris, 75005 Paris, France

*e-mail: f.ruppin@ip2i.in2p3.fr

[20]Department of Physics and Astronomy, University of Pennsylvania, 209 South 33rd Street, Philadelphia, PA, 19104, USA
[21]Institut d'Astrophysique de Paris, Sorbonne Université, CNRS (UMR 7095), 75014 Paris, France
[22]Caltech, Pasadena, CA 91125, USA
[23]Department of Physics, University of California, One Shields Avenue, Davis, CA 95616, USA

Abstract. We present preliminary results from an on-going program that aims at mapping the intracluster medium (ICM) temperature of high redshift galaxy clusters from the MaDCoWS sample using a joint analysis of shallow X-ray data obtained by *Chandra* and high angular resolution Sunyaev-Zel'dovich (SZ) observations realized with the NIKA2 and MUSTANG-2 cameras. We also present preliminary results from an on-going Open Time program within the NIKA2 collaboration that aims at mapping the ICM temperature of a galaxy cluster at $z = 0.45$ from the resolved detection of the relativistic corrections to the SZ spectrum. These studies demonstrate how high angular resolution SZ observations will play a major role in the coming decade to push the investigation of ICM dynamics and non-gravitational processes to high redshift before the next generation X-ray observatories come into play.

1 Introduction

Mapping the temperature of the hot plasma within galaxy clusters from the spectroscopic information embedded in X-ray data has enabled tremendous advances in the understanding of merger dynamics, sloshing, and AGN feedback at low redshift [*e.g.* 1–3]. However, producing such temperature maps is currently unfeasible for most clusters at $z > 0.5$ because of the limited effective area of current X-ray observatories. Unfortunately, the next generation of X-ray observatories such as *Athena* and *Lynx* will not be operational until the 2030s. In the meantime it is essential to push the investigation of ICM dynamics and thermodynamic properties to higher redshift in order to understand how galaxy clusters formed and evolved. Fortunately, during the last decade, we have witnessed the arrival of instruments capable of mapping the SZ effect at high angular resolution on scales of few arcmin [*e.g.* 4–6]. The amplitude of the SZ signal being redshift independent and proportional to the ICM pressure content, the joint analysis of SZ and X-ray data offers a powerful way of mapping the ICM temperature distribution at high redshift [*e.g.* 7, 8]. Another challenging method that enables mapping the ICM temperature at high redshift consist in detecting the relativistic corrections to the SZ effect (rSZ). This method has the advantage of being independent of X-ray observations. It could therefore allow us to unveil potential calibration biases affecting the temperature measurements realized with current X-ray observatories. In this paper, we present preliminary results from two on-going programs using respectively joint analyses of X-ray and SZ data and the detection of the rSZ effect in order to map the ICM temperature of high redshift clusters.

2 Mapping the ICM temperature with joint X-ray / SZ analyses

Simulations have shown that the most active period in cluster formation history is expected at redshifts $1 < z < 2$ [*e.g.* 9]. Studying merger dynamics and non-gravitational processes such as AGN feedback through temperature mapping thus requires high angular resolution X-ray and SZ observations at $z > 1$.

The *Massive and Distant Clusters of WISE Survey* (MaDCoWS) is an Infra-Red (IR) survey that has enabled detecting hundreds of massive clusters at $z > 1$ [10] observable from the

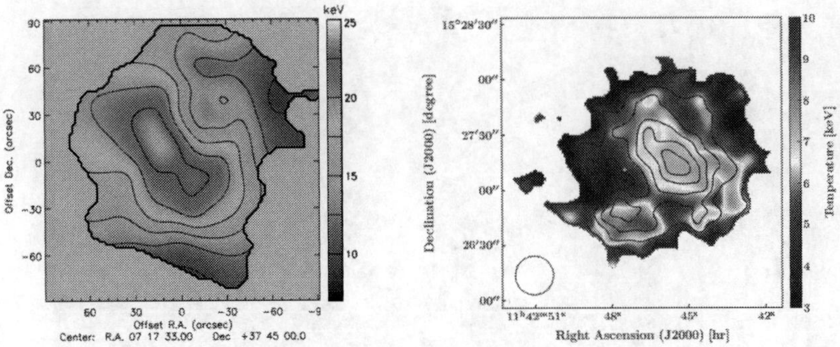

Figure 1. Left: Spectroscopic temperature map derived from *Chandra* for a cluster at $z = 0.55$. **Right:** Temperature map obtained from the joint analysis of *Chandra* X-ray and NIKA2 SZ observations for a cluster at $z = 1.2$. The relative uncertainties on temperature are similar in both panels. Figures extracted from [7, 8].

northern hemisphere, where the NIKA2 and MUSTANG-2 SZ cameras are operational. We have selected 10 clusters from the MaDCoWS catalog in order to realize high angular resolution follow-up observations in X-ray and SZ with *Chandra*, NIKA2, and MUSTANG-2. This sample, dubbed MOO-X/SZ, covers a redshift range from $z = 0.9$ to $z = 1.75$. The joint analysis of the SZ and X-ray data obtained for each MOO-X/SZ cluster will allow us to investigate ICM physics with a level of details and precision that is not achievable with current X-ray observatories on their own at such high redshifts.
We have already completed the analysis of the most massive cluster in this sample, *i.e.* MOO J1142+1527 at $z = 1.2$ [8]. As shown in Fig. 1, the quality of the temperature map of this high-z cluster that we realized from the joint analysis of *Chandra* and NIKA2 data with a combined exposure of only 23.2 hours (right) is similar to the one achieved from the analysis of the X-ray spectroscopic data obtained with a 42.5 hour *Chandra* exposure of the $z = 0.55$ cluster MACS J0717.5+3745 (left) [7]. Obtaining a similar map using only X-ray information from *Chandra* at such high redshift would have required an exposure at least 30 times greater. This highlights the strong complementarity between X-ray and high angular resolution SZ observations to go beyond the characterization of individual systems at $z > 1$ and start investigating the ICM properties in samples of high redshift clusters.
Building upon the success of this first study, a NIKA2 Open Time (OT) program aiming at mapping the SZ signal of 5 MOO-X/SZ clusters already observed by *Chandra* as well as a Large Program (LP) dedicated to the *Chandra* observation of 4 MOO-X/SZ clusters already observed by MUSTANG-2 have been accepted. The NIKA2 OT observations were completed in March 2021 and the *Chandra* LP observations are still on-going. In Fig. 2, we present preliminary maps of 4 MOO-X/SZ clusters obtained in SZ (top) and X-ray (bottom). The first SZ map on the left has been obtained with MUSTANG-2 at 90 GHz and the three other maps present NIKA2 data at 150 GHz [11]. The SZ signal is detected at $> 7\sigma$ in all maps. We observe similar cluster morphologies in SZ and X-ray and we identify ICM disturbances in each cluster. For example, MOO J0319−0025 at $z = 1.19$ displays a clear bi-modal morphology while MOO J1014+0038 at $z = 1.23$ presents ICM substructures in the south-west region. These preliminary results tend to confirm our expectation that ICM disturbances due to merging events are more frequently observed at $z > 1$ than in low redshift cluster samples such as REXCESS [12] or CCCP [13]. Based on the tools that we developed for the analysis of MOO J1142+1527, we are currently analyzing jointly our X-ray and SZ data sets in order to provide a detailed characterization of the ICM properties in a sample of $z > 0.9$ clusters for the first time. In particular, we are producing ICM temperature and entropy maps that

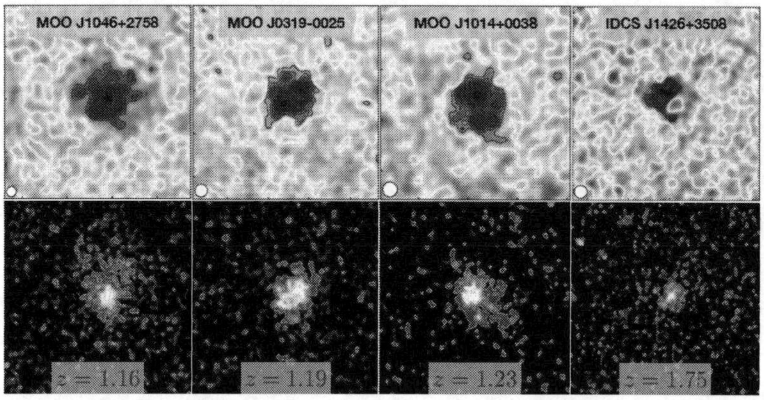

Figure 2. Preliminary SZ (top) and X-ray (bottom) maps obtained with the MUSTANG-2 (far left) and NIKA2 camera at 90 GHz and 150 GHz and the ACIS-I instrument on-board *Chandra* for four clusters in the MOO-X/SZ sample. Contours in the MUSTANG-2 and NIKA2 maps start at 3σ with 1σ steps. Photon count iso-contours are shown in the X-ray maps for visual purposes. We indicate the name and redshift of each cluster in the upper and lower panels, respectively. All maps have a side length of about ~4 arcmin.

we cross correlate with the galaxy distribution of each cluster in order to understand their merger dynamics. We also aim at exploring the connection between the ICM thermodynamic properties of all MOO-X/SZ clusters and the potential radio emission of the brightest cluster galaxy in order to better understand AGN feedback at high redshift. The analysis of the spectroscopic information embedded within our *Chandra* data using prior information on the ICM temperature obtained from the joint X-ray / SZ analysis will allow us to constrain the metallicity content of these cluster in order to understand metal enrichment at high redshift. Finally, the hydrostatic masses obtained from the estimated density and pressure profiles of these clusters will be compared to the richness-derived masses obtained with *Spitzer* in order to unveil potential systematic effects on the mass-richness scaling relation derived at high redshift. The full characterization of the MOO-X/SZ sample will thus provide valuable information that will pave the way to an in-depth understanding of cluster formation at $z > 1$ and to precise cluster cosmological constraints from the forthcoming optical/IR surveys such as *Euclid* [14] and *Rubin*-LSST [15].

3 Mapping the ICM temperature from the detection of the rSZ effect

Assuming that the contributions to the SZ effect from sub-dominant electron populations such as ultra-relativistic electrons accelerated in AGN jets or electrons arising from the annihilation of dark matter are negligible, the total relative variation of the CMB specific intensity at frequency ν induced from the SZ effect is given by:

$$\frac{\Delta I_\nu}{I_0} = y_{tSZ}\, f(\nu, T_e) + y_{kSZ}\, g(\nu, T_e, v_z), \tag{1}$$

where the amplitudes of the thermal and kinetic SZ signals are given by:

$$y_{tSZ} = \frac{k_B T_e}{m_e c^2}\tau \quad \text{and} \quad y_{kSZ} = -\frac{v_z}{c}\tau \quad \text{with} \quad \tau = \sigma_T \int n_e\, dl. \tag{2}$$

The SZ signal thus depends on three quantities: the density of ICM electrons n_e along the line of sight l, the ICM temperature T_e, and the bulk velocity of the ICM electrons along the line of sight v_z. The relativistic corrections inducing variations of the shape of the tSZ and

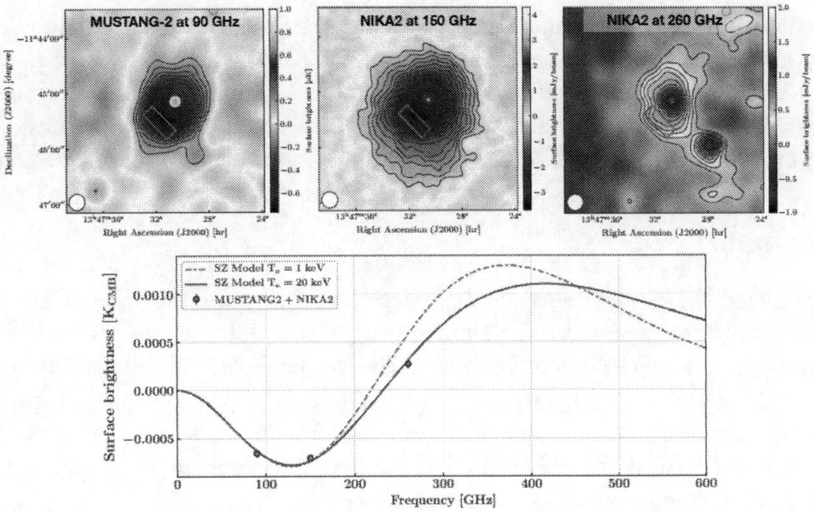

Figure 3. Top: Preliminary SZ maps obtained with MUSTANG-2 and NIKA2 at 90, 150, and 260 GHz, respectively. Signal-to-noise contours start at 3σ with 1σ increments. We highlight the region considered for the SZ surface brightness measurements shown in the bottom panel with red rectangles. **Bottom:** Preliminary measurements of the SZ surface brightness in the shock region of the cluster at the three frequencies available with NIKA2 and MUSTANG-2 (purple points) along with the expected tSZ spectra for ICM temperatures of 1 and 20 keV.

kSZ spectra $f(\nu, T_e)$ and $g(\nu, T_e, v_z)$ are often referred to as the rSZ effect and depend directly on the value of the ICM temperature. Sampling the total SZ spectrum with at least three independent data points on a large enough frequency range thus enables estimating the ICM temperature independently from any X-ray observation. The detection of the rSZ effect has already been made using millimeter spectroscopy with Z-Spec [16] and by stacking analyses [*e.g.* 17]. These studies enable measuring the mean ICM temperature of galaxy clusters but they were unable to provide resolved detection of the rSZ effect which is a prerequisite to the production of ICM temperature maps.

To this end, we proposed to use both MUSTANG-2 and NIKA2 in order to map the SZ signal of the most SZ luminous cluster in the sky, RXJ1347−1145 at $z = 0.45$, at an angular resolution < 18 arcsec and on scales of few arcmin at 90, 150, and 260 GHz. The MUSTANG-2 observations were realized at the Green Bank Telescope during the instrument commissioning phase for 2 hours. The NIKA2 observations have been conducted in the context of an Open Time program for a total of 24 hours. The top panels of Fig. 3 show the MUSTANG-2 map (left) obtained with the MIDAS software [18] and the NIKA2 maps (middle and right) realized from a preliminary analysis based on the procedure described in [8]. All three maps have been smoothed at an effective resolution of 20 arcsec. The extended SZ signal is detected at $> 3\sigma$ on a sky area of about 2×1.5 arcmin^2. We detect a strong positive signal in the cluster core at 260 GHz due to a radio AGN. This signal compensates the negative SZ signal at 90 GHz and 150 GHz. We also detect a sub-millimeter galaxy to the south-west of the cluster, which also alters its morphology at 150 GHz. As a preliminary test, we measured the average SZ surface brightness in a region (red rectangle in Fig. 3) where a shock has been detected in previous analyses [19] and far enough from the detected point sources to minimize their contamination. The measurements are shown as purple points in the lower panel of Fig. 3. We use the SZpack software [20] in order to compute the expected SZ spectrum for ICM temperatures of 1 keV (dashed blue line) and 20 keV (red line), assuming a null bulk velocity along the line of sight[1]. While changing the value of the Compton parameter y_{tSZ}

[1]Note that having three SZ constraints will also enable fitting the kinetic SZ component in the total SZ signal.

allows us to fit the data at 90 GHz and 150 GHz in both cases, we are able to exclude an ICM temperature of 1 keV at more than 5σ in the shock region based on the measurement realized at 260 GHz. A careful characterization of the point source, cosmic infrared background, and kSZ contamination using *Herschel* and spectroscopic data will allow us to produce the first ICM temperature map of a galaxy cluster based on the resolved detection of the rSZ effect.

4 Conclusions

We have completed the observations of two independent SZ programs aiming at mapping the ICM temperature of high redshift galaxy clusters. The joint analysis of our SZ data and *Chandra* observations will allow us to characterize the ICM properties of a sample of MaDCoWS clusters at $z > 1$ with a level of details that cannot be achieved with current X-ray observatories on their own. Furthermore, the resolved detection of the rSZ effect with MUSTANG-2 and NIKA2 will enable providing the first ICM temperature map of a merging cluster without considering any X-ray information.

Acknowledgements

We would like to thank the IRAM staff for their support during the campaigns. The NIKA2 dilution cryostat has been designed and built at the Institut Néel. In particular, we acknowledge the crucial contribution of the Cryogenics Group, and in particular Gregory Garde, Henri Rodenas, Jean Paul Leggeri, Philippe Camus. This work has been partially funded by the Foundation Nanoscience Grenoble and the LabEx FOCUS ANR-11-LABX-0013. This work is supported by the French National Research Agency under the contracts "MKIDS", "NIKA" and ANR-15-CE31-0017 and in the framework of the "Investissements d'avenir" program (ANR-15-IDEX-02). This work has benefited from the support of the European Research Council Advanced Grant ORISTARS under the European Union's Seventh Framework Programme (Grant Agreement no. 291294). F.R. acknowledges financial supports provided by NASA through SAO Award Number SV2-82023 issued by the Chandra X-Ray Observatory Center, which is operated by the Smithsonian Astrophysical Observatory for and on behalf of NASA under contract NAS8-03060.

References

[1] H. R. Russell *et al.*, MNRAS **423**, 236–255 (2012)
[2] M. Rossetti *et al.*, Astron. Astrophys. **556**, A44 (2013)
[3] D. A. Rafferty *et al.*, MNRAS **428**, 58–70 (2012)
[4] A. Monfardini *et al.*, Astron. Astrophys. **521**, A29 (2010)
[5] R. Adam *et al.*, Astron. Astrophys. **609**, A115 (2018)
[6] S. R. Dicker *et al.*, J. Low Temp. Phys. **176**, 808–814 (2014)
[7] R. Adam *et al.*, Astron. Astrophys. **606**, A64 (2017)
[8] F. Ruppin *et al.*, ApJ **893**, 74 (2020)
[9] G. B. Poole *et al.* MNRAS **380**, 437-454 (2007)
[10] A. H. Gonzalez *et al.*, ApJS **240**, 33 (2019)
[11] L. Perotto *et al.*, Astron. Astrophys. **637**, A71 (2020)
[12] H. Böhringer *et al.*, Astron. Astrophys. **469**, 363-377 (2007)
[13] H. Hoekstra *et al.*, MNRAS **427**, 1298–1311 (2012)
[14] Euclid Collaboration *et al.*, Astron. Astrophys. **627**, A23 (2019)
[15] H.-Y. Wu *et al.*, ApJ **910**, 28 (2021)
[16] M. Zemcov *et al.*, ApJ **749**, 114 (2012)
[17] J. Erler *et al.*, MNRAS **476**, 3360-3381 (2018)
[18] C. Romero *et al.*, ApJ **891**, 90 (2020)
[19] S. W. Allen *et al.*, MNRAS **335**, 256 (2002)
[20] J. Chluba *et al.*, MNRAS **426**, 510–530 (2012)

Cosmology with galaxy clusters: Impact of theoretical and observational systematic uncertainties

Laura Salvati[1,2,*]

[1]Université Paris-Saclay, CNRS, Institut d'Astrophysique Spatiale, 91405, Orsay, France
[2]INAF – Osservatorio Astronomico di Trieste, Via G. B. Tiepolo 11, 34143 Trieste, Italy

Abstract. In this talk I focus on how the modelling of the mass-observable relation and the halo mass function can impact the accuracy and precision of cosmological constraints inferred from galaxy clusters. I present a new analysis of clusters detected in mm wavelengths by the Planck satellite, highlighting the need of an improved description and calibration for the mass-observable relation. I also discuss how to improve our analysis in view of future cluster surveys, with a particular focus on the impact of the halo mass function calibration and the need for a universal definition.

1 Introduction

Galaxy clusters are the largest, gravitationally bound structures in the Universe. They are associated with peaks in the matter density field on megaparsec scales. The abundance of clusters depends on the underlying cosmological model, tracking the growth of structure and matter density, and therefore it provides constraints on cosmological parameters; see e.g. [1].

In general, when studying galaxy clusters for the cosmological analysis, number counts are used as a cosmological probe. Cluster number counts should coincide with the halo mass function, i.e. the number distribution of clusters in bins of redshift and mass. However, cluster masses cannot be measured directly. Thus, we rely on observables that act as mass-proxies and that tightly correlate with the underlying cluster mass, via some statistical scaling relation. We use these scaling relations, with a model for the cluster selection process, to transform the theoretical mass function into a prediction for the distribution of clusters in the space of survey observables. The calibration of the scaling relations represents the current limiting systematic in cluster cosmology studies and it is usually referred to as the "mass-calibration problem"

In this analysis, we consider at first clusters detected in the mm-wavelengths, through the thermal Sunyaev-Zeldovich (tSZ) effect [2], by the Planck satellite [3, 4]. We then consider clusters observed in the optical and near-infrared wavelengths, performing a forecasts analysis for future experiments. We focus on how different modelling and assumptions for scaling relations and mass function can largely impact the constraints on cosmological parameters inferred from cluster number counts. We also comment on how these differences affect the possible tension with other cosmological probe, e.g. the σ_8 discrepancy between tSZ clusters and cosmic microwave background (CMB) primary anisotropies.

*e-mail: laura.salvati@universite-paris-saclay.fr

2 Mass-observable relation calibration

Planck cluster catalog for the cosmological analysis is composed by 439 clusters, in the redshift range $z = [0, 1]$ and above signal-to-noise ratio $q_{\rm min} = 6$. For the full description of the number counts modelling, we refer the reader to [4]. We recall here the model for the formulation of the scaling relations: under the assumption of self-similarity and hydrostatic equilibrium (HE), we find a relation between tSZ observable and cluster mass, estimated from the tSZ effect, $M_{\rm SZ}$. However, since this evaluation is based on HE, it provides biased-low masses. In order to take into account this effect, we add in the analysis the mass bias parameter, defined as the ratio between the tSZ mass and the real cluster mass (with the cluster being defined as an over-density of 500 times the critical density of the Universe), $(1-b) = M_{\rm SZ}/M_{500}$. Recent hydrodynamical simulations provide the estimation $(1-b) \sim 0.8$, see e.g. collection of results in [5, 6].

The calibration of Planck scaling relations is based on a multi-wavelength analysis, using X-ray observations and weak lensing (WL) measurements. In particular, the mass bias calibration is obtained from the CCCP WL analysis [7], $(1 - b) = 0.780 \pm 0.092$. As discussed in [4], a possible way to reconcile cosmological constraints from clusters and CMB anisotropies points towards a low value of the mass bias, $(1 - b) = 0.58 \pm 0.04$. However, this value is not in agreement with results from hydrodynamical simulations, WL evaluations and measurements of cluster gas fraction, as discussed e.g. in the X-COP analysis [8].

2.1 Combining tSZ observables

In order to fully understand the CMB-cluster tension and the impact of the mass calibration, we perform a new cosmological analysis of the Planck tSZ signal: we combine the cluster counts ($\rm NC^{tSZ}$) with the power spectrum of the full tSZ signal [9], $C_\ell^{\rm tSZ}$, for which we consider also detections from the South Pole Telescope (SPT) [10]. The tSZ power spectrum is modelled with the same ingredients as number counts, in particular the scaling relations and the mass function. Combining these two probes would ideally break degeneracy between cosmological and scaling relation parameters, since we find: $\rm NC^{tSZ} \propto \sigma_8^9 \Omega_m^3 (1 - b)^{3.6}$ and $C_\ell^{\rm tSZ} \propto \sigma_8^{8.1} \Omega_m^{3.2} (1 - b)^{3.2}$. We start analysing the ΛCDM scenario and we then consider the impact of varying the total neutrino masses. The full analysis is presented in [5].

We show the main results in figure 1. On the left panel, we report the two-dimensional probability distributions in the (Ω_m, σ_8) parameter plane for different dataset combinations, when considering the standard cosmological model. First of all, we notice that results for $\rm NC^{tSZ}$ + $C_\ell^{\rm tSZ}$ (green curves) combination are dominated by the $\rm NC^{tSZ}$ constraining power (orange curves). We then stress that results from tSZ probes and CMB primary anisotropies (red curves) are now fully in agreement, in particular the σ_8 constraints agree within $2\,\sigma$. This is also due to the further improvement in the Planck CMB analysis [11–13] which now prefers lower values of σ_8. In the right panel we analyse results when considering the sum of neutrino masses as a varying parameter. In this case, we find that $C_\ell^{\rm tSZ}$ contribution strongly improves the constraints, moving the 95% upper limit on massive neutrinos down to $\sum m_\nu < 1.88\,\rm eV$. This is mainly due to the ability of $C_\ell^{\rm tSZ}$ to constrain scales at which massive neutrinos start to have an impact on the evolution of the matter power spectrum (see e.g. [14] for a full discussion of the impact of massive neutrinos on the matter power spectrum). This effect is even stronger when combined with CMB data: we improve the results from $\sum m_\nu < 0.68\,\rm eV$ (for CMB alone, red curves) to $\sum m_\nu < 0.23\,\rm eV$ (for CMB+tSZ probes, blue curves).

We now discuss results for the mass bias. We focus on constraints obtained for the CMB+tSZ probes combination, for the standard model of cosmology, $(1 - b) = 0.65 \pm 0.04$, and when considering a varying neutrino mass, $(1 - b) = 0.67 \pm 0.04$. These constraints are

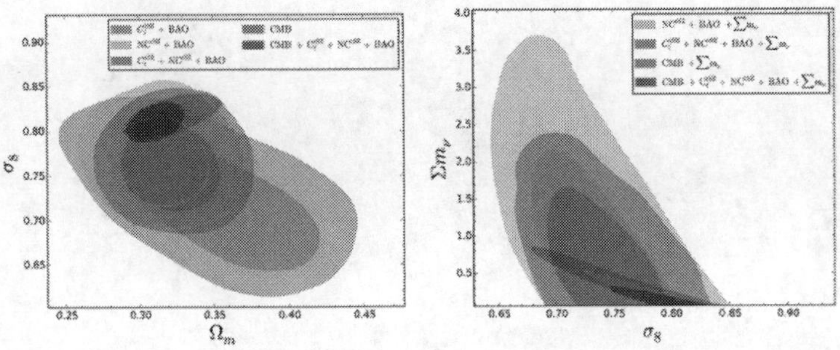

Figure 1. Two-dimensional probability distributions for Ω_m, σ_8 and $\sum m_\nu$ for different dataset combinations, as discussed in the text. These plots are taken from [5].

higher than the Planck 2015 analysis [4], due to the shift on the σ_8 parameter for CMB analysis. Nevertheless, they are still lower than the expected 0.8 value, pointing to a remaining inconsistency between CMB primary anisotropies and tSZ probes. We stress that this tension is not solved when exploring extensions to ΛCDM: we tested also the impact of a varying equation of state (EoS) parameter for dark energy, for which we find $(1 - b) = 0.63 \pm 0.04$.

2.2 Improving the mass bias definition

In order to further understand this remaining discrepancy on the scaling relation calibration, we analyse a new definition of the mass bias, considering a mass and redshift dependence. We stress that, in general, WL calibrations of the mass bias are based on very small subsamples of the full Planck cosmological cluster sample, considering few tens of objects in a limited redshift range. It is therefore possible that the subsample used for the calibration is not representative of the full cluster sample and a more complex definition for the mass bias is needed. We adopt the following definition:

$$(1 - b)_{\rm var} = (1 - \mathscr{B}) \left(\frac{M}{M_*} \right)^{\alpha_b} \left(\frac{1 + z}{1 + z_*} \right)^{\beta_b} . \qquad (1)$$

We consider the tSZ probes defined above ($\mathrm{NC}^{\rm tSZ} + C_\ell^{\rm tSZ}$) and compare with results obtained when adding also CMB primary anisotropies. The full analysis is presented in [15], we discuss here the main results. On the one hand, when focusing on the tSZ+CMB dataset combination, we find that CMB data are driving the constraining power, not allowing for any evidence of mass and redshift evolution for the mass bias, and therefore finding consistent results with the ones discussed in section 2.1. On the other hand, when considering tSZ probes alone, we find a mild evidence for a redshift evolution of the mass bias, with $\beta_b = 0.24 \pm 0.2$. We show this evolution for the mass bias in figure 2. In the left panel, the green and blue curves represent the mass bias for fixed minimum and maximum values of mass respectively, $M_{\rm min} = 0.83 \cdot 10^{14} M_\odot$ and $M_{\rm max} = 14.69 \cdot 10^{14} M_\odot$. We find therefore a trend for an increasing mass bias with redshift. However, this result is the opposite of what we would expect: as clusters evolve with cosmic time, we expect them to reach the HE condition towards $z = 0$, reaching therefore $(1 - b) \simeq 1$.

In order to further understand our findings, we perform different consistency tests. We refer again the reader to the full analysis in [15] and report here the main results. We select a subsample of the full Planck cosmological sample, considering only clusters with $z \geq 0.2$. We choose this lower limit in redshift to be consistent with the range used for the CCCP calibration and to highlight the behaviour of Planck sample when it starts to show a mass-limited selection. We show the new results in the right panel of figure 2. The red and blue curves represent the evolution of the mass bias for fixed $M_{\rm min}$ and $M_{\rm max}$ respectively, pointing to an opposite trend with respect to the previous results.

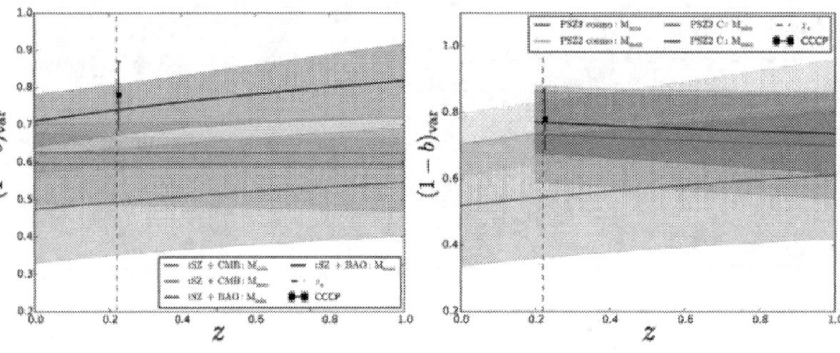

Figure 2. Evolution of the mass bias with respect to redshift, at fixed values of mass, for different dataset combination. We compare results for the full Planck cosmological sample (left panel), with results obtained when selecting clusters with $z \geq 0.2$ (right panel). These plots are taken from [15].

We stress therefore that these results are completely dominated by the adopted sample selection. This result is confirmed when we compare similar analysis performed in recent years (see a list of references in [15]): a possibile mass and redshift evolution for the mass bias produces different results depending on the selected cluster subsample.

We conclude with a general discussion for the results presented up to now: the mass bias parameter is introduced into the analysis to generally take into account the incomplete HE assumption. In practice, it may encompass different sources of uncertainties entering into the analysis, such as the model assumed for the cluster detection or the impact of the detection noise. Our results can also point to a general need of a more complete description for the scaling relations, improving the modelling of the interplay between astrophysics and cosmology in cluster formation and evolution and moving on from the simplistic assumptions of self-similar evolution and hydrostatic equilibrium.

We further improve this analysis, by combining Planck cluster sample with the SPT cluster observations [16, 17]. We exploit SPT cluster cosmological constraining power to provide an independent calibration for Planck scaling relations. Among other results, we find a mass and redshift evolution for the mass bias that is consistent with the above analysis. This analysis is developed within the SPT collaboration and will be presented in [18].

3 Halo mass function calibration

Next generation experiments will provide samples of thousands of clusters suitable for the cosmological analysis. We need therefore to quantify the impact on cosmological constraints of the other assumptions we perform in the analysis, apart from the mass calibration. We focus here on the impact of the halo mass function, with a forecasts analysis.

When modelling the cluster number counts, for the mass function we adopt fitting formulas calibrated from numerical simulations. In recent years, many analysis have provided different formulations and calibrations that can be used in the cosmological analysis (see e.g. the discussion in [19] and references therein). It has been shown that the mass function calibration can impact the final results on cosmological parameters up to $\sim$ 10%, see e.g. the discussion in [20, 21]. Indeed, fitting formulas obtained from different numerical simulations may reflect different initial conditions and assumptions performed during the simulations (e.g. assumed initial cosmology, definition of the cluster mass and detection, resolution of the simulation).

 https://doi.org/10.1051/epjconf/202225700044

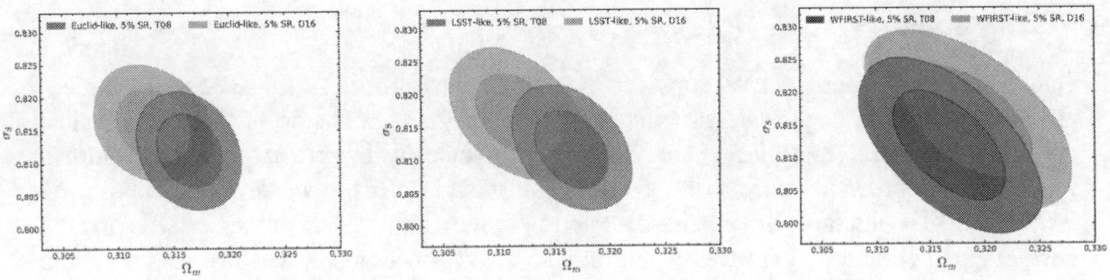

Figure 3. Two-dimensional probability distributions in $(\Omega_m \sigma_8)$ parameter plane for the Euclid-like, LSST-like and WFIRST-like experiments. These plots are taken from [28].

In our investigation, we compare two different formulations for the mass function, both widely used in the cosmological community: [22] (hereafter T08) and [23] (hereafter D16). We choose these two formulations since they represent two main approaches in evaluating the mass function (see e.g. discussion in [24]). We forecast results for three future surveys: the Euclid satellite [25], the LSST - Vera Rubin telescope [26] and the WFIRST - Nancy Grace Roman telescope [27]. The full analysis is discussed in [28], we report here the main results.

Assuming a 5% accuracy on the calibration of the scaling relations, we compare the results we obtain on the cosmological parameters when adopting the two mass function formulations. We show the results in figure 3, for the ΛCDM scenario. The two mass function evaluations produce different constraints, especially for the Euclid-like and LSST-like experiments, which are expected to produce larger cluster catalogs (due to the larger observed sky area). As discussed in [23], the two mass functions show consistent behaviour in the intermediate mass and redshift range, while D16 predicts more cluster at high redshift. This difference in the predicted number counts seems to be the cause for the discrepancy we find in the cosmological parameters. In our analysis we adopt an intrinsic scatter for the scaling relations with a redshift dependence, following [29]:

$$\sigma^2_{\ln M}(z) = \sigma^2_{\ln M,0} - 1 + (1+z)^\beta . \qquad (2)$$

When adopting the two mass functions, for the redshift evolution of the scatter, we find $\beta = 0.1250 \pm 0.0025$ for T08 and $\beta = 0.1056 \pm 0.0025$ for D16, for the Euclid-like experiment (for which we have the larger impact), showing therefore a $\sim 8\,\sigma$ discrepancy, which compensates for the high-redshift difference between the two mass functions. To further understand the impact of this different redshift evolution, we consider a non-constant EoS for dark energy, following the parametrisation $w = w_0 + (1-a)w_a$ [30, 31]. Still focusing on the Euclid-like experiment, for the two mass functions we find $w_0 = -1.000 \pm 0.010$ and $w_a = 0.35^{+0.32}_{-0.13}$ for T08, $w_0 = -1.095 \pm 0.012$ and $w_a = 0.13^{+0.51}_{-0.26}$ for D16, while having consistent results for the β parameter. In this case, the different redshift evolution between the two mass functions produces the same $\sim 8\,\sigma$ discrepancy on the w_0 parameter: a different EoS for dark energy implies a different expansion rate for the background, consistent with the different redshift evolution for the mass functions.

From this analysis, the halo mass function emerges as a non-negligible source of systematic uncertainties. It is therefore necessary to find a "universal" calibration, independent on the adopted simulation.

4 Conclusions

Galaxy clusters are emerging as a powerful cosmological probe, useful to describe the latest evolution of the large scale structure. The accuracy and precision of the cosmological constraints inferred from cluster analysis are dominated by theoretical and observational modelling. In order to improve the analysis, we would need further developments in hydrodynamical simulations and precise description of the astrophysical processes affecting the cosmological evolution of clusters. The exquisite quality of future cluster observations (such as NIKA2 follow-up program) will help in characterising the cluster properties.

References

[1] S. W. Allen *et al.* , Ann. Rev. Astron. Astrophys. **49** (2011), 409-470
[2] R.A. Sunyaev and Y.B. Zeldovich, Astrophys. Space Sci. **7** (1970), 3-19
[3] P. A. R. Ade *et al.* [Planck], Astron. Astrophys. **594** (2016), A27
[4] P. A. R. Ade *et al.* [Planck], Astron. Astrophys. **594** (2016), A24
[5] L. Salvati, M. Douspis and N. Aghanim, Astron. Astrophys. **614** (2018), A13
[6] G. Gianfagna *et al.* Mon. Not. Roy. Astron. Soc. **502** (2021) no.4, 5115-5133
[7] H. Hoekstra *et al.*, Mon. Not. Roy. Astron. Soc. **449** (2015) no.1, 685-714
[8] D. Eckert *et al.* Astron. Astrophys. **621** (2019), A40
[9] N. Aghanim *et al.* [Planck], Astron. Astrophys. **594** (2016), A22
[10] E. M. George, *et al.* Astrophys. J. **799** (2015) no.2, 177
[11] N. Aghanim *et al.* [Planck], Astron. Astrophys. **596** (2016), A107
[12] R. Adam *et al.* [Planck], Astron. Astrophys. **596** (2016), A108
[13] N. Aghanim *et al.* [Planck], Astron. Astrophys. **641** (2020), A6 [erratum: Astron. Astrophys. **652** (2021), C4]
[14] J. Lesgourgues and S. Pastor, New J. Phys. **16** (2014), 065002
[15] L. Salvati *et al.*, Astron. Astrophys. **626** (2019), A27
[16] L. E. Bleem *et al.* [SPT], Astrophys. J. Suppl. **216** (2015) no.2, 27
[17] S. Bocquet *et al.* [SPT], Astrophys. J. **878** (2019) no.1, 55
[18] L. Salvati *et al.* [SPT], in preparation
[19] P. Monaco, Galaxies **4** (2016) no.4, 53
[20] A. Paranjape, Phys. Rev. D **90** (2014) no.2, 023520
[21] S. Bocquet *et al.*, Mon. Not. Roy. Astron. Soc. **456** (2016) no.3, 2361-2373
[22] J. L. Tinker *et al.*, Astrophys. J. **688** (2008), 709-728
[23] G. Despali, *et al.* Mon. Not. Roy. Astron. Soc. **456** (2016) no.3, 2486-2504
[24] Z. Sakr, S. Ilić, A. Blanchard, J. Bittar and W. Farah, Astron. Astrophys. **620** (2018), A78
[25] R. Laureijs *et al.* [EUCLID], [arXiv:1110.3193 [astro-ph.CO]].
[26] P. A. Abell *et al.* [LSST Science and LSST Project], [arXiv:0912.0201 [astro-ph.IM]].
[27] D. Spergel, *et al.* [arXiv:1503.03757 [astro-ph.IM]].
[28] L. Salvati, M. Douspis and N. Aghanim, Astron. Astrophys. **643** (2020), A20
[29] B. Sartoris, *et al.* Mon. Not. Roy. Astron. Soc. **459** (2016) no.2, 1764-1780
[30] M. Chevallier and D. Polarski, Int. J. Mod. Phys. D **10** (2001), 213-224
[31] E. V. Linder, Phys. Rev. Lett. **90** (2003), 091301

Constraining AGN feedback model with SZ profile

Hideki Tanimura[1,*], *Gary* Hinshaw[2,3,4], *Ian G.* McCarthy[5], *Ludovic* Van Waerbeke[2], *Nabila* Aghanim[1], *Yin-Zhe* Ma[6,7], *Alexander* Mead[8], *Tilman* Tröster[8], *Alireza* Hojjati[2], and *Bruno* Moraes[9,10]

[1]Université Paris-Saclay, CNRS, Institut d'Astrophysique Spatiale, Bâtiment 121, 91405 Orsay, France
[2]Department of Physics and Astronomy, University of British Columbia, Vancouver, BC V6T 1Z1, Canada
[3]Canadian Institute for Advanced Research, 180 Dundas St W, Toronto, ON M5G 1Z8, Canada
[4]Canada Research Chair in Observational Cosmology
[5]Astrophysics Research Institute, Liverpool John Moores University, Liverpool, L3 5RF, United Kingdom
[6]Astrophysics and Cosmology Research Unit, School of Chemistry and Physics University of KwaZulu-Natal, Durban, South Africa
[7]NAOC-UKZN Computational Astrophysics Center (NUCAC), University of KwaZulu-Natal, Durban, 4000
[8]Institute for Astronomy, University of Edinburgh, Royal Observatory, Blackford Hill, Edinburgh, EH9 3HJ, UK
[9]Department of Physics and Astronomy, University College London, Gower Street, London, WC1E 6BT, UK
[10]Instituto de Fisica, Universidade Federal do Rio de Janeiro, 21941-972, Rio de Janeiro, RJ, Brazil

Abstract. Relativistic jets from AGN have a wide range of impacts on galaxy groups and clusters and are key for understanding their formation and physical properties. However, this non-gravitational process is not well understood. Galaxy groups with shallow gravitational potentials are ideal laboratories to study and constrain the AGN feedback model. We studied hot gas in ~66,000 SDSS LRG halos with an average halo mass of 3×10^{13} $M_\odot$ using the *Planck* tSZ map. We have detected their average tSZ radial profile at ~17σ and compared it with the cosmo-OWLS cosmological hydrodynamical simulations with different AGN feedback models. The best agreement has been obtained for the AGN 8.0 model in the simulations. We have also compared our measured tSZ profile with the prediction from the universal pressure profile assuming the self-similar relation and found them consistent if the model accounts for the clustering of neighboring haloes via a two-halo term.

1 Introduction

A power-law relation between the electron (or gas) pressure and mass, $P - M^{2/3}$, is valid under the assumption that the galaxy-formation process is dominated by gravity, called self-similar relation. Any deviation from this relation points to the presence of more complex processes such as baryonic feedback effects. X-ray and SZ observations demonstrate that

*e-mail: hideki.tanimura@ias.u-psud.fr

massive galaxy clusters almost follow the self-similar scaling relation [1]. On the other hand, numerous X-ray observations find a deficit of baryonic gas inside low-mass halos compared to the cosmic average baryon fraction [2–5], implying the deviation from the self-similar scaling relation. The gas deficit may be caused by AGN feedback: the response of the intracluster medium (ICM) to the relativistic plasma ejected by the active galactic nuclei (AGN) [6]. The AGN feedback is key for understanding the evolution of galaxies and the formation of large-scale structures; however, not well understood.

2 Stacking y map centered on LRGs

We performed a stacking analysis using the *Planck* y map [7] for Sloan Digital Sky Survey (SDSS) data release 7 (DR7) luminous red galaxies (LRGs) at $0.16 < z < 0.47$ [8], which are considered to be mostly central galaxies in dark matter halos [9]. We constructed an average y profile centered on the 66,479 LRGs with the stellar mass[1] of $10^{11.2} \leq M_*/M_\odot \leq 10^{11.7}$, roughly corresponding to halo masses of $10^{13} \leq M_{500}/M_\odot \leq 10^{14}$. We detected a significant signal at $\sim 17\sigma$ out to ~30 arcmins well beyond the extent of the 10 arcmin beam of the *Planck* y map in Fig. 1.

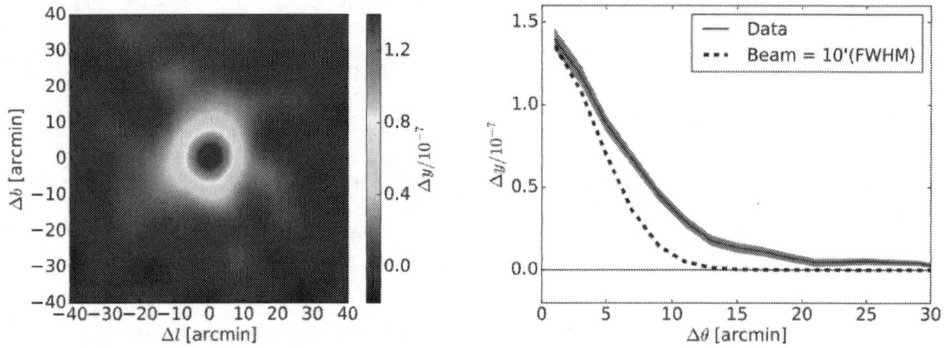

Figure 1. *Left*: Average y map centered on 66,479 LRGs. *Right*: Average y profile around the LRGs. The 1σ statistical uncertainty is represented via the width of the blue line. The FWHM = 10 arcmin Gaussian beam of the *Planck* y map is shown as a black dashed line for comparison, the peak of which is normalized to the center of the LRGs' y profile.

3 Comparison with the cosmo-OWLS hydrodynamic simulations

We compared the average y profile around LRGs with predictions from the cosmo-OWLS cosmological hydrodynamical simulations [10–12]. The simulations are designed with cluster cosmology and large-scale structure surveys in mind. They consist of box-periodic hydrodynamical simulations and have volumes of $400h^{-1}\text{Mpc}^3$ with 1024^3 baryonic and dark matter particles. They simulated five different models (Table. 1), which are compared with our measured y profile in Fig. 2.

In the comparison, a clear difference between the data and REF model can be seen. In general, energy released from the center of a halo heats cluster gas, this in turn prevents

[1]The stellar masses of the LRGs are provided in the New York University Value-Added catalogue (NYU-VAGC) at http://sdss.physics.nyu.edu/vagc/

cooling and thus the star formation around the central region. Therefore, if we consider halos of the same total mass, the stellar mass of the central galaxy is decreased as the power of the central AGN is increased. Since we select central galaxies based on stellar mass, lower-mass halos are selected in the REF model compared to the models that include AGN feedback. This is apparent as the lower central peak value of the simulated y profiles in the REF model compared to the AGN models. We also see a visible trend in that the higher the power of AGN feedback, the lower the peak of y profile. This is due to the fact that the AGN feedback ejects gas from the center of halos outward and the overall gas density is lowered.

In summary, this comparison agrees best with simulations that include AGN feedback (AGN 8.0) but not with simulations that do not include it (REF) or with simulations with very violent AGN feedback (AGN 8.7). The reduced χ^2 values are 1.0 (AGN 8.0), 1.3 (AGN 8.5), 9.9 (AGN 8.7), and 75.9 (REF), respectively. This result is consistent with other studies showing that the AGN 8.0 model reproduces a variety of observed gas features in optical and X-ray data (e.g., [10]).

Table 1. The baryon feedback models in the cosmo-OWLS simulations.

Simulation	Cooling	Star formation	SN feedback	AGN feedback	$\Delta T_{\rm heat}$
NOCOOL	No	No	No	No	...
REF	Yes	Yes	Yes	No	...
AGN 8.0	Yes	Yes	Yes	Yes	$10^{8.0}$ K
AGN 8.5	Yes	Yes	Yes	Yes	$10^{8.5}$ K
AGN 8.7	Yes	Yes	Yes	Yes	$10^{8.7}$ K

4 Halo model prediction with the UPP

The average y profile around the LRGs is also compared with a prediction using the halo model with the halo mass function and halo bias [13] and a universal pressure profile (UPP) [14] in Fig. 3. The stellar masses of the LRGs are converted to halo masses using the stellar-halo mass relation from either [15] (C15-SHM) or [16] (W16-SHM), estimated using gravitational lensing measurements. Then, the UPP is scaled to different masses of halos, assuming a self-similar relation. The predicted y profile is consistent with the data, but only if we account for the two-halo clustering term in the model and if we assume the stellar-halo mass relation from C15-SHM or W16-SHM. This consistency may imply that the UPP, estimated for massive galaxy clusters in the mass range of $10^{14.4} - 10^{15.3}$ $M_\odot$, can be applied even in low-mass halos down to $\sim 10^{13.5}$ $M_\odot$ in a self-similar manner.

5 Conclusion

According to the study with the cosmo-OWLS simulations [17], the AGN 8.0 simulation predicts more extended pressure profiles around low-mass halos than the UPP. However, the AGN 8.0 model also reproduces our observed y profile, supporting a self-similar assumption. This apparent inconsistency is explained by the fact that the deviations from the self-similar relation in the AGN 8.0 simulation are mainly confined to halo masses below $M_{500} \sim 10^{13.5}$ $M_\odot$, which roughly corresponds to the average mass of our sample. Furthermore, the impact of coarse angular resolution of the *Planck* y map is not negligible in our analysis: the UPP and AGN 8.0 pressure distributions only differ significantly on scales of $r < r_{500}$ in the simulations, in which they are compared without the beam. Thus, data from higher-resolution tSZ maps (such as those from ACT or SPT with FWHM of order an arcminute) would be necessary in this regard.

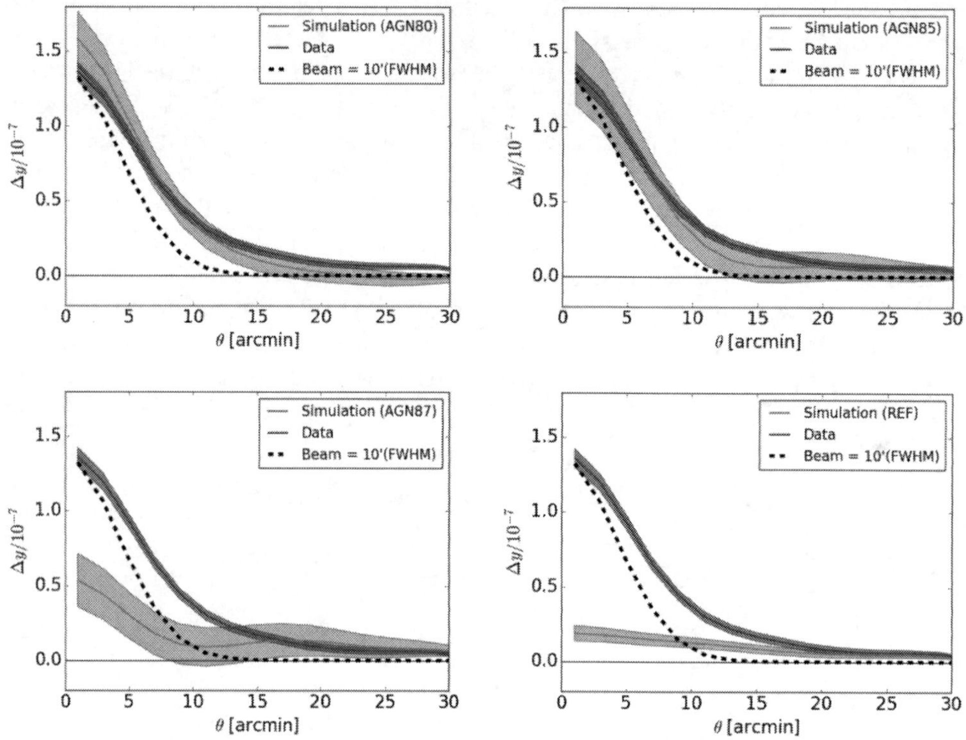

Figure 2. The average y profile around LRGs (blue) is compared to the y profiles of the simulated central galaxies (gray) in different AGN feedback models respectively. In each case we have matched the stellar mass and redshift distributions from the simulations to be the same as those in the data. *Top left*: AGN 8.0 model, *Top right*: AGN 8.5 model, *Bottom left*: AGN 8.7 model, and *Bottom right*: REF model.

Acknowledgement

This research has been supported by the funding for the ByoPiC project from the European Research Council (ERC) under the European Union's Horizon 2020 research and innovation programme grant agreement ERC-2015-AdG 695561.

References

[1] Planck Collaboration, A&A, **557**, A52 (2013)
[2] Gastaldello, F., et al., ApJ, **669**, 158 (2007)
[3] Pratt, G. W., et al., A&A, **498**, 361 (2009)
[4] Sun, M., et al., ApJ, **693**, 1142 (2009)
[5] Gonzalez, A. H., et al., The Astrophysical Journal, **778**, 14 (2013)
[6] Gitti, Myriam, et al., Advances in Astronomy, **2012**, 950641 (2012)
[7] Planck Collaboration, A&A, **594**, A22 (2016)
[8] Blanton, Michael R., et al., AJ, **129**, 2562-2578 (2005)
[9] Reid, B. A., et al., AJ, **698**, 143-154 (2009)
[10] Le Brun, Amandine M. C., et al., MNRAS, **440**, 1270-1290 (2014)

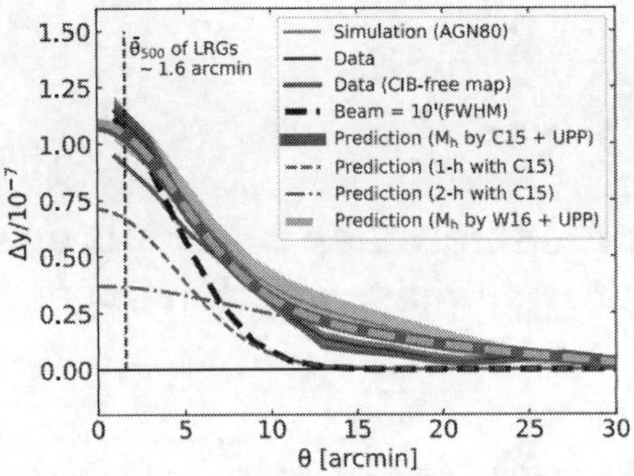

Figure 3. The average y profile around 66,479 LRGs (blue) is compared to the predictions using a halo model with the halo mass function and halo bias [13] and UPP [14]. The halo masses of the LRGs are estimated using either the SHM relation of [15] (C15-SHM: magenta) and [16] (W16-SHM: yellow). The one-halo (dashed line in magenta) and two-halo (dash-dotted line in magenta) terms are shown separately for the model prediction using the C15-SHM. The y profile of the simulated central galaxies in the AGN 8.0 simulation is shown in grey. To show an impact of beam, the average angular size of the LRGs ($\bar{\theta}_{500} \sim 1.6$ arcmin) is shown in vertical black dashed line.

[11] van Daalen, M. P., et al., MNRAS, **441**, 2997-3010 (2014)
[12] McCarthy, I. G., et al., MNRAS, **440**, 3645-3657 (2014)
[13] Tinker, Jeremy L., et al., ApJ, **724**, 878-886 (2010)
[14] Planck Collaboration, A&A, **550**, A131 (2013)
[15] Coupon, J., et al., MNRAS, **449**, 1352-1379 (2015)
[16] Wang, W., et al., MNRAS, **456**, 2301-2320 (2016)
[17] Le Brun, Amandine M. C., et al., MNRAS, **451**, 3868-3881 (2015)

Constraining hydrostatic mass bias and cosmological parameters with the gas mass fraction in galaxy clusters

Raphaël Wicker[1,*], *Marian* Douspis[1], *Laura* Salvati[1], and *Nabila* Aghanim[1]

[1]Université Paris-Saclay, CNRS, Institut d'Astrophysique Spatiale, 91405, Orsay, France

Abstract. The gas mass fraction in galaxy clusters is a convenient tool to use in the context of cosmological studies. Indeed this quantity allows to constrain the universal baryon fraction Ω_b/Ω_m, as well as other parameters like the matter density Ω_m, the Hubble parameter h or the Equation of State of Dark Energy w. This gas mass fraction is also sensitive to baryonic effects that need to be taken into account, and that translate into nuisance parameters. Two of them are the depletion factor Υ, and the hydrostatic mass bias $B = (1 - b)$. The first one describes how baryons are depleted in clusters with respect to the universal baryon fraction, while the other encodes the bias coming from the fact that the mass is deduced from X-ray observations under the hypothesis of hydrostatic equilibrium.

We will show preliminary results, obtained using the *Planck*-ESZ clusters observed by XMM-*Newton*, on both cosmological and cluster parameters. We will notably discuss the investigation on a possible redshift dependence of the mass bias, which is considered to be non-existent in hydrodynamic simulations based on Λ-CDM, and compare our results with other studies.

1 Introduction

Being the most massive gravitationally bound systems of our universe, galaxy clusters carry a lot of information. They can notably be used as powerful cosmological probes [1], or as astrophysical objects of study, to better understand the physics of the intra-cluster medium. Their baryonic component being mainly under the form of hot gas [2], of which the fraction is assumed to be relatively well known and understood, the gas mass fraction of galaxy clusters can be used as a robust cosmological probe (see e.g. [3] and references therein). Indeed this gas fraction is considered to be a good proxy for the universal baryon fraction [4].

Clusters being the siege of astrophysical phenomena, the gas mass fraction is also sensitive to the baryonic physics inside these objects. Baryonic physics are encoded for one part in the depletion factor Υ, which describes how the gas is depleted with respect to the universal baryon fraction [5], and for the other part in the hydrostatic mass bias $B = (1 - b) = \frac{M_{measured}}{M_{true}}$.This bias comes into play when measuring the cluster masses from X-ray or SZ observations. Indeed such measures assume that clusters are at the hydrostatic equilibrium, which is not entirely true due to a fraction of non-thermal pressure support inside these objects. Several works have been studying the hydrostatic mass bias (see e.g. [6], [7]), and notably its evolution with the redshift. Our goal in this work is to try and compare the findings of these studies with results obtained using the gas mass fraction in clusters.

*e-mail: raphael.wicker@ias.u-psud.fr

2 Data

We use gas masses and total masses at R_{500} from 120 clusters of the *Planck*-ESZ sample [8] as seen in follow-up XMM-*Newton* observations analysed in [9]. These clusters span the redshift range [0.059; 0.546], with their total mass derived from the hydrostatic equilibrium equation given below :

$$M_{tot}(< r) = -\frac{rk_BT(r)}{G\mu m_p}\left(\frac{d\ln\rho(r)}{d\ln r} + \frac{d\ln T(r)}{d\ln r}\right) \quad (1)$$

From the gas masses and total masses, we can compute the gas fraction of these clusters, $f_{gas} = M_{gas}/M_{tot}$, displayed in Figure 1 below.

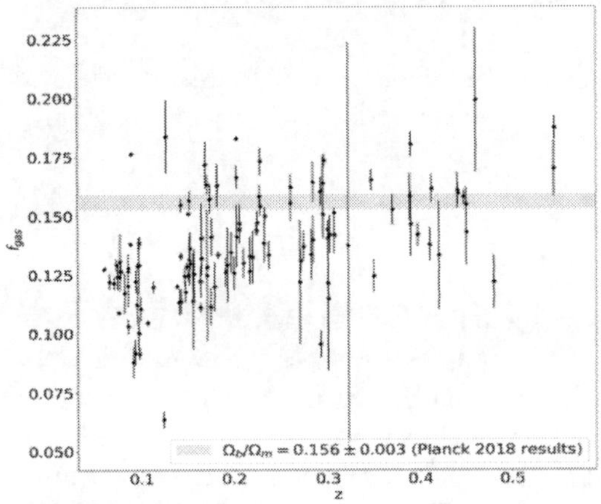

Figure 1. Observed gas fraction of the *Planck*-ESZ sample, compared to Ω_b/Ω_m from [13], with errors at 1σ.

3 Modelling the redshift evolution of the bias

The redshift evolution of the gas mass fraction depends on several instrumental, astrophysical, and cosmological contributions, which can be summed up in equation 2 below, from [5]:

$$f_{gas}(z) = K \times \frac{\Upsilon(z)}{B(z)} \times A(z) \times \left(\frac{\Omega_b}{\Omega_m}\right) \times \left(\frac{D_A^{fid}(z)}{D_A(z)}\right)^{3/2} \quad (2)$$

where K is an instrumental calibration constant (assumed to be 1 in this work), Υ is the baryon depletion factor, and B is the hydrostatic mass bias $B = (1 - b)$. $A(z)$ is an angular correction parameter, constant and equal to 1 if we are in the fiducial cosmology considered. (Ω_b/Ω_m) is the universal baryon fraction, and D_A is the angular diameter distance. In this analysis we consider as fiducial cosmology a flat Λ-CDM with $h = 0.7$, $\Omega_m = 0.3$ and $\Omega_\Lambda = 0.7$.

As shown in equation 2, what we actually constrain when looking at the redshift evolution of f_{gas} is the evolution of the ratio $\Upsilon(z)/B(z)$. Based on hydrodynamical simulations from [10], we assume a constant depletion factor $\Upsilon(z) = \Upsilon_0$. For a study of the evolution of $\Upsilon(z)$ in galaxy cluster data, see [11]. We therefore assume that all the evolution comes from the bias. Following [12], we assume a linear evolution [1] with a constant term B_0 (the bias at $z = z_{pivot}$) and a slope B_1 :

$$B(z) = B_0 + B_1 \times (z - z_{pivot}) \quad (3)$$

with $z_{pivot} = \langle z \rangle = 0.218$. In a first part of the study we consider a cosmology fixed at the values of [13]. We thus fit the measured f_{gas} with a MCMC using the python package `emcee`, with the set of priors given in Table 1 and adding a term accounting for the intrinsic scatter in the data, σ_f.

[1]In the full study from Wicker et al. *in prep*, we will look at other possible parametrizations of the bias evolution.

Table 1. Set of priors used in the first part of the analysis. $\mathcal{U}(l, u)$ means a uniform prior of lower bound l and upper bound u, while $\mathcal{N}(\mu, \sigma)$ means a normal prior of mean μ and standard deviation σ. The prior on Υ_0 comes from [10].

B_0	$\mathcal{U}(0.3, 1.7)$
B_1	$\mathcal{U}(-1.5, 1.5)$
Υ_0	$\mathcal{N}(0.85, 0.03)$
σ_f	$\mathcal{U}(0, 1)$

4 Results

Matching the selection from weak lensing studies like the CoMaLit [14] or LoCuSS [15] studies, we chose to study the subsample $z > 0.2$, as well as our full sample. This choice was also motivated by the results from [6] showing that the trends in the mass bias depended on the considered redshift range, with results changing when considering only clusters with $z > 0.2$.

4.1 Full sample

We first carried out our analysis on the full sample, with redshifts in the range [0.059, 0.546]. We kept all the cosmological parameters fixed at their values from [13], leaving free the constant term of the bias B_0, the slope of the bias B_1, and the intrinsic scatter σ_f. The depletion factor Υ is treated as a nuisance parameter. Using this model, we obtained the plot given in Figure 2. In particular with $B_1 = -0.60 \pm 0.11$ we obtain a slope very different from

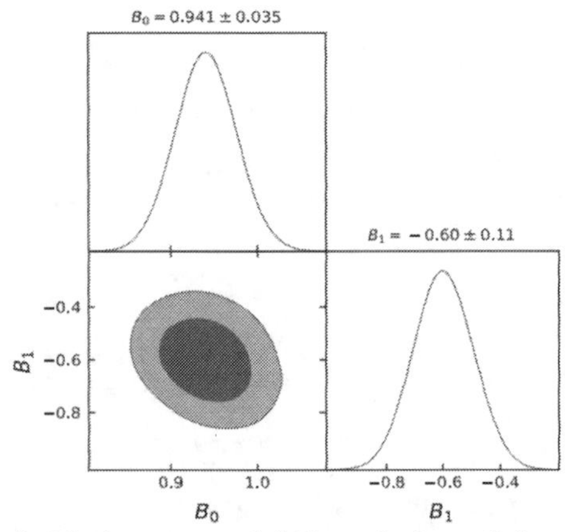

Figure 2. Constraints on B_1 for the full sample. Shaded areas mark 1 and 2σ regions while the numeric values give the uncertainties at 1σ.

0, hinting at a redshift evolution of the mass bias at 5.5σ, with a bias increasing with the redshift. Indeed with $B_1 < 0$, we have a parameter $B(z) = (1 - b)$ decreasing with the redshift, meaning that the masses are more biased at higher redshift. These trends are in agreement with [14] and [15] based on weak lensing studies, but in contradiction with [6] based on tSZ number counts with almost the same clusters as this work.

4.2 High redshift clusters

Following [6] we now focus on the clusters for which z > 0.2, using the same model as in Sect. 4.1. We obtain the contours from Figure 3.

We find this time $B_1 = -0.24 \pm 0.15$ which means a slope much closer to 0, with this time a mild 1.7σ discrepancy with no evolution of the bias. This behaviour had already been noted

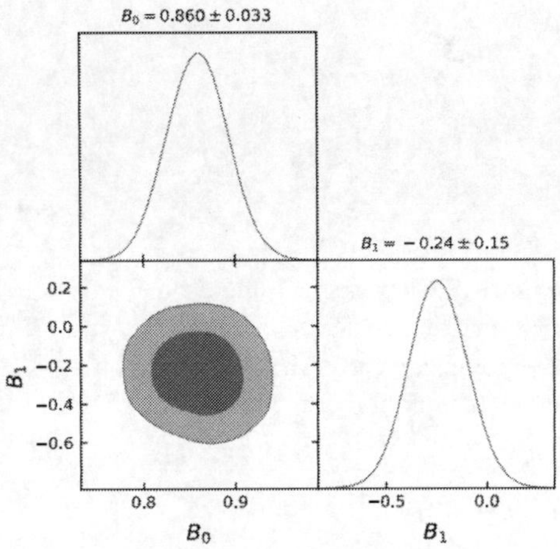

Figure 3. Constraints on B_1 for clusters with $z > 0.2$. Shaded areas mark 1 and 2σ regions.

in [6], which found no evolution of the bias within 1σ when focusing only on the clusters with z > 0.2. In addition, the value of $B_0 = 0.860 \pm 0.035$ is also fully compatible with other values of $B = (1 - b)$ from weak-lensing studies, like [16] or obtained from hydrodynamical simulations, like [17].

5 Impact on cosmological analysis

In a second part of the analysis we let some of the cosmological parameters free. In this preliminary work we focused only on the high redshift clusters. A more complete study focused on different subsamples and selections in mass and redshift are proposed in Wicker et al. *in prep.* We first considered no evolution of the hydrostatic mass bias, then applied a non-zero B_1, based on the values derived in the previous part of this work. The set of priors considered for this work is the following:

Table 2. Set of priors used in the second part of the analysis. $\mathcal{U}(l, u)$ means a uniform prior of lower bound l and upper bound u, while $\mathcal{N}(\mu, \sigma)$ means a normal prior of mean μ and standard deviation σ. The prior on B_0 comes from [18], and the prior on h comes from [13]. The priors on Υ_0 and σ_f are the same as in Table 1.

B_0	$\mathcal{N}(0.780, 0.092)$
B_1	$\mathcal{N}(0, 0.001)$ (Sect. 5.1) then $\mathcal{N}(-0.24, 0.15)$ (Sect. 5.2)
Ω_b/Ω_m	$\mathcal{U}(0.05, 0.3)$
Ω_m	$\mathcal{U}(0.01, 1.0)$
h	$\mathcal{N}(0.674, 0.005)$

5.1 Considering no evolution of the bias

We show in Figure 4 that when we consider no evolution of the bias, the derived value of $\Omega_b/\Omega_m = 0.133^{+0.017}_{-0.019}$ is compatible with the value from Planck (2σ orange bands), yet peaks −0.023 lower. We can also see the degeneracy between B_0 and Ω_b/Ω_m, as expected from equation 2. On the other hand we can see that the posterior for Ω_m is very wide and peaks high above the Planck value, as we obtain $\Omega_m = 0.59^{+0.16}_{-0.20}$.

5.2 Considering an evolution of the bias

When we consider an evolution of the bias with the redshift, keeping the same prior on B_0, we get back to values that are much more compatible with the *Planck* values, as we can see in

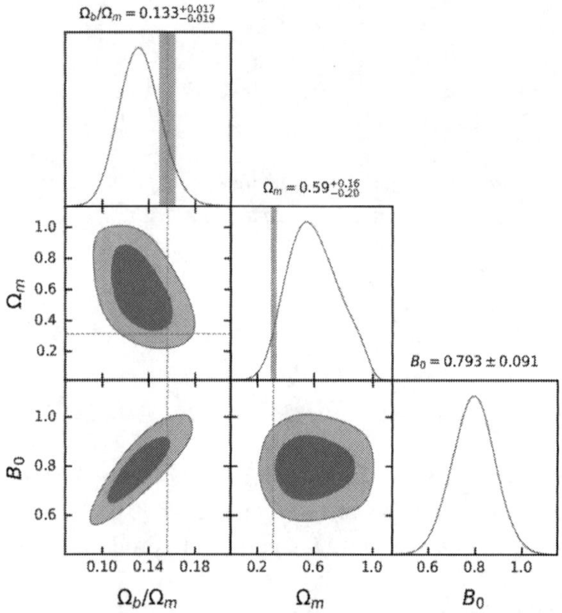

Figure 4. Contours at 1 and 2σ and marginalized posteriors for the cosmological study when assuming no redshift evolution of the bias.

Figure 5. Indeed we find $\Omega_b/\Omega_m = 0.144^{+0.018}_{-0.021}$ and $\Omega_m = 0.35^{+0.17}_{-0.25}$, although the latter peaks slightly below the *Planck* value of 0.315 [13]. We can also see that just like B_0 and Ω_b/Ω_m, Ω_m and B_1 are strongly degenerated, explaining the strong impact on the constraints of this parameter when we go from a constant bias to a varying bias.

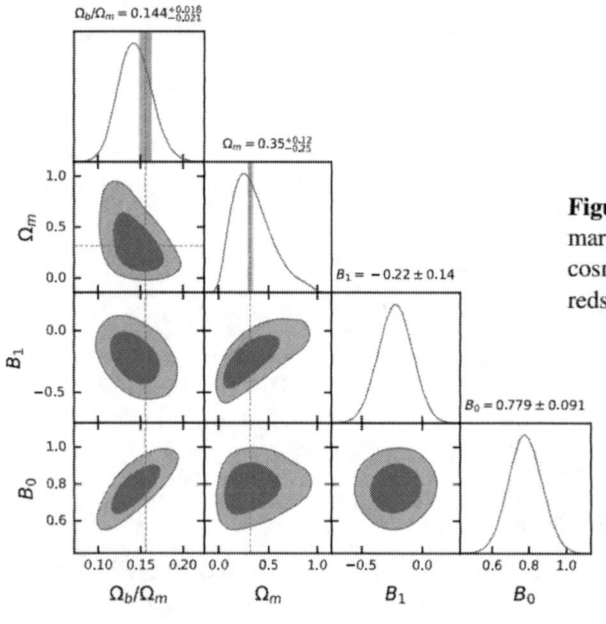

Figure 5. Contours at 1 and 2σ and marginalized posteriors for the cosmological study when assuming a redshift evolution of the bias.

6 Conclusion

As we have shown, our results on the redshift evolution of the mass bias are strongly dependent on the redshift range at which we chose to study our clusters, hinting at a strong sample dependence. To try and mitigate this effect, a similar work could be carried out on larger samples, like PSZ2 or eFEDS [19] clusters. We would also need to take into account possible mass selection effects, as cluster at higher redshifts in the sample are also of higher mass, and the hydrostatic mass bias may also be mass dependent [6]. The paper from Wicker et al.

in prep will propose a complete study of the mass and redshift evolution of fgas in galaxy clusters, and as such of the hydrostatic mass bias, based on the method described in these proceedings. Finally, we have shown that these results on the evolution of B with redshift absolutely need to be taken into account when carrying out a cosmological analysis, as disregarding such evolution could strongly bias the quality of the constraints obtained in the end. As it turns out, most cosmological studies using f_{gas} are carried out at R_{2500} instead of R_{500}. Using high spatial resolution measurements in X-rays *and* in SZ, which could be allowed by NIKA2, could help to constrain better the evolution of the mass bias at these radii.

References

[1] S.D.M. White, J.F. Navarro, A.E. Evrard, C.S. Frenk, Nature, **366**, 429 (1993)

[2] Y.T. Lin, J.J. Mohr, ApJ, **617**, 879 (2004), `astro-ph/0408557`

[3] R.F.L. Holanda, G. Pordeus-da-Silva, S.H. Pereira, JCAP, **2020**, 053 (2020), `2006.06712`

[4] S. Borgani, A. Kravtsov, Advanced Science Letters **4**, 204 (2011), `0906.4370`

[5] S.W. Allen, D.A. Rapetti, R.W. Schmidt, H. Ebeling, R.G. Morris, A.C. Fabian, MNRAS, **383**, 879 (2008), `0706.0033`

[6] L. Salvati, M. Douspis, A. Ritz, N. Aghanim, A. Babul, A&A, **626**, A27 (2019), `1901.03096`

[7] G.W. Pratt, M. Arnaud, A. Biviano, D. Eckert, S. Ettori, D. Nagai, N. Okabe, T.H. Reiprich, Space Science Reviews, **215**, 25 (2019), `1902.10837`

[8] Planck Collaboration, P.A.R. Ade, N. Aghanim, M. Arnaud, M. Ashdown, J. Aumont, C. Baccigalupi, A. Balbi, A.J. Banday, R.B. Barreiro et al., A&A, **536**, A8 (2011), `1101.2024`

[9] L. Lovisari, G. Schellenberger, M. Sereno, S. Ettori, G.W. Pratt, W.R. Forman, C. Jones, F. Andrade-Santos, S. Randall, R. Kraft, ApJ, **892**, 102 (2020), `2002.11740`

[10] S. Planelles, S. Borgani, K. Dolag, S. Ettori, D. Fabjan, G. Murante, L. Tornatore, MNRAS, **431**, 1487 (2013), `1209.5058`

[11] K. Bora, S. Desai, European Physical Journal C, **81**, 296 (2021), `2103.12695`

[12] A.B. Mantz, S.W. Allen, R.G. Morris, D.A. Rapetti, D.E. Applegate, P.L. Kelly, A. von der Linden, R.W. Schmidt, MNRAS, **440**, 2077 (2014), `1402.6212`

[13] Planck Collaboration, N. Aghanim, Y. Akrami, M. Ashdown, J. Aumont, C. Baccigalupi, M. Ballardini, A.J. Banday, R.B. Barreiro, N. Bartolo et al., A&A, **641**, A6 (2020), `1807.06209`

[14] M. Sereno, S. Ettori, MNRAS, **468**, 3322 (2017), `1603.06581`

[15] G.P. Smith, P. Mazzotta, N. Okabe, F. Ziparo, S.L. Mulroy, A. Babul, A. Finoguenov, I.G. McCarthy, M. Lieu, Y.M. Bahé et al., MNRAS, **456**, L74 (2016), `1511.01919`

[16] R. Herbonnet, C. Sifón, H. Hoekstra, Y. Bahé, R.F.J. van der Burg, J.B. Melin, A. von der Linden, D. Sand, S. Kay, D. Barnes, MNRAS, **497**, 4684 (2020), `1912.04414`

[17] V. Biffi, S. Borgani, G. Murante, E. Rasia, S. Planelles, G.L. Granato, C. Ragone-Figueroa, A.M. Beck, M. Gaspari, K. Dolag, ApJ, **827**, 112 (2016), `1606.02293`

[18] H. Hoekstra, R. Herbonnet, A. Muzzin, A. Babul, A. Mahdavi, M. Viola, M. Cacciato, MNRAS, **449**, 685 (2015), `1502.01883`

[19] Y.E. Bahar, E. Bulbul, N. Clerc, V. Ghirardini, A. Liu, K. Nandra, F. Pacaud, I.N. Chiu, J. Comparat, J. Ider-Chitham et al., arXiv e-prints arXiv:2110.09534 (2021), `2110.09534`

AUTHOR INDEX

AUTHOR INDEX

AUTHOR INDEX